Advances in Soil Science

Advances in Soil Science

B.A. Stewart, Editor

Advances in Soil Science

Volume 16

Edited by B.A. Stewart

With Contributions by
D.L. Brakensiek S.K. De Datta T.J. Gish
W.J. Rawls S.K. Sanyal C.P. Schulthess S.R. Smith P.N. Soltanpour
D.L. Sparks

With 45 Illustrations

Springer-Verlag
New York Berlin Heidelberg London
Paris Tokyo Hong Kong Barcelona

B.A. Stewart
USDA Conservation & Production Research Laboratory
Bushland, TX 79012, USA

ISSN: 0176-9340

Printed on acid-free paper.

Softcover reprint of the hardcover 1st edittiion 1991

Typeset by Asco Trade Typesetting Ltd., Hong Kong.

9 8 7 6 5 4 3 2 1

ISBN-13:978-1-4612-7812-2 e-ISBN-13:978-1-4612-3144-8

DOI:10.1007/978-1-4612-3144-8

Preface

The International Society of Soil Science was organized in 1924 and the 1st International Congress of the Society was held in Washington, D.C., United States of America, in 1927. The 14th Congress was held August 12–18, 1990 in Kyoto, Japan. Dr. Akira Tanaka, Society President, stated "It is a serious concern of people in the world to sustain high farming productivity for feeding the ever growing population, and also to protect the environment for human habitation. For this purpose, it is necessary to maximize crop yield per unit field area through intensive management of the soil and crops on lands which are best suited for crop production, and to leave lands which are suboptimal for crops under natural conditions insofar as possible. Soil science is responsible for planning the optimum land-use system, for managing soils and crops with maximum efficiency, and for fighting intelligently against soil deterioration." Dr. Tanaka clearly stated the challenges facing soil scientists, and they are indeed challenges. Even though the rate of world population growth has decreased somewhat in recent years, the consensus is that total population will double—to some ten billion—before it stabilizes. The increased production of food, fiber, and fuelwood necessary for the people will require a continued supply of new technologies and management practices.

The consensus of many scientists is that the greatest challenge is not increasing production, but preventing serious deterioration of the soil and water resource base so the production level can be sustained. There are many historical examples where civilizations were successful for extended periods only to see soil degradation lead to their collapse. There is sufficient knowledge today, and it is being added to daily, that the mistakes of the past need not occur in the future. However, success will require a blend of science and government policy to prevent short-term gains at the expense of long-term sustainability. This will be particularly difficult in areas where population pressures are so great that crop production is being extended to fragile lands.

Advances in Soil Science was establised to provide a forum for leading scientists to analyze and summarize the available scientific information on a subject, assessing its importance and identifying additional research needs. Most importantly, contributors are asked to develop principles that have practical applications to both developing and developed agriculture. It is not the purpose of the series to report new research findings because there are many excellent scientific journals for that need. Communications in scientific journals, however, are generally restricted to short and technical presentations. Therefore, *Advances in Soil Science* fills a gap between the scientific journals and the comprehensive reference books in which scientists can delve in depth on a particular subject relating to soil science. While intended primarily for scientists and students of soil science, the series provide technical information for anyone interested in our natural resources and man's influence on these resources. The reviews are written by leading scientists from many countries to provide the readers with information from a wide array of conditions.

The acceptance of *Advances in Soil Science* by both authors and readers has been very gratifying and confirms our perception that a need did exist for a medium to publish soil science reviews. I want to thank the authors for their excellent contributions, the Editorial Board for their help in selecting such competent contributors, and the Springer-Verlag staff for their kind assistance and counsel. Last, and most important, I want to thank the readers for their acceptance and use of *Advances in Soil Science.*

B.A. Stewart

Contents

Contributors

D.L. Brakensiek, USDA-ARS, Hydrology Laboratory, Beltsville, Maryland 20705, USA
S.K. De Datta, Agronomy-Physiology-Agroecology Division, International Rice Research Institute, 1099 Manila, Philippines
T.J. Gish, Department of Agricultural Engineering, University of Maryland, Beltsville, Maryland 20705, USA
W.J. Rawls, U.S. Department of Agriculture, Agricultural Research Service, Beltsville, Maryland 20705, USA
S.K. Sanyal, Agronomy-Physiology-Agroecology Division, International Rice Research Institute, 1099 Manila, Philippines
C.P. Schulthess, National Chemical Laboratory for Industry, I-I Higashi, Tsukuba, Ibaraki 305, Japan
S.R. Smith, WRc Medmenham, Marlow, Buckinghamshire SL7 2HD, UK
P.N. Soltanpour, Department of Agronomy and the Experiment Station, Colorado State University, Fort Collins, Colorado 80523, USA
D.L. Sparks, Department of Plant and Soil Sciences, University of Delaware, Newark, Delaware 19717-1303, USA

Chemistry of Phosphorus Transformations in Soil

S.K. Sanyal and S.K. De Datta

I. Introduction

Phosphorus (P) is essential for plants and animals because of its role in vital life processes, such as in photosynthesis in plants and energy transformations in all forms of life. It also has a significant role in sustaining and building up soil fertility, particularly under intensive systems of agriculture.

Soils are known to vary widely in their capacities to supply P to crops because only a small fraction of the total P in soil is in a form available to crops. Thus, unless the soil contains adequate amount of plant-available P, or is supplied with readily available-(inorganic)-P fertilizers, crop growth will suffer.

A large proportion of P depleted by the agricultural crops comes from the native P content of the soil, and this should be replenished primarily through inorganic P fertilizers. Furthermore, unlike nitrogen (N), which can be recycled to the soil by fixation from air, P once removed from the soil by the crop or by erosion, runoff, or leaching cannot be replenished except from external sources.

There is also concern for a rapid depletion of high-quality phosphate rock (PR) ores for production of soluble fertilizers, which is further complicated by the high manufacturing costs involved. Reactive phosphate rocks and their partially acidulated products are therefore often used as direct-application fertilizer materials, mainly in acid soils.

The dynamics of P transformations in soils, and the fixation and release characteristics of P are reported in this chapter. The importance of various physicochemical processes in governing P concentration in soil solution has been highlighted. In particular, the distinct nature of P transformation processes in flooded soils as compared with those in upland soils has been discussed. An area of more recent investigations is the transformation of organic P and the buildup of microbial biomass P in soil (Tate, 1984; Stewart and Tiessen, 1987). Recent developments in methodology have made it possible to estimate better the pool size of organic P, and its contribution to available P in the soil. Key processes of interaction of P with carbon (C), N, and sulfur (S) have also been identified, and incorporated into models of P cycles (Parton et al., 1988). Research data have been generated to validate these models of P dynamics in a cropping system for medium- or long-term fertility trials in various environments.

II. Physicochemical Processes Governing Phosphorus Concentration in Soil Solution

A. Sorption–Desorption

Phosphorus (P) adsorption by soils is a widely researched subject. This is primarily because of the widespread P deficiency reported for agricultural soils, and the fact that P adsorption by soils is a process mainly responsible for rendering soluble phosphate in soil solution unavailable to plants. The adsorption process, which refers to surface P accumulation on soil components, may, in some cases, be accompanied by penetration of the adsorbed P by diffusion into the adsorbent body, leading to further absorption of the adsorbed species. The general term *sorption* is sometimes used to denote both of these processes taking place simultaneously.

In acid soils, P adsorption is generally attributed to hydrous oxides of iron (Fe) and aluminum (Al), and to (1:1) layer lattic clays, particularly in tropical soils with low pH. The possibility that some natural phosphates of aluminum and/or iron (such as variscite and strengite) are formed in these soils was discounted by some authors (Ryden and Pratt, 1980). Phosphate adsorption in acid soils was also considered (Hsu, 1965) to be a special case of precipitation wherein aluminum (or iron) remained as the constituent of the original phase while reacting with P by the use of residual force on the surface. Furthermore, a more fundamental approach to study P immobilization in acid soils suggested would be to examine the development of reactive aluminum hydroxides and iron oxides rather than the solubility of some P compounds. However, in recently fertilized soils, the local conditions of low pH and high-phosphate concentrations in the vicinity of the P fertilizer granules are conducive to the dissolution of clays and reprecipitation of a wide variety of P compounds by reaction with soil components (Sample et al., 1980; Ryden and Pratt, 1980). Such a process of crystallization of P compounds, having definite values of solubility products, is slow, and the attainment of equilibrium is likely to be disturbed by P uptake by plants, diffusion, and more importantly, by rapid adsorption of the soluble P by surface-reactive aluminum hydroxides and iron oxides, especially if the latter is predominantly amorphous in nature. As a consequence, persistence of fertilizer reaction products in soils is unlikely. Details on P precipitation–dissolution equilibrium in soils will be discussed in a subsequent section (Section IIB).

In neutral and alkaline soils, various forms of calcium phosphates are the stable minerals that govern P concentration in soil solutions. In calcareous soils, treatments with a high level of P tend to converge to a common concentration of P in solution of 1–2 ppm P after a long period of contact (Al-Khateeb et al., 1986). This behavior can best be explained by the slow formation of a calcium (Ca)–P compound (Barrow, 1987). More direct evidence of formation of calcium phosphates on a calcite surface, reacting with P solutions, was provided by Freeman and Rowell (1981) who demonstrated by scanning electron microscopy and X-ray diffraction technique the formation on a calcite surface of dicalcium phosphate (DCP) that slowly changed to octacalcium phosphate.

1. Equilibrium Systems

a. Adsorption Isotherms

The relationship between the amount of P adsorbed per unit weight of soil (x) and the equilibrium P concentration in solution (c) bathing the soil at

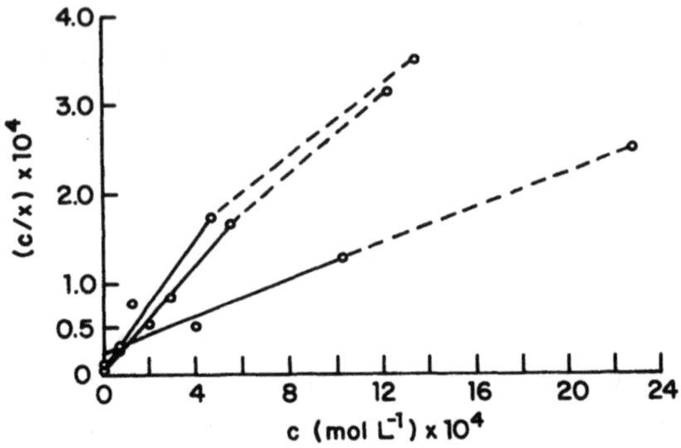

Figure 1. Phosphorus sorption isotherm data plotted according to the Langmuir isotherm. (*Source*: Bennoah, E.O. and D.K. Acquaye. Phosphate sorption characteristics of selected major Ghanaian Soils, *Soil Science* 148, 114–123. © Williams & Wilkens 1989.)

a constant temperature has been described by several adsorption isotherms. The main motivations for describing adsorption curves were to (1) identify the soil constituents involved in adsorption (Adams et al., 1987; Loganathan et al., 1987), (2) predict the amount of fertilizer needs of soils to meet the demand of plant uptake for an optimum yield (Fox and Kamprath, 1970; Fox 1974; Fox and Kang, 1978; Roy and De Datta, 1985; Greenland and De Datta, 1985; Klages et al., 1988), and (3) study the nature of the adsorption process to learn more about the mechanism of the process (Barrow, 1984, 1987).

Langmuir Adsorption Isotherm. There have been many attempts to fit the results of P adsorption studies on soils, clays, and sediments to the simple Langmuir equation (Bache and Williams, 1971; Sims and Ellis, 1983a; Vig and Dev, 1984; Bennoah and Acquaye, 1989). A close fit to the simple Langmuir equation is generally obtained at low concentrations (< 15 mg/l) over a limited range. In fact, the adsorption maximum calculated for a lower concentration range is often exceeded at higher concentrations (Barrow, 1978; Harter, 1984). A typical plot of the simple Langmuir equation for P adsorption by soils is shown in Figure 1, which illustrates the rather narrow range of linearity of the plot. A curved relationship between (c/x) and c over a wide range of concentrations implies that the bonding energy is not a constant but rather a function of adsorption, and that there is no well-defined maximum. A probable reason for these deviations may be the migration of sorbed P to subsurface layers and crystalline hydrous Fe oxides. The restriction to a monomolecular layer assumed in the Langmuir model also seems unlikely, especially at higher concentrations where some kind of surface structure may start forming (Olsen and Khasawneh, 1980). In conclusion, the assumption of no lateral interaction among the sorbed P species and constant free energy of adsorption does not fit well with the present knowledge that the sorbed P species carries a charge and that surface charge and potential decrease as more P is sorbed on oxide minerals or in the soil system (Sposito, 1981; Kuo and McNeal, 1984).

To account for such nonlinearity of the simple Langmuir plots, researchers have proposed several modifications to the simple Langmuir equation. The equation proposed by Gunary (1970) includes an additional square-root term in concentration. In some cases, the data can best be fitted with two intersecting straight lines, and this has been taken to indicate the existence of more than one kind of P adsorption site on the surface. Syers et al. (1973), Ram et al. (1987), and Mehadi and Taylor (1988) used a Langmuir two-surface equation, and obtained satisfactory agreement with experimental data. These workers proposed that the two surfaces are characterized by different bonding energy and adsorption maximum.

Ryden et al. (1977a), while studying P adsorption on soils of contrasting properties and an iron oxide gel, proposed the existence of three types of adsorption sites—regions I, II, and III—of widely different reactivities. For regions I and II, chemisorption was the suggested mechanism, whereas for region III, Ryden et al. (1977a) proposed a physical adsorption mechanism. Table 1 reproduces the relevant sorption parameters for regions I, II, and III. The data reveal that the thermodynamics of adsorption for different soils and the iron oxide gel were independent of the adsorbing surface in all regions, even though the Langmuir adsorption maximum (*b*) varied with sorbent surface and experimental conditions. This suggested that the experimental conditions affected only the degree rather than the nature of the P adsorption process. The contention of Ryden et al. (1977a) that the adsorption in region III is more physical than chemical was contested by Parfitt (1978) on the grounds that such weak adsorption could well be the result of ligand exchange on a nearly saturated surface where the surface charge is much reduced. Nanzyo (1984), from studies of infrared absorption spectra of P sorbed on alumina gel, suggested that the reaction of P with alumina gel is one of ligand exchange, and the adsorbed P is readily converted to a state similar to aluminum phosphate gel.

An important point in favor of the two- or three-surface Langmuir model (Ryden and Pratt, 1980) is that sorption, especially at higher concentrations, were very similar. This similarity for a wide range of surface soils, subsoils, and hydrous ferric oxides provided the basis of a useful approach for assessing soils for P adsorption. Barrow (1978), on the other hand, questioned the necessity of postulating more than one kind of sorption site for soil P. He surmized, instead, that the observed deviations from the simple Langmuir equation could well be because the fact that the bonding energy is a decreasing function of the increasing surface-saturation was discounted.

A more serious criticism on the validity of the two- (or three-) surface Langmuir equation as a physical model for P sorption by soils was offered by Sposito (1982). He demonstrated that, if the distribution coefficient for an ion sorbed by a soil is a finite decreasing function of the amount adsorbed (x), and extrapolates to zero at some finite value of x, then the sorption isotherm

Table 1. Sorption constants describing the three regions (I, II, and III) of P sorption by four soils and Fe gel using different experimental conditions

Sorbent	ΔG_I	kJ/mol ΔG_{II}	ΔG_{III}	b_I	b_{II} μ mol/g	b_{III}
		A. Equilibrium, 10^{-1} *M* NaCl				
Egmont soil	−38.5	−29.9	−21.8	39.3	48.4	104
Okaihau soil	−39.1	−30.2	−21.0	21.5	33.2	55.8
Porirua soil	−37.0	−29.4	−19.3	4.2	9.2	17.1
Waikakahi soil	−36.8	−29.6	−22.5	1.4	2.6	10.5
Fe gel[a]	ND	−29.1	−18.4	590	445	1020
		B. 40 h, 10^{-1} *M* NaCl				
Egmont soil	−36.5	−29.1	−21.2	20.3	26.8	94.8
Okaihau soil	−39.5	−29.7	−20.1	9.7	29.4	48.4
Porirua soil	−39.3	−28.3	−19.9	1.6	7.4	15.5
Waikakahi soil	−38.3	−28.7	−20.3	0.6	2.5	9.1
Fe gel	ND	−29.7	−18.4	480	450	1018
		C. 40 h, 10^{-4} *M* NaCl				
Egmont soil	−41.3	−31.0	−21.2	8.0	18.1	37.4
Okaihau soil	−43.6	−29.6	−20.1	4.2	14.7	22.4
Porirua soil	−40.8	−31.1	−21.9	0.7	2.8	7.2
Waikakahi soil	−38.6	−29.8	−20.8	0.6	2.3	8.0
Fe gel	ND	−36.0	−19.0	226	242	643

Note: ΔG is the free energy of sorption (derived from the Langmuir Sorption Energy Constant.
a Constants relate to sorption during 690 h; final pH = 7.0
b Sorption maximum
ND, Not determined.
Source: Ryden et al., 1977a.

can always be represented mathematically by a two-surface Langmuir equation, independent of the chemical mechanism of ion sorption. It thus follows that the adjustable parameters in the two-surface Langmuir equation cannot be interpreted in terms of surface reactions without additional independent evidence that adsorption on two kinds of surface site actually is involved in the ion sorption reaction.

The curvature of the simple Langmuir plot was attributed by Lin et al. (1983a) to heterogeneous adsorption energies of soil minerals, implying that a homogeneous surface provides a sufficient condition but not a necessary condition for a linear Langmuir plot. These authors also opposed the idea that curvilinear Langmuir plots indicate the existence of two or more different adsorption sites.

An alternative to Langmuir sorption was suggested where ion product considerations were combined with estimates of solid-phase P activity, based on regular solid-solution concepts, to predict P solubility as a function of sorbed P (Blanchar and Stearman, 1984, 1985).

Sibbesen (1981) proposed an extended Langmuir equation, wherein the bonding energy term was replaced by one, the value of which decreased with increasing P concentration. The extended Langmuir equation reads,

$$x = \frac{ABc^{-D}c}{1 + Bc^{-D}c} \tag{1}$$

where A, B, and D are coefficients. Eq. (1) was better able to explain the P adsorption data in the experimental soils than did the simple, or the two-surface, Langmuir equations.

To improve the applicability of the simple Langmuir equation for P sorption studies in soils, Kuo (1988) used a modified version developed through statistical mechanics. The modified equation takes into account the electrostatic interactions among the surface-sorbed P species. The equations used by Kuo (1988) are the following. Eq. (2) and (3) show the classical Langmuir isotherm, and the modified isotherm, respectively.

$$\theta/(1 - \theta) = ac(Q/RT) \tag{2}$$

$$\theta/(1 - \theta) = ac\exp[(Q - zW\theta)/RT], \tag{3}$$

where θ is the amount of P sorbed (x) at a P concentration (c) over the maximum sorption capacity (b), R is the gas constant, T is the absolute temperature, Q is the energy derived from the chemical component, and W is the interaction energy of which the net contribution to the total energy of sorption can increase or decrease depending on the number of the nearest neighbors (z) surrounding a central P surface species, and on the fraction of sites covered; a is a coefficient. Equation 3 differs from the Tempkin equation (see below) in that the interaction energy (W) is independent of the chemical energy.

Equation (3) contains only three adjustable parameters if the actual determination of z is not necessary, considering that zW is the interaction energy for all neighbors surrounding a central sorbed P species. Introducing $\alpha = zW$, and $K = a\exp(Q/RT)$, Eq. (3) yields

$$\theta/(1 - \theta) = Kc\exp(-\alpha\theta/RT) \tag{4}$$

Solution for b, K, and α in Eq. (4) was made using a nonlinear least square program (Kuo, 1988), and the accuracy of the estimated parameters b and K by this model, compared with the same parameters estimated by the classical Langmuir isotherm, has been evaluated based on the goodness of fit of the experimental results. Figure 2 demonstrates the tendency of the classical Langmuir isotherm to underestimate considerably the capacity of the soils to sorb P, particularly at higher P concentrations.

Tempkin Equation. This equation is based on the assumption that the bonding energy of adsorption decreases linearly with increasing surface coverage. The

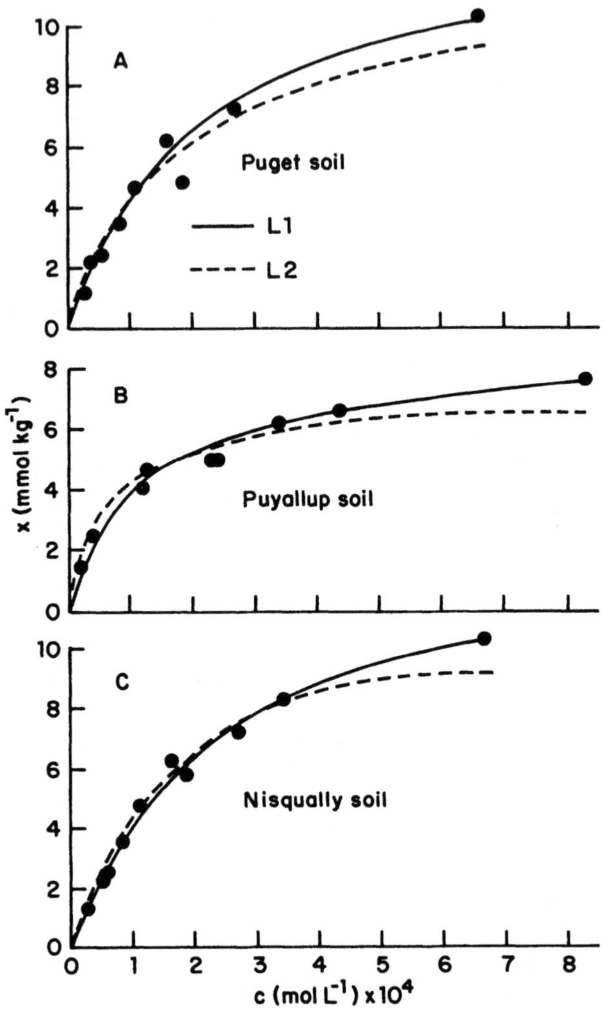

Figure 2. Phosphate sorption (x) as a function of P concentration in solution (c) and the predicted phosphate sorption based on the modified Langmuir isotherm (L1), and the classical Langmuir isotherm (L2). (*Source*: Adapted from Kuo, 1988.)

resulting concentration dependence is quite complex, but for middle range of surface coverages, the equation reads (Bache and Williams, 1971),

$$\frac{x}{b} = \frac{RT}{a} \ln Ac, \tag{5}$$

where A and a are coefficients, R is the universal gas constant, and b is the Langmuir adsorption maximum. According to Eq. (5), a plot of x against $\log c$ should yield a straight line. Such plots for soils in many cases, however,

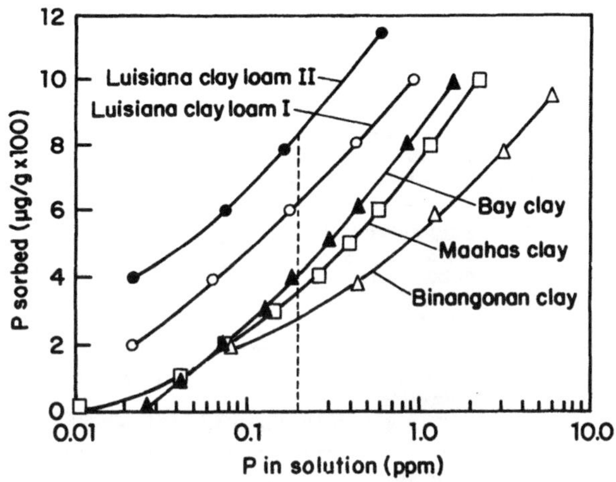

Figure 3. Phosphate sorption isotherm for five Philippine soils (nonreduced). (*Source*: Roy and De Datta, 1985. With permission from Kluwer Academic publishers.)

yielded gentle curves, rather than straight lines (Fox and Kamprath, 1970; Bache and Williams, 1971; Le Mare, 1982; Rao et al., 1983; Roy and De Datta, 1985; Russell et al., 1988), but the agreement with the experimental data was, in general, better over a wider range of concentrations than that with the Langmuir plots. Figure 3 shows the Tempkin's plots for P sorption by five Philippine soils in a study conducted by Roy and De Datta (1985).

For P adsorption in soil, the intensity factor is, in fact, the chemical potential of P in equilibrium solution. The latter is a logarithmic function of P activity in solution. Furthermore, the Tempkin plots enables one to bring into one plot the P adsorption data over a very wide range of concentrations, and also to compare the relative P sorption capacity of various soils (Fig. 3) corresponding to empirically observed equilibrium P concentrations that lead to optimum growth of lowland rice (Roy and De Datta, 1985; Sanyal et al., 1990). As mentioned earlier, this technique of estimating P requirement was used in many investigations of P needs of crops. Indeed, Klages et al. (1988) showed that P fertilizer requirement for dryland winter wheat was predicted more accurately by such methods based on sorption isotherm than by those based on the Olsen test for soil P. The effects of various soil properties on P requirement estimated from sorption isotherms are illustrated in Figure 4.

Some caution, however, must be exercised while using such adsorption isotherms to estimate P fertilizer requirements from the amounts required to raise P in solution to a predetermined level (Kamprath and Watson, 1980; Diamond, 1985). Thus, this approach considers only the intensity factor for solution P while disregarding other factors that affect P uptake by roots (Olsen and Khasawneh, 1980). One example is the buffer capacity of soils and the transport of solution P to the root zone. In addition, this method is of less

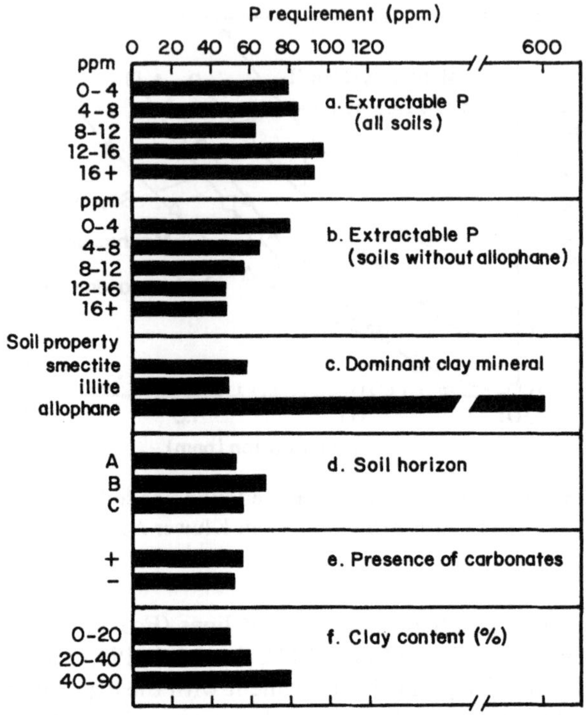

Figure 4. Average isotherm phosphorus requirement of soils as affected by soil properties. (*Source*: Klages N.G., R.A. Olsen and V.A. Haby, Relationship of phosphorus isotherms to $NaHCO_3$ = extractable phosphorus as affected by soil properties. Soil Science 146, 85–91. © by Williams & Wilkens 1988.)

value when the amount of P already present in soils is large relative to the amount adsorbed. As suggested by Barrow (1978), one way of overcoming this problem is to measure the slope of the adsorption isotherm at the required concentration, rather than the amount adsorbed. The slope will be independent of the amount of P originally present. Also, taking the slope provides important information on the P adsorption buffer power of the soil (Holford, 1979, 1988; Bowman and Olsen, 1985) that governs the availability of a plant nutrient. The diffusive mobility of an adsorbable solute is also diminished by high buffer power (Gregory, 1988), thereby affecting adversely the plant availability of the nutrient.

Bache and Williams (1971) measured the slope at 10^{-4} *M* P concentration for their soils, and found that the isotherm slope was strongly correlated with that at 10^{-3} and 10^{-5} *M* concentrations. Bache and Williams (1971) also sought correlations of their isotherm slope, taken as a reference index, with several sorption parameters for 42 soil samples. They established that the Langmuir adsorption maximum, sorption at 10^{-3} M P in solution, and (x/log c) for one addition of 50 μmol P/g soil gave the best correlations. Of these, the first two involve a number of determinations, whereas the third is a single-point method. The latter was therefore suggested as a simple routine index for P sorption.

The P sorptivity of a large number of soil samples from nine orders and 26 great groups of the soil taxonomy was satisfactorily estimated in terms of single-point sorption index of Bache and Williams (1971). Some statistical evidence was presented to prove that the variation in sorption index within a soil series usually is considerably less than the variations in a more diverse collection of soil samples (Burnham and Lopez-Hernandez, 1982; Lopez-Hernandez, 1987). This index was also used by Lajtha and Bloomer (1988) for desert soils.

Freundlich Adsorption Isotherm. Freundlich equation was widely used to describe P adsorption in soils (Mead, 1981; Le Mare, 1982; Polyzopoulos et al., 1985; Torrent, 1987; Shaviv and Shachar, 1989; Buchter et al., 1989). Some authors tend to attach less significance to the Freundlich equation because it does not provide any measure of an "adsorption maximum" in which soil scientists may primarily be interested. Nevertheless, the Freundlich coefficient, k, may be regarded as a hypothetical index of P sorbed from a solution having unit equilibrium concentration. In a recent study on P sorption by a number of acid and acid sulfate soils of South and Southeast Asia, a high degree of correlation was found between the Freundlich k and the Langmuir adsorption maximum, or the Tempkin adsorption parameter (Sanyal et al., 1990).

More importantly, the Freundlich equation, although orginally empirical, was since derived rigorously by Sposito (1980). Moreover, it implies that the affinity (bonding energy) decreases exponentially with increasing surface coverage, a condition which is perhaps nearer to the reality than the assumption of constant bonding energy as in the simple Langmuir equation.

One difficulty of using the simple Freundlich equation is how best to overcome the difficulties of distinguishing the adsorption by soils of added P from that already present, about which a brief mention was made above. Thus, denoting the native P as Q, the Freundlich equation may be written as (Barrow, 1978),

$$x + Q = kc^{1/n} \qquad (n > 1) \tag{6}$$

or

$$x = kc^{1/n} - Q. \tag{7}$$

Deviations that one may expect in the plot of the simple Freundlich equation (e.g., plot of log x against log c) due to a neglect of Q, expecially if the latter is relatively large, were demonstrated by Barrow (1978) using data from an experiment (Barrow, 1974) that studied the effect of previous P additions on adsorption of fresh P by soils (Fig. 5).

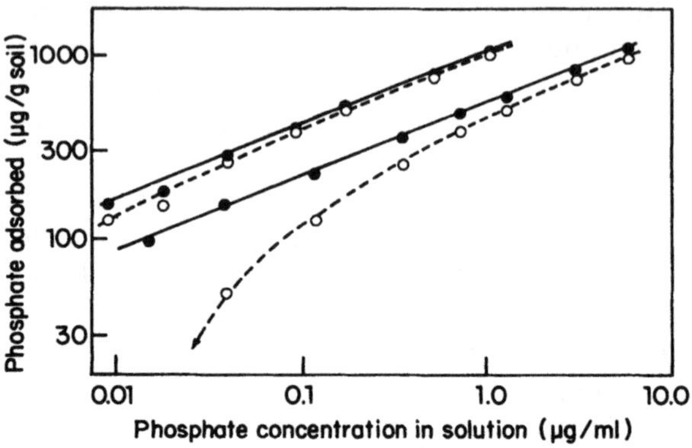

Figure 5. Plot of the data of Barrow (1974) on a double logarithmic scale. Values are for a soil without added phosphate (upper pair of lines) and after incubation at 25°C for 1 year with 400 μg P/g soil (lower pair). Open symbols and broken lines, plots of adsorption (x) against concentration (c) according to Eq. (7) but with (1/n) fixed at 0.4; closed symbols and solid lines, adjusted values of adsorption according to Eq. (6) with $Q = 0$. (*Source*: Barrow, N.J., *Soil Science* 1978 by permission of Oxford University Press.)

Q may be estimated from isotopically exchangeable P (Olsen and Khasawneh, 1980), or its effect on x may be eliminated by taking the derivative (dx/dc) at any concentration independent of Q. Thus, from Eq. (6),

$$\frac{d(x+Q)}{dc} = \frac{dx}{dc} = \frac{d}{dc}(kc^{1/n}). \tag{8}$$

However, Barrow (1978) points out that previous P application affects the slope of adsorption isotherm, presumably by increasing the negative charge on the surface of colloids, thereby affecting x.

In a recent study, Sahrawat and Warren (1989) used the simple Freundlich equation to study the comparative P sorption behavior of Vertisol and Alfisol in India. Vertisol was found to have a higher capacity and buffer power for P adsorption than was Alfisol, implying a lower response of the former to fertilizer P. Vertisol maintained a greater level of dissolved and labile P.

Sibbesen (1981) proposed a modified version of the Freundlich equation, namely

$$x = kc^{(c^{-D/n})}, \tag{9}$$

in which the original shape governing parameter $1/n$ of the Freundlich equation (Eq. 6) was replaced by $(c)^{-D}/n$, D being a coefficient. Out of several sorption equations used (e.g., the simple and two-surface Langmuir equations, extended Freundlich and Langmuir equations, Gunary equation, and Fitter–Sutton equation), Eq. (9) was found to yield the best fit to the experimental data on P sorption by soils. Moreover, coefficients of Eq. (9) were also found to be the least correlated.

In a recent study, Polyzopoulos et al. (1985) demonstrated that P sorption data from 14 representative Alfisols of Greece may be described by the simple and two-surface Langmuir equations, and Freundlich and Tempkin equations, with the Freundlich and the two-surface Langmuir equation proving slightly superior. The Freundlich equation, being simpler, was preferred by the authors.

Ratkowsky (1986) compared statistically seven nonlinear mathematical equations used to describe P sorption by soils, including the Freundlich equation, and its extended form (Sibbesen, 1981), Langmuir equation, two-surface Langmuir and the extended Langmuir equations (Sibbesen, 1981), and Gunary equation (1970). Freundlich and the extended Freundlich equations performed the best, with Gunary equation also having acceptable statistical properties.

Mechanistic Models. An adherence of experimental sorption data to an adsorption isotherm equation does not indicate the actual mechanism of a sorption process in soils. A mechanistic model for describing P sorption is thus often preferred over an empirical equation.

Advances have been made using mechanistic models in the description of P sorption by Al and Fe oxides, and soils. These models were reviewed by Morel et al. (1981) and Sposito (1984). The variable-charge models developed by Bowden et al. (1977, 1980), and extended by Barrow (1983a) described the reaction between divalent phosphate ions and a variable-charge surface, such as goethite (Bowden et al., 1980; Barrow et al., 1980, Posner and Barrow, 1982; Bolan and Barrow, 1984), or between P and soils (Barrow, 1983a). The latter could provide a close description of the effects on sorption of P concentration, pH, time of contact, and temperature. The model explains the apparent lack of reversibility of the P sorption process (see below) in terms of a penetration of the sorbed P into the soil particles, and the continuing slow reaction over a prolonged period of time.

These mechanistic models agreed better with the experimental data and with the observations that earlier led to postulating two or three kinds of P adsorption sites on surfaces, than did the simple or multiple-surface Langmuir equations. The main point of distinction of these models with the simple adsorption isotherms is that in these models, the change of surface charge and potential accompanying P adsorption by soils is given due consideration. Moreover, the distribution pattern of the adsorbed species at the interface region is also considered (Bowden et al., 1977, 1980). However, one disadvantage is the presence of many adjustable parameters in these models. For instance, the variable-charge model of Bowden et al. (1977, 1980) contains seven adjustable parameters.

The constant capacitance model of Stumm et al. (1980), as applied by Goldberg and Sposito (1984a,b), requires three assigned values for protonation and deprotonation of surface hydroxyl (OH) groups and capacitance, as well as three adjustable variables for describing pH-dependent P sorption by oxide minerals. This model was also capable of predicting the pH-dependence of P sorption and the P surface species in oxide minerals and in soils (Goldberg and Sposito, 1984a,b). However, these models are not often suited to the estimation of P sorption capacities of soils. The maximum P sorption has to be assigned (Bowden et al., 1980) or obtained by extrapolation (Goldberg and Sposito, 1984a,b). In contrast, the classical Langmuir, Tempkin, or Freundlich isotherms, each with only two parameters, are simple and have been extensively used in describing P sorption by soils in terms of some relative sorption parameters.

b. Factors Affecting Phosphate Adsorption

Many soil properties influence P adsorption by soils, soil minerals, and sediments. These include the nature and amount of soil components (e.g. clay, organic matter, and hydrous oxides of iron and aluminum); background electrolyte—its concentration, and valency of the constituent cation; and pH of the adsorption system.

Iron and Aluminum Oxides and Hydroxides. Of the soil properties tested, acid ammonium oxalate extractable (amorphous) iron and aluminum proved to be an important criterion for P adsorption in several soils (Juo, 1981; Borggaard, 1983; Araki et al., 1986; Loganathan et al., 1987; Adams et al., 1987; Wada et al., 1989; Buchter et al., 1989). Oxalate extraction is known to dissolve amorphous and poorly crystalline oxides of iron and aluminum while having little or no effect on crystalline iron and aluminum minerals. Crystalline iron oxides are relatively inactive in P sorption (Ryden and Pratt, 1980). The difference in P adsorption capacity (PAC) between fresh soil, and soil after extraction with oxalate showed that a high proportion of P adsorption by soils from 11 horizons of an acidic soil was attributable to poorly ordered minerals (Adams et al., 1987). Ethylenediaminetetraacetate (EDTA) which is also known to dissolve amorphous iron oxides, has also been used in place of oxalate, and the decrease in PAC following EDTA extraction of several soils from Denmark and Tanzania were attributed to the removal of amorphous iron oxides (Borggaard, 1983). Figure 6 demonstrates the close correlation between such a decrease in the PAC of soils (relative to acetate-extracted soil as control), subjected to EDTA extraction, and EDTA-extractable iron.

The role of oxides and hydroxides of iron and aluminum in retaining inorganic P, when the latter was added in concentrations greater than those present in interstitial waters, was also emphasized by several authors (Ryden and Pratt, 1980). Notably, short-range (amorphous) iron oxide gels, in general, sorb 10–100 times more P than do crystalline iron and aluminum hydrous oxides (e.g., hematite and gibbsite), and approaching 1000 times more than do crystalline aluminosilicates and calcium carbonate (Ryden and Pratt, 1980).

Clay. A significant correlation of P sorption parameters with clay content has been reported by several workers (Fox and Kamprath, 1970; Ayodele and

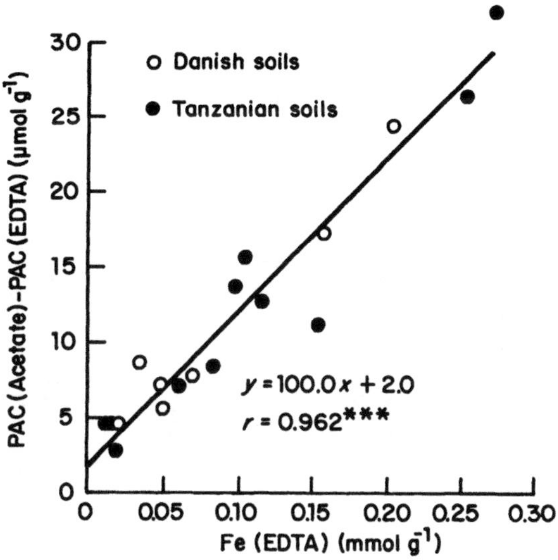

Figure 6. The decrease in phosphate adsorption capacity (PAC) following EDTA extraction of soils as a function of EDTA-extractable iron (amorphous iron oxides). x denotes Fe (EDTA), y denotes PAC (acetate)—PAC (EDTA), r denotes the simple linear correlation coefficient, whereas *** denotes statistical significance at 0.1% level. (*Source*: Borggaard, 1983.)

Agboola, 1981; Dolui and Gangopadhyay, 1984; Loganathan et al., 1987; Solis and Torrent, 1989; Wada et al., 1989; Bennoah and Acquaye, 1989), and this may be a mere reflection of the effect of specific surface area on P adsorption. Clays, particularly (1 : 1) lattice clays, may contribute to P sorption in tropical soils, especially at low pH, when the activity of iron and aluminum is also expected to be higher. Table 2 shows the simple linear correlation coefficients between measured P sorption parameters and several soil properties for acid and acid sulfate soils of South and Southeast Asia with high P-fixing capacity (Sanyal et al., 1990). A high degree of correlation between the Langmuir adsorption maximum or Freundlich k and the clay content, organic carbon, and cation exchange capacity of the soils was observed. Also, significant correlation was obtained with "active iron" (free iron oxide) content of the soils. Freundlich k, but not Langmuir maximum, was correlated with the exchangeable aluminum content.

Hydrous oxides of iron and aluminum have been found to occur as fine coatings on surfaces of clay minerals in soil (Greenland et al., 1968; Haynes, 1983). These coatings, characterized by large surface areas, hold an appreciable quantity of P, thereby implying a secondary role of crystalline aluminosilicates (clays) in P sorption (Ryden and Pratt, 1980).

Agboola and Ayodele (1983), however, reported no correlation between P sorption maximum and clay content of soils.

Table 2. Simple linear correlation coefficients at 25°C between phosphate sorption–desorption parameters and relevant soil properties of some acid and acid sulfate soils of South and Southeast Asia

Sorption parameters	Soil properties: pH	Clay	Organic C	Active Fe	CEC	Exchangeable Al
Freundlich k	−0.803***	0.868***	0.866***	0.563*	0.838***	0.675*
Langmuir adsorption maximum	−0.701**	0.850***	0.730**	0.687**	0.848***	0.507 NS
Bonding energy	−0.640*	0.726**	0.786**	0.183 NS	0.614*	0.636*
P sorption at 0.12 ppm in solution	−0.806**	0.714*	0.659**	0.607*	0.700*	0.565*
Adsorption rate coefficient (ka) (at 35°C)	−0.810***	0.920***	0.773**	0.569*	0.828***	0.628*
Desorption rate constants (at 35°C)						
k'_d	0.604 NS	−0.489 NS	−0.667*	−0.224 NS	−0.603 NS	−0.593 NS
k''_d	0.779*	−0.828**	−0.563 NS	−0.388 NS	−0.769*	−0.475 NS

*, **, *** indicate statistical significance at 5%, 1%, and 0.1% level, respectively.

NS, not significant.

Source: Sanyal et al., 1990.

pH and Supporting Electrolytes. A strongly negative correlation noted between Freundlich *k* or Langmuir maximum, and soil pH (Table 2) agrees with several observations (Adams and Odom, 1985; Bolan et al., 1988; Mehadi and Taylor, 1988), and corroborates with what has been stated regarding the activity of iron and aluminum at low pH.

However, an opposite effect of pH had also been observed on P sorption in soils. Thus, a considerable controversy regarding the pH effect exists (Sanchez and Uehara, 1980; Probert, 1980; Haynes, 1982, 1984; Barrow, 1985, 1987). A decrease in adsorption with pH was reported when dilute solutions of NaCl or KCl was used to measure P adsorption (Barrow, 1984). But when dilute $CaCl_2$ solution was used, P retention by a limed soil increased (Amarasiri and Olsen, 1973; Barrow, 1984; Naidu et al., 1990).

It was suggested that the freshly precipitated iron and aluminum oxides (at pH 6.5) were responsible for the increased P sorption. Phosphorus adsorption on goethite was also found to be less dependent on pH when $CaCl_2$ rather than NaCl was used as the indifferent electrolyte (Parfitt, 1978). Mokwunye (1975) reported a positive effect of pH on P retention by some soils from the savannah zones of Nigeria. He surmised that this could be associated with the P retention by an activated hydroxy–aluminum species (at pH 5.0) that held P through the surface-exchange of a hydroxyl group. Naidu et al. (1990) reported an increase in P sorption by strongly acidic soils on liming after an initial fall, and beyond pH 5.5–6.0. They attributed this rise to the formation of insoluble Ca–P compounds. Smillie et al. (1987) also suggested a chemical association between sorbed P and Ca in soils, having adequate exchangeable Ca, as an important P retention mechanism.

According to Barrow (1984, 1987), these observations may be explained by keeping in view the effect of pH on the P ionic species that is adsorbed, as well as the charge and electrostatic potential of the variable-charge surfaces that retain P in soil. These effects are also dependent on the nature of the supporting electrolyte, thus complicating the issue even more.

With an increase in pH, the charge and the electrostatic potential of positive sites on variable-charge materials in soil decrease, causing a fall in P retention. However, at higher pH (>5.0), the concentration of deprotonated $H_2PO_4^-$ (i.e., of $HPO_4^=$ ion) increases, and the existing theories (Bowden et al., 1980; Barrow, 1984, 1987) treat $HPO_4^=$ ion as the dominant adsorbable species (e.g., on goethite). This encourages the P adsorption by goethite at higher pH (up to a pH of 7.0, which is the value of pk_2 of H_3PO_4). Thus, P adsorption by goethite (Bowden et al., 1980) and amorphous hydroxy-Al surfaces (Kwong et al., 1979) decrease relatively slowly up to pH 7.0, and rapidly decreases beyond pH 7.0.

The net result of these opposing tendencies is usually governed by the nature of the supporting electrolyte in which pH effects on P adsorption are measured. Electrolyte concentration and valency of cation affect the electrostatic potential on the surface. An increase in concentration of the electrolyte, especially if the cation is polyvalent, renders the potential on the

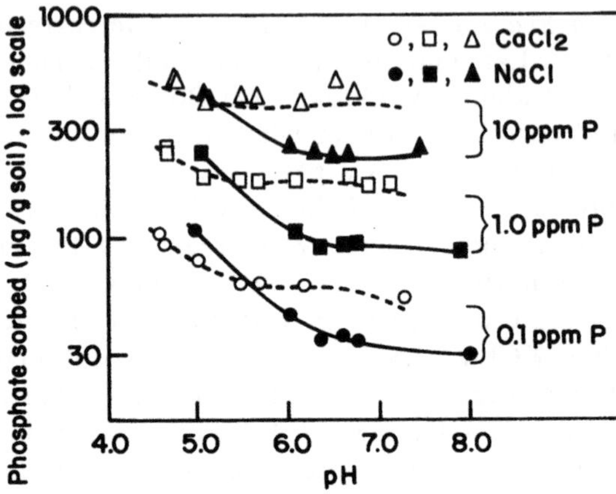

Figure 7. Effect of background electrolyte on phosphate sorption at the indicated concentrations in solution. (*Source*: Barrow, N.J., 1987. With permission from Kluwer academic publishers.)

surface less negative (Bowden et al., 1980; Barrow, 1987), and the effect of higher pH in bringing down P adsorption is less marked. This also causes desorption of the previously added/adsorbed P to decrease (Barrow and Shaw, 1979).

For a dilute solution of NaCl or KCl, however, the effect of diminishing potential at higher pH dominates, and P sorption decreases with pH. Figure 7 illustrates the effect of background electrolyte on P sorption at varying pH. In the case of a divalent cation exhibiting specific adsorption (e.g., Zn^{2+} or Ca^{2+}) on the surface, the fall of P adsorption with pH is even more gradual (Barrow et al., 1980). As noted by Haynes (1984), liming may partially offset the inverse pH effect on P adsorption by soils by increasing the Ca^{2+} concentration and ionic strength of the soil solution.

Organic Matter. Several authors have reported significant positive correlations between soil organic matter content and P sorption (Singh and Tabatabai, 1977; Mizota et al., 1982; Bennoah and Acquaye, 1989; Sanyal et al., 1990). The role of organic matter in augmenting P sorption in soil has often been attributed to the association with and possible stabilization of the soil organic matter by the "free" sesquioxides. Thus, Wada and Gunjigake (1979) observed that P adsorption by volcanic soils was correlated with organically bound aluminum, and to a lesser extent, with iron extracted by sodium pyrophosphate.

The possibility that a gel complex (of the type proposed by Mattson et al., 1950) consisting largely of hydrated iron oxide along with a smaller amount of organic matter, aluminum, associated $Si(OH)_4$, and inorganic phosphates is the major contributor to P sorption by soils and lake sediments had been considered by some workers (Saunders, 1965). The results of Bennoah and Acquaye (1989) also seemed to indicate that iron and aluminum intimately associated with organic matter can sorb much more P than can the same amount of free Fe_2O_3 and Al_2O_3. Harter (1969), however, disagreed with the idea that organic matter and P were adsorbed in soil by the same mechanism. He suggested that P may even be directly bonded to organic matter by replacing the organic hydroxyl groups.

Vijayachandran and Harter (1975) also proposed that the anion exchange sites on soil organic matter were responsible for the correlation between organic carbon and the P sorption maximum.

Reduction of P sorption by organic matter in soils has also been observed (Yuan, 1980; Sibanda and Young, 1986; Anderegg and Naylor, 1988). This can be explained by a possible competitive action between P and organic matter for sorption sites on, for instance, hydrous oxides of iron and aluminum. Earl et al. (1979) noted that citrate, tartrate, and acetate were effective (in this order) in reducing P sorption by soils, synthetic iron, and aluminum oxide gels in contrast with the behavior of inorganic anions. Thus, Ryden et al. (1987) have shown that except for OH^- ion, inorganic anions have a limited ability to compete with P for sites on hydrous ferric oxide gel.

Yuan (1980) also supported the idea that some of the adsorption sites for P and organic matter in soils are common. This leads to a competitive

Table 3. Correlation between clay content and phosphate adsorption by some acid and acid sulfate soils of South and Southeast Asia

Location of soil sample	Clay (%)	Langmuir adsorption maximum/ % clay
Quezon, Philippines	59	29.5
Bentre, Vietnam	54	20.3
My Trung, Vietnam	50	33.4
Port Blair, India	46	38.6
Cox Bazar acid sulfate soil, Bangladesh	37	48.6
	$r =$	-0.863 NS
Pathumthani, Thailand	54	22.4
Sukamandi, Indonesia	41	25.0
Satkhira, Bangladesh	30	24.2
Giridih, India	25	33.0
Muktagacha, Bangladesh	20	30.2
Jamirdia Bhaluka, Bangladesh	19	41.5
Anandapur, India	16	38.3
	$r =$	-0.814*

Note: Abbreviation as in Table 2. r is the simple linear correlation coefficient and * denotes statistical significance at 5% level.

Source: Sanyal et al., 1990.

effect. In agreement with this, Sibanda and Young (1986) demonstrated that humic acid and fulvic acid competed strongly with P for adsorption sites on goethite and gibbsite at low pH values. Further, soils of their study also showed a similar effect with a reduction in P adsorption resulting from the adsorption of humic acid at the pH of soils. The effect was attributed to the strong reactions between humic or fulvic acid and hydrous oxides of Fe and Al. Evans (1985) found that phytic acid, present in soil solutions of a coarse-textured soil, was capable of strongly reducing P sorption by soils. Identification of phytic acid in soil solution suggests that the phytic acid inhibition of P adsorption may accelerate P leaching in coarse-textured soil.

In a study conducted by Sanyal et al. (1990), P sorption per unit weight of clay or organic matter increased with a decrease in clay or soil organic matter content. Tables 3 and 4 show the Langmuir adsorption maximum values per unit weight of clay or organic carbon. Trends became clearer when the strongly P-fixing soils were considered separately from the moderately and weakly sorbing soils (Tables 3 and 4). Fox and Kamprath (1970) observed this trend for clay, and proposed an easier access of P to clay surfaces when the clay is dispersed through media like sand. It is possible that intimate clay–organic complexes, formed through cationic bridges, render some of the active surfaces of both the colloidal components inaccessible for P sorption.

Table 4. Correlation between organic carbon content and phosphate adsorption by some acid and acid sulfate soils of South and Southeast Asia

Location of soil sample	Organic C (%)	Langmuir adsorption maximum/ % organic C
My Trung, Vietnam	3.87	432
Bentre, Vietnam	2.95	371
Cox Bazar acid sulfate soil, Bangladesh	2.10	856
Quezon, Philippines	2.09	832
Port Blair, India	1.76	1009
	$r = -0.891^{*}$	
Satkhira, Bangladesh	1.22	596
Sukamandi, Indonesia	0.888	1153
Muktagacha, Bangladesh	0.778	766
Jamirdia Bhaluka, Bangladesh	0.713	1107
Pathumthani, Thailand	0.667	1817
Anandapur, India	0.240	2554
Giridih, India	0.194	4247
	$r = -0.876^{**}$	

Note: Significance of *r* value and asterisk as in Table 4. ** denotes statistical significance at 1% level.

Source: Sanyal et al., 1990.

Free Iron. A correlation between initial P adsorption and free iron oxides was reported by Ayodele and Agboola (1981), Pena and Torrent (1984), Torrent (1987), Mehadi and Taylor (1988), Manikandan and Sastry (1988), and Solis and Torrent (1989). However, Vijayachandran and Harter (1975) noted no such correlation for six great soil groups in the U.S.A. Sanyal et al. (1990) obtained significant correlation of free iron oxide (active iron) with the Langmuir adsorption maximum and the Freundlich k (Table 2). The bonding energy, on the other hand, was not correlated. It was also less strongly correlated with the clay and the organic matter contents of the soils as compared with the above two indexes of P sorption (Table 2). This perhaps indicates that the amount of P sorbed is a function of the number of sorption sites rather than of their bonding energy.

c. Phosphate Adsorption by Amorphous Soils

Allophanic soils, such as Andepts, are known to bind large amounts of P. Gebhardt and Coleman (1974) reported on the P sorption capacities of some Andepts from Mexico and Hawaii, which ranged from 300 to 700 μmol/g, whereas a value of the order of 50 μmol/g was reported for Oxisols (Parfitt, 1977). Evidently, such high adsorption capacities are related to large specific surface areas of these soils.

Wada (1985) suggested that the aluminum- and iron-humus complexes, allophane, imogolite, and related soil materials are related to P sorption. The pH dependence of P sorption was quite appreciable among the Andepts containing allophane and imogolite, whereas it was less so among those containing iron- and aluminum-humus complexes for which P concentration in solution was the governing factor (Gunjigake and Wada, 1981).

Phosphate retention by allophanic soils was also shown to increase with the degree of weathering (Fox, 1974). In Andepts from New Guinea, allophane, aluminum hydrous oxides, and humus-complexed iron were said to be responsible for providing sorption sites for P (Moody and Radcliffe, 1986).

The relative P sorption power of allophane was compared with that of inorganic components, containing iron and aluminum, by McLaughlin et al. (1981). The P sorbed decreased in the order: allophane > fresh Al oxide gel > Fe oxide gel > pseudoboehmite > aged Al oxide gel > dried Fe oxide gel > Fe-coated kaolinite > hematite > goethite > akaganeite > gibbsite = ground kaolinite > dispersed kaolinite. The shape of the adsorption isotherm (Freundlich) was similar for each adsorbent, and the differences in the extent of P adsorption were attributed to the number of functional groups, M-OH (M = Fe or Al), exposed to the soil/solution interface.

d. Phosphate Desorption

Desorption of once-adsorbed P from soils and clays had often been shown to be irreversible leading to a large hysteresis effect (Madrid and Posner, 1979; Olsen and Khasawneh, 1980; Okajima et al., 1983; Mouat, 1983). The desorption isotherm was thus displaced to the left of the sorption isotherm. Such hysteresis effect leads to an overestimation of the replenishing ability of soils to supply P to the soil solution, when P sorption isotherms are used for the purpose (Okajima et al., 1983). Although Munns and Fox (1976) and Madrid and Posner (1979) found that the resulting degree of hysteresis decreases with longer time allowed after P addition, Ryden and Syers (1977) found the irreversibility of P desorption from soils and hydrous ferric oxide gel to increase with increasing time above 30 h. The predominant view at present is to treat such irreversibility as arising from the incomplete attainment of equilibrium during the slower reaction phase of adsorption (Okajima et al., 1983; Barrow, 1983b, 1985).

Moreover, it has also been suggested that because adsorption equilibrium was slow, an apparent readsorption during the desorption step is possible (Barrow, 1983b). The diffusive migration of initially adsorbed P beneath the adsorbing surface has also been cited as a probable reason for apparent irreversibility of P desorption (Barrow, 1983b, 1985). In the event of the latter happening, a part of the adsorbed P would no longer be in equilibrium with the solution, hence, the irreversibility. Indeed, both sorption and desorption continue for long periods, although the rate of change may become too slow (Barrow, 1979a; Bache and Ireland, 1980). Thus, it was shown that the plots of desorbed P against concentration were continuous with those of newly adsorbed P, but that the plots for originally added P did not coincide with either of these (Barrow, 1983b). Madrid and Posner (1979) also demonstrated that when the total of adsorption and desorption time is long enough, both adsorption and desorption points tend to lie on a single curve that corresponds to the isotherm calculated according to the Stern model of double layer theory.

From goethite, however, most of the adsorbed P was found to be isotopically exchangeable implying that the adsorbed P was in equilibrium with the solution phase (Atkinson et al., 1972).

In a recent study, desorption of P adsorbed by five limed soils was followed (Le Mare and Leon, 1989), and the labile P pool was found to be enriched by exchangeable P released on two desorptions. Such exchangeable P was strongly correlated with oxalate-extractable aluminum. Liming was also found to decrease the buffer power for desorbing P in the soils, thereby causing a specific amount of exchangeable P to maintain a higher concentration in solution.

Kuo et al. (1988) observed that recovery of sorbed P from several soils, having contrasting properties, by using $NaHCO_3$, NaOAc or $NH_4F{-}HCl$ extraction, was strongly dependent on P sorption capacity, but not on the buffer capacity of the soils or the bonding energy estimated by the application of Langmuir equation.

2. Kinetics of the Processes

a. Kinetic Equations

The study of the kinetics of P sorption and release by soils is of considerable interest in soil and environmental science. The time factor is certainly of relevance for P uptake by plants. However, comparisons among soils on the basis of rate constants alone do not seem to be of great practical value, since amounts of P desorbing during any time interval would also depend on the reserve of desorbable P present (Pavlatou and Polyzopoulos, 1988). On the other hand, a better understanding of the energetics of P sorption, based on kinetic studies, may help elucidate mechanisms of P adsorption–desorption in soils.

The reaction between P and soils is rapid at first. It then becomes slow, and continues for a very long time (Barrow, 1983c; Bolan et al., 1985; Al-Khateeb et al., 1986). In many cases, it is doubtful whether a true equilibrium is reached within a practicable reaction period, although an apparent equilibrium is possible within a reasonable period. This has led several authors to use a variety of equilibrium isotherms (reviewed above) to describe P adsorption and desorption by soils.

A number of kinetic equations have been used by several investigators and presented in Table 5. Extensive (Barrow, 1983c; Sparks, 1986), as well as brief (Bolan et al., 1985; Pavlatou and Polyzopoulos, 1988) reviews of these kinetic equations, used to describe P sorption and desorption rates, have been published. Broadly, two approaches have been employed in these kinetic studies. In one, data have been fitted to some selected standard kinetic equations, whereas in the second, some empirical equations (e.g., modified Freundlich and Elovich equations) have been used (Table 5).

As an experimental technique, the anion-exchange resin-extractable P from soils was widely used to provide a measure of the rate of P release in soils as well as index of quantity of available P. However, Dalal (1985) suggested that P desorbed from soils by cation-anion-exchange resin (CAER-P) may be a better measure of available P than that by anion exchange resin due to the capacity of the former to take up cations as well as P, thus better simulating cations and P uptake by plants.

In many studies (reviewed by Sparks, 1986), the experimental data were inadequately described by a single first-order kinetic reaction, and were often interpreted as a combination of two or three simultaneous first-order reactions, corresponding to conceptually distinct transformations of different forms of Ca–P compounds (in soil) having varying degrees of crystallinites and solubilities.

The temperature dependence of P sorption and release by several soils and soil minerals has been found to be generally small in these earlier studies, leading to low activation energies of the processes. This tends to show that adsorption–desorption processes are diffusion-controlled (Sparks, 1986).

Table 5. Summary of kinetic equations used for phosphate reaction with soil constituents

Equation	Assumptions
First order kinetics $\log C = \log C_0 - Kt$	Rate of change in concentration is proportional either to the concentration in solution or to the number of empty sites
Second order kinetics $(1/C_0) - (1/C) = Kt$	Rate of change in concentration is proportional to both the concentration in solution and the number of empty sites
Diffusion equation $X = R\sqrt{t} + b$	Rate limiting step is the diffusion of phosphate ions either from the solution to the surface or from the surface to the interior of the particle
Modified Langmuir equation $X = b_1 + \Delta b_1 + \frac{b_{11}K_{11}C}{1 + K_{11}C} + \frac{b_{111}K_{111}C}{1 + K_{111}C}$	Rate of adsorption is proportional to the concentration in the solution and the number of empty sites: with time, adsorbed phosphate redistribution in region 1
Modified Freundlich equation[a] $X = KC^b t^{b_1}$	Phosphate reaction in soil system contained three compartments, A, B, and C and reacts according to $A \rightleftharpoons B \rightleftharpoons C$: rate limiting step is $B \rightarrow C$
Elovich equation $X = (1/\beta)\ln(\alpha\beta) + (1/\beta)\ln t$	Activation energy of adsorptions increases linearly with surface coverage

Note: Where X is the amount of phosphate sorbed, C_0 and C are the initial and final concentrations of phosphate in solution, t is the time, and all others are parameters.

Source: Adapted from Bolan et al., 1985.

[a] The two-constant kinetic equation, due to Kuo and Lotse (1974a), also comes under this category.

More recently, a low heat of reaction for P desorption from acid soils has been reported (Chien et al., 1982). Also, Sharpley and Ahuja (1983) further demonstrated that P desorption from soils can be described by a diffusion-controlled mechanism in which the amount desorbed in a given time was linearly related to the initial desorbable P on soil (Sharpley et al., 1981; Sharpley and Ahuja, 1983).

Barrow and Shaw (1975a) and Barrow (1979b), on the other hand, reported a much higher activation energy and a surface-reaction as the controlling step, for P adsorption by soils by fitting their experimental data to an empirical equation (modified Freundlich equation in Table 5) they developed. These authors pointed out an apparent discrepancy in an earlier treatment (Kuo and Lotse, 1974a, b) in identifying the temperature-sensitive parameter, and attributed the low activation energy to this. Torrent (1987) and Mendoza and Barrow (1987a) successfully used Barrow and Shaw's (1975a) equation to describe the rapid and slow P sorption by the Mediterranean soil clays and the Argentinian soils, respectively. Recently, Kato and Owa (1989) also found an equation of the same type to describe P sorption kinetics well in soils of their study.

The kinetics and heats of adsorption for interaction between P and sodium (Na)-saturated fractions of allophanic-rich soils were investigated by Imai et al. (1981). The findings were interpreted in terms of three simultaneous reactions, namely, an "instantaneous" adsorption due to exposed sites, and two inverse exponential rates of adsorption on internal and freshly forming external sites.

A parabolic diffusion equation has also been used to fit the P desorption data in soils. The equation reads (Sparks, 1986)

$$\frac{1}{t}\left(\frac{C_t}{C_\infty}\right) = \left(\frac{4}{\pi^{1/2}}\right)\left(\frac{D}{r^2}\right)^{1/2}\left(\frac{1}{t^{1/2}}\right) - \frac{D}{r^2}, \tag{10}$$

where C_t = amount of ion adsorbed at time t, C_∞ = amount of ion adsorbed at equilibrium, r = average radius of soil or clay particle, t = time, and D = diffusion coefficient.

Eq. (10) may also be transformed into a simple form, Eq. (11)

$$\frac{C_t}{C_\infty} = Rt^{1/2} + \text{constant}, \tag{11}$$

where R is the overall diffusion coefficient.

The form of Eq. (11) is, in fact, identical with the diffusion equation given in Table 5, as C_∞ for a given experiment is constant.

A square-root relation of the type of Eq. (11) was also used by Tambe and Savant (1978) and Novak and Petschauer (1979) to denote the effect of time on P sorption. Bolan et al. (1985) argued that in either case, the source concentration varies with time, and hence the simple diffusion kinetics leading to a linear dependence of P sorption on square root of time (e.g., Eq. 11) should not be expected to yield a good fit to the experimental data.

The Elovich (or the Roginsky–Zeldovich) equation or its modified forms often have been used to treat the kinetics of P adsorption and release in soils, and soil minerals (Chien and Clayton, 1980; Chien et al., 1980a; Hingston, 1982; Mouat, 1983). Chien and Clayton (1980) have shown that the Elovich equation can yield a simple linear plot between the amount adsorbed or desorbed (q) and time (on a logarithmic scale, see below) in cases where the experimental data fail to conform to a simple first-order kinetic equation and are needed to be explained in terms of a combination of two or three simultaneous first-order reactions (Sparks, 1986). This form of Elovich equation is

$$\frac{dq}{dt} = \alpha \exp(-\beta q), \tag{12}$$

which, on transformations (with the assumption that $\alpha\beta t \gg 1$), becomes

$$q = \frac{1}{\beta}\ln(\alpha\beta) + \frac{1}{\beta}\ln t, \tag{13}$$

where q is the amount adsorbed in time t, and α and β are constants for a given sorption process. α may be regarded as the initial reaction rate because $dq/dt \to \alpha$ as $q \to 0$, that is, a rapid adsorption rate not governed by the exponential law. A plot of q versus *ln t* should yield a straight line, the slope and intercept of which lead to α and β. On replacing q in Eq. (13) by $(C_0 - C_t)$, where C_0 and C_t are the P concentrations at the initial time and at any arbitrary time (t), one obtains (Sparks, 1986),

$$C_0 - C_t = \frac{1}{\beta}\ln(\alpha\beta) + \frac{1}{\beta}\ln t \tag{14}$$

with the initial condition that $C_0 - C_t = 0$ at $t = 0$. Figure 8 gives the plots of data of Ryden et al. (1977b) on P sorption by soils according to the first-order kinetics ($\ln C = \ln C_0 - kt$) and Eq. (14). Obviously, the deviations of the plots from the simple first-order kinetics disappear in the plots according to the Elovich equation. It was thus not necessary to postulate different regions (e.g., regions I, II, and III) for the overall adsorption isotherm (Ryden et al., 1977b).

According to Chien and Clayton (1980), the value of the constant β is a function of the P source and P adsorbent, whereas the degree of dependence of α on soil types varies with the type of the process investigated. The value of α for dissolution of phosphate rocks is independent of rock sources in a given soil, although its variations were observed with different soil types treated with a given phosphate rock (Chien et al., 1980a). The constants α and β render a comparison possible between the P sorption or release rates in different soils. A decrease in β and/or an increase in α would enhance the reaction rate (Eq. 12) as demonstrated by Chien and Clayton (1980). When applying the Elovich equation, however, it is necessary to obtain accurate sorption–desorption data at short intervals (Havlin and Westfall, 1984).

The modified Elovich equation (Eq. 14) was used by other investigators to describe the kinetics of P adsorption and desorption in soil (Ayodele and Agboola, 1981; Onken and Matheson, 1982). Ayodele and Agboola (1981), while studying P fixation capacity of tropical Savannah soils of Nigeria, observed that clay and organic matter contents of the soils were most active during the initial rapid phase of P fixation. They further noted that up to 1 day of reaction time, the clay content was most highly correlated with P sorption, while as the time progressed, the influence of clay and organic matter declined and that of iron and aluminum oxides became more significant. In the kinetic study conducted by Sanyal et al. (1990), the sorption rate coefficient (ka) for P adsorption by some acid and acid sulfate soils was also found to be

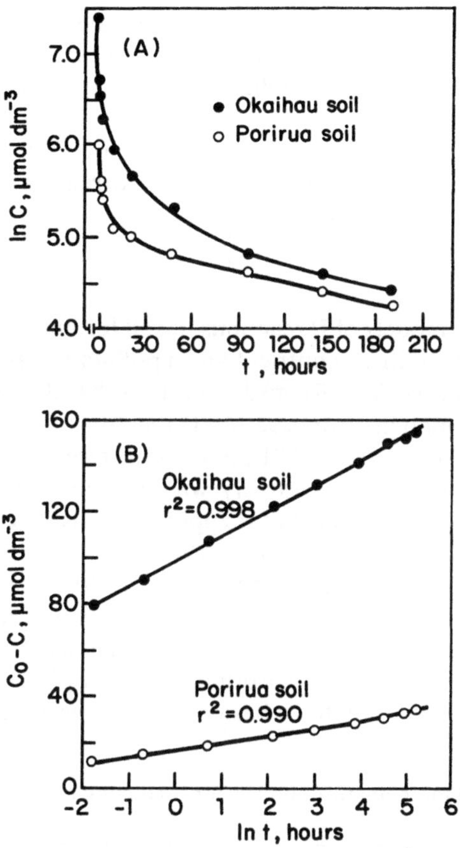

Figure 8. A, Plot of first-order kinetic equation for phosphate sorption in two soils; B, Plot of Elovich equation for phosphate sorption in two soils. Data for A and B from Ryden et al. (1977b). (*Source*: Adapted from Chien and Clayton, 1980.)

strongly dependent on clay, and progressively less so on organic matter and on the free iron oxide content of the soils (Table 2). Because the kinetic experiments were continued up to 21 h, these correlations are considered to agree with the observations of Ayodele and Agboola (1981).

Polyzopoulos et al. (1986) recently questioned the applicability of the simplified Elovich equation in describing the kinetics of P sorption in soils. They identified situations wherein an appreciable part of the P is sorbed or released at rates different from those characterizing the sorption or release of the remaining part for which the simple Elovich equation may be applied. In these situations, comparison among soils on the basis of β values does not seem possible. Barrow (1983c) also demonstrated that the slopes ($1/\beta$) of Elovich plots vary with the level of P addition and lead to apparently different sorption characteristics for the same soil at different initial additions of P.

Enfield et al. (1976) fitted the P adsorption kinetic data to five kinetic models, namely, a linearized first-order sorption, a first-order Freundlich sorption, an empirical function, a diffusion-limited Langmuir sorption, and a diffusion-limited Freundlich sorption. The best fit was obtained for a diffusion-limited model based on the Langmuir or Freundlich equation. The regression coefficients of the models were not correlated with physical and chemical properties of the soils.

b. Kinetic Models

The need for kinetic models of P reactions in soil for modeling transport and transformations of P, both as a nutrient and a pollutant, has been emphasized by various investigators (Van Riemsdijk et al., 1984; Barrow, 1985; Sparks, 1986). The various models proposed for P reactions in soils have been reviewed by Mansell and Selim (1981), and more recently by Sparks (1986). Mansell and Selim (1981) dealt largely with the mathematical models for predicting reactions and transport of P applied to soils. Their classification of these models is given in Table 6 (Sparks, 1986).

Enfield and Ellis (cited in Shaviv and Sachar, 1989) classified soil inorganic P into three groups: those containing iron and aluminum P, those containing calcium P, and those combining with silicate minerals. They also proposed an approximate correlation of P sorption, with soil pH according to which Ca–P compounds predominate in basic soils. For various calcareous soils,

Table 6. Summary of mathematical models for predicting phosphorus reactions in soil

Type of model
I. Mathematical models that assume chemical nonequilibrium
a. Transport models that assume reversible kinetic reactions for applied phosphate
b. Transport models that assume irreversible kinetic reactions for applied phosphate
c. Transport models that assume both reversible and irreversible reactions for applied phosphate
d. Nontransport sorption models that assume both reversible and irreversible kinetic reactions for applied phosphate
II. Mathematical models that assume phosphate removal from solution to occur simultaneously by equilibrium and nonequilibrium reactions
a. Transport models that assume two types of phosphate sorption sites
III. Mechanistic multiphase models for reactions and transport of phosphate applied to soils

Source: Reprinted with permission from Sparks, D.L., 1986. Kinetics of reactions in pure and in mixed systems. Soil Physical Chemistry. © CRC Press, Inc. Boca Raton, Florida.

Shaviv and Shachar (1989) showed that the rate constants obtained for each of the kinetic reactions proposed were similar. This implies that the time-dependent reactions in calcareous soils can be modeled by considering the precipitation of Ca–P compounds in an environment in which conditions are dominated by the Ca–P–CO_2 system.

Lin et al. (1983a,b, 1986) developed a multifactor kinetic model involving a reversible two-step reaction mechanism between the solution and nonlabile P phase via the labile phase. The model was used to simulate experimental results obtained on reactions of P with minerals in acidic soils under a variety of conditions determined by three factors—solution pH, solution concentration, and mineral-specific surface area. The experimental data as well as the kinetic models showed that high pH values, low concentrations of P in the reacting solution, and small specific surface area will reduce retention of P. For the minerals studied in this experiment, gibbsite was found to be most sensitive to variations of these three factors. Goethite was found to rank next in sensitivity, although its P retention capacity was less than that of kaolinite, due presumably to its greater tendency for desorption.

A simplified soil and plant P model by Jones et al. (1984a,b) and Sharpley et al. (1984) was designed to provide long-term estimates of labile and organic P, their changes for various soils under different management practices, and also of crop response to soil test P and fertilizer P requirements for several crops and soils. The model has been tested for calcareous and highly weathered and tropical soils. Thus, accurate simulations were obtained of P transformations, during 60 years of cultivation of calcareous Houston Black clay, a fine, montmorillonitic, thermic Udic Pellustert, and yield of maize (*Zea mays* L.) on two highly weathered Hawaiian soils receiving varying amounts of fertilizer P (Sharpley et al., 1989).

c. Slow Reactions

As briefly mentioned earlier, the reaction between P and soil continues for a long period, and it is often doubtful whether the reaction ever reaches equilibrium. These slower reactions may not be useful for indexes of labile P and may affect the plant-availability of P (Barekzai and Mengel, 1985; Parfitt et al., 1989).

Rates of these slow reactions are strongly temperature-dependent (Barrow and Shaw, 1975a,b; Barrow, 1980). However, the kinetics of these slow reactions were found to be insensitive to soil moisture content or soil type such that reactions in solution may not be the rate-limiting processes (Parfitt, 1978), and that the slow reactions perhaps occur at the original sorption site (Ryden and Pratt, 1980). It is thus possible that the diffusive migration of adsorbed P inside the adsorbing surface, and/or precipitation on the surface, followed by occlusion may be responsible for such slow reaction rates (Ibrahim and Pratt, 1982; Barrow, 1983a; Van Riemsdijk et al., 1984; Bolan et al., 1985; Mendoza and Barrow, 1987a,b).

Table 7. Effects of temperature and time of reaction on quantity of phosphate sorbed by ferrihydrite, and its extractability

Temperature (°C)	Time (d)	P sorbed (mmol/g)	Sorbed P extracted by NaOH (%)
25	6	0.93	99
25	30	1.15	97
25	90	1.19	107
35	30	1.18	103
35	90	1.20	106
50	30	1.20	95
50	90	1.20	103

Source: Willett et al., 1988.

Observation based on fractionation of sorbed P that the amount of surface-bound P (0.1 *M* NaOH-extractable) decreases with time (Williams et al., 1967) implies an absorption of the initially adsorbed P. This process can also account for a decrease in the ease of isotopic exchange and P desorption (Ryden and Pratt, 1980).

In a recent study, such slow P fixation reaction is also shown to be affected by the calcium carbonate content (Solis and Torrent, 1989). On the other hand, by studying the slow reaction of P with aggregated particles of ferrihydrite, Willett et al. (1988) ruled out the possibility of penetration of P into the crystal lattice, and instead attributed the slow reaction to the migration of P to surface sorption sites of decreasing accessibility within the aggregates. Table 7 shows the effects of temperature and time on P sorption by ferrihydrite, and its extractibility by NaOH. It is thus apparent that P continued to react with ferrihydrite for at least 90 days at 25°C, but was completely recovered by extraction with 0.1 *M* NaOH. This indeed implies that sorbed P remains primarily at the surface of the oxide.

B. Precipitation–Dissolution

The adsorption process may be regarded as that leading to net accumulation at an interface, whereas precipitation is the process that causes an accumulation of a substance to form a new bulk-solid phase. Both these processes cause loss of materials from a solution phase. However, the former is essentially two-dimensional, whereas the latter is inherently three-dimensional (Sposito, 1984).

Lin et al. (1983a) observed that the P adsorption in soils must involve the same mechanism at a molecular level, as that involved in the formation of discrete phase of iron, aluminum, or calcium phosphates, each being the result of the attractive forces between P and, for example, Fe or Al. These authors believed that adsorption mechanisms prevail at low P concentrations and precipitation at higher concentrations in soil.

Ryden and Pratt (1980) pointed out a distinction between sorption and precipitation. Sorption requires the structure of sorbent to remain essentially unchanged as the process progresses even though its surface activity decreases. This leads to a higher concentration maintained in solution at a greater surface-saturation than that at a lower saturation. In the precipitation process, however, the surface activity remains constant.

Where mixed precipitates are inhomogeneous solids, with one component restricted to a thin outer layer because of poor diffusion (Sposito, 1984), it is difficult to distinguish between the two processes. For soils, a further complication arises from the precipitation of new solid phases homogeneously onto the surfaces of existing solid phases, and from the weathering solids providing host surfaces for more stable phases into which the former transform chemically (Sposito, 1984).

A number of early research investigations on P chemistry in soil postulated the formation of a variety of insoluble inorganic phosphates of Fe, Al, and Ca through precipitation reaction (Kittrick and Jackson, 1955a,b, 1956).

Phosphate activities in aqueous soil extracts were also correlated with different solubility isotherms constructed from solubility product constants of various iron, aluminum, and calcium phosphates (Lindsay et al., 1959a,b; Murrmann and Peech, 1968; Jensen, 1971; Talibudeen, 1974). These solubility isotherms indicate that in acid soil, variscite and strengite are the probable P minerals, whereas in alkaline soils, several calcium phosphates govern the P concentration in soil solution. Whereas these calcium P minerals (e.g., apatite) are unstable in acid condition, variscite and strengite start dissolving at higher pH. Although at a pH >6.5, reprecipitation of P (e.g., on liming) as calcium phosphates begins (Haynes, 1984).

However, with the extremely slow dissolution rate of Fe, Al, and Ca phosphates in soil (Murrman and Peech, 1969), it is unlikely that variscite and fluorapatite, although the ultimate reaction products of the applied P in acid and limed (or alkaline) soils, respectively, can control the P concentrations in soil solution. McLaughlin and Syers (1978) while comparing the P concentrations maintained by hydrous ferric oxide gel, known for its P sorption capacity, and strengite, or short-range order ferric phosphate (amorphous), also concluded that such iron phosphate compounds are unlikely to persist in soil, and that their influence to control the P concentration in soil solution was limited.

Bache (1964) demonstrated for reactions between P solutions and gibbsite or hydrous ferric oxide that the ionic activity products for the precipitation reaction do not correspond with the solubility constants of variscite or strengite. Bache (1963) also suggested that the thermodynamic solutions to the P equilibria in soil systems are more likely to be found in terms of free energies of surface reactions (i.e., sorption–desorption). More recently, Harrison and Adams (1987) suggested that ionic activity products could not be used to predict, even indirectly, that solid-phase minerals were or were not controlling solution P in soils.

Further limitation of the solubility isotherm approach was illustrated by Ryden and Pratt (1980). Thus, the pH values and P concentrations of soil extracts over the pH range 4.0–7.5 were shown to be apparently independent, without conforming to any specific solubility isotherm, with the soil extracts remaining undersaturated with respect to all the phosphates (likely to be formed in soil) over the pH range 5.5–6.5.

In farmers' fields, highly water-soluble fertilizer P or concentrated fluid fertilizers are often applied to soil. This leads to the formation of strongly concentrated P solutions and often a low pH in the vicinity of the fertilizer granule. Depending on the P source used, P concentration may range from 1.5 to more than 6 *M* with the concentrations of the accompanying cations reaching as high as 10–12 *M*. The solution pH may be anywhere in the range from 1 to 10 (Sample et al., 1980). The concentrated P solutions, leaving the application site, contact successive soil increments and induce the dissolution and/or exchange of appreciable quantities of cations in the solution phase (Fixen and Ludwick, 1982a, b; Blanchar and Stearman, 1984).

For instance, with monocalcium phosphate monohydrate (MCPM), a granular P fertilizer added to soil, a concentrated phosphoric acid solution is generated that reacts with soil constituents causing the dissolution (or exchange) of appreciable quantities of Fe, Al, and several other metallic cations. This may be followed by recrystallization of new products. Ammonium taranakite and variscite were reportedly formed in a laterite soil treated with monoammonium phosphate (Prabhudesai and Kadrekar, 1984). Several of these initial reaction products may also be amorphous, having much higher solubility products than their crystalline analogues, and are thus capable of maintaining higher P concentrations (Ryden and Pratt, 1980).

Moreover, the formation of X-ray amorphous analogues of variscite and sodium montebrasite was demonstrated (Veith and Sposito, 1977; Veith, 1978) when aluminum hydrous oxides, aluminum oxides, synthetic aluminosilicates, and allophanic soils were reacted with sodium orthophosphate. The evidence for such a formation of aluminum phosphate by precipitation, as compared with surface-adsorption of P, was indicated by the slowness of the reaction, the intermediate and significant increase of silicon in solution when aluminosilicates were reacted, and by the very large amounts of phosphates reacted (Veith and Sposito, 1977).

On treating soils with KH_2PO_4, Al-Khateeb et al. (1986) found that β-tricalcium phosphate dominated the reaction products. Precipitated forms were dicalcium phosphate dihydrate, octacalcium phosphate, and β-tricalcium phosphate. Out of the magnesium phosphates, bobiernite was the dominant form. Calcium phosphates, rather than magnesium phosphate, controlled P solubility in the given soils. It was also reported by Fixen and Ludwick (1982a) that P-fertilized soils appear buffered very near the solubility isotherm of β-tricalcium phosphate, which lies between octacalcium phosphate and hydroxyapatite on a solubility diagram.

Based on their studies, Havlin and Westfall (1984) suggested that under field conditions, octacalcium phosphate or a mineral of similar solubility was the metastable mineral phase accumulating on the surface of the P-fertilized soils. At the Olsen P level, of <32 mg/kg, however, β-tricalcium phosphate apparently controlled P solubility in soils. Despite the fact that hydroxyapatite (HA) is thermodynamically predicted to control P activity in many soil solutions, equilibrium P values are often oversaturated with respect to HA, and are controlled by other Ca solid phases metastable to HA (Fixen et al., 1983; O'Connor et al., 1986). The kinetic rates of HA precipitation being slow, a role of kinetic inhibitors has been suggested to control the precipitation rate of HA in soils (Aoba and Moreno, 1985). In another study, Inskeep and Silvertooth (1988) noted that humic, fulvic, and tannic acids can inhibit HA precipitation, and this was suggested to cause increases in P concentrations in many agricultural soils with sludge or organic manure.

A recent study on the interaction of naturally occurring goethite surfaces with dilute KH_2PO_4 solution (Martin et al., 1988) tended to show that P retention in the iron oxide system is due primarily to precipitation of, for instance, the mineral griphite rather than to adsorption, either alone, or in conjunction with surface penetration of the surface-sorbed P. Nanzyo (1986) concluded from infrared spectral studies that at high pH, P was sorbed on iron hydroxide gel as a binuclear surface complex similar to that on goethite.

The initial reaction products of fertilizer P in soil, as discussed above, are likely to be metastable, and may redissolve in soil solution on changing conditions. A further reprecipitation of the more stable reaction products is also possible. Even if such more stable products are formed, they are unlikely to govern the P concentration in soil solution (Ryden and Pratt, 1980).

On liming soils, which are initially high in exchangeable aluminum, freshly precipitated amorphous polymeric hydroxy aluminum cations form independently or coat surfaces of the colloidal soil particles and thereby form a new, highly active adsorbing surface (Haynes, 1983). Hence, liming up to pH 6.5 has been found to increase P adsorption (Freisen et al., 1980; Haynes and Ludecke, 1981; Naidu et al., 1990). Le Mare and Leon (1989) proposed that the effect of aluminum associated with organic matter in an acid soil having a high content of the latter must be considered in assessing the effect of liming on P retention by soil. Thus, in unlimed soil, added P exchanges with organic anions associated with Al (Bloom, 1981). Liming may enhance organic matter decomposition, releasing Al which combines with P, and also with hydroxyl ions to form aluminum hydroxide precipitate. The latter adsorbs P. However, Haynes (1983) and Freisen et al. (1980) observed that air-drying the limed soil decreased its capacity for subsequent P sorption. Evidently, air-drying modified the surface-charge characteristics (Haynes, 1983).

Formation of amorphous hydroxy aluminum phosphates, obtained by reaction of aluminum-saturated montmorillonite and phosphoric acid, which are important soil–fertilizer reaction products and often a good source of P for plants in neutral soil, was also reported by Webber (1978) and Kodama and Webber (1975). Liming an acid soil (high in exchangeable Al) may accelerate their formation (White and Taylor, 1977). Sims and Ellis (1983b) showed that the Al–OH fraction in highly weathered acid soils has widely varying capacities to adsorb and retain (fix) P, depending on the degree of polymerization/crystallization of the Al–OH compounds.

Although the precipitation–dissolution concept is often used to describe P interaction with soil and sediment components (Syers and Ru-Kun, 1989), the surface-reactive amorphous aluminum hydroxides and iron oxides (and to a lesser extent, their crystalline analogues) tend to dominate the process of P fixation (through adsorption), rather than Al^{3+} and Fe^{3+} ions in solution (Ram and Rai, 1987). In alkaline soils, the various calcium phosphates are more likely to govern the P concentration through sorption–desorption processes.

Furthermore, even if the precipitation of various P compounds are the controlling factor in soils, the solubility product isotherms are only useful in recognizing the kind of P compound expected to control the P concentration in soil solution at equilibrium (Olsen and Khasawneh, 1980). They are unable to provide the quantity of such compounds and hence, of the reserves of solution P, a factor that is of paramount importance for crop uptake studies.

III. Reactions of Phosphorus Fertilizers in Soil

A. Reactions of Soluble Fertilizers

Reactions of soluble phosphatic fertilizers in soils are widely studied and documented (Sample et al., 1980). Discussion in this paper, however, will be restricted mainly to the reactions of superphosphates, the most widely used among the soluble phosphatic fertilizers. These reactions also help to understand the changes that phosphate rocks and partially acidulated phosphate rocks undergo in soils, as directly applied fertilizers.

Monocalcium phosphate (MCP) is the essential phosphate component in superphosphates. In soil, MCP of single or triple superphosphate undergoes hydrolysis to form an acid (pH 1.48) metastable triple point solution (MTPS) of MCP, dicalcium phosphate dihydrate (DCPD), and phosphoric acid. The solution remains near this point for at least 24 h, and over a period of 7 days, dicalcium phosphate dihydrate gradually dissolves, and less soluble dicalcium phosphate (DCP) precipitates out (Stephen and Condron, 1986). The incongruent dissolution of MCP leaves the solid phase partially depleted with respect to P. The solution, leaving the site of fertilizer granule by diffusion down the gradient of chemical potential becomes highly acidic and enriched in P. This hydrolytic dissolution of MCP may be represented as

$$Ca(H_2PO_4)_2 \cdot H_2O + xH_2O \rightarrow CaHPO_4 + H_3PO_4 + (x + 1)H_2O. \quad (a)$$

The dicalcium phosphate that remains in the fertilizer granule dissolves incongruently, and the highly insoluble hydroxyapatite precipitates out to form phosphoric acid in solution.

$$10CaHPO_4 + 2H_2O \rightarrow Ca_{10}(PO_4)_6(OH)_2 + 4H_3PO_4. \quad (b)$$

With the solubility of DCP being lower than that of MCP, the diffusion gradient between the fertilizer granule site and the bulk soil solution in reaction (b) is much lower than for reaction (a). As a consequence, P contribution to water-soluble P pool in soil from reaction (b) becomes very low.

All water-soluble P fertilizers produce nearly saturated solutions in soil, even under nonsaturated moisture conditions (Sample et al., 1980). Moisture may be drawn by capillary or even vapor transport. More water flows into the saturated fertilizer solution from the adjoining soil solution by osmotic transport, whereas the fertilizer solution itself moves out of the application zone into the bulk solution by diffusion (Hagin, 1985). To determine the soluble P concentration at the application site, the composition of the corresponding saturated aqueous solutions should be considered. Soluble P concentrations for MCP, DCP, hydroxyapatite, and diammonium phosphate are given in Table 8 (Sample et al., 1980; Hagin, 1985).

As discussed in section IIB, the concentrated phosphoric acid solution, moving out of the fertilizer granule, enters into several reactions with soil minerals. Thereafter, it induces the dissolution and/or exchange of appreciable quantities of cations such as Fe, Al, (manganese) Mn, Ca, and (magnesium) Mg (Sample et al., 1980). After several reactions take place, the soil solution gets

Table 8. Composition of saturated solutions of mono- and dicalcium phosphates, hydroxyapatite, and diammonium phosphate

			Composition of saturated solution		
				Accompanying cation	
Compound	Formula	pH	P mol/l	Cation	Solubility mol/l
Monocalcium phosphate	$Ca(H_2PO_4)_2 \cdot H_2O$	1.0	4.5	Ca^{2+}	1.3
Dicalcium phosphate	$CaHPO_4$ $CaHPO_4 \cdot 2H_2O$	6.5	~0.002	Ca^{2+}	0.001
Hydroxyapatite	$Ca_{10}(PO_4)_6(OH)_2$	6.5	$\sim 10^{-5}$	Ca^{2+}	0.001
Diammonium phosphate	$(NH_4)_2HPO_4$	8.0	3.8	NH_4^+	7.6

Source: Adapted from Sample et al., 1980, and Hagin, 1985.

Table 9. Solubility products of some soil-phosphorus fertilizer reaction products

Compound	Formula	pK_{sp}[a]
Variscite	$AlPO_4 \cdot 2H_2O$	21.5–22.5
Ammonium taranakite	$H_6(NH_4)_3Al_5(PO_4)_8 \cdot 18H_2O$	175.5
Potassium taranakite	$H_6K_3Al_5(PO_4)_8 \cdot 18H_2O$	178.7
Strengite	$FePO_4 \cdot 2H_2O$	35.3
Dicalcium phosphate	$CaHPO_4$	6.66
Dicalcium phosphate dihydrate	$CaHPO_4 \cdot 2H_2O$	6.56
Octacalcium phosphate pentahydrate	$Ca_8H_2(PO_4)_6 \cdot 5H_2O$	93.8
Hydroxyapatite	$Ca_{10}(PO_4)_6(OH)_2$	111.8
Flourapatite	$Ca_{10}(PO_4)_6F_2$	120.8
Dimagnesium phosphate trihydrate	$MgHPO_4 \cdot 3H_2O$	5.82
Magnesium ammonium phosphate hexahydrate	$MgNH_4PO_4 \cdot 6H_2O$	13.2

Note: [a] pK_{sp} is the negative logarithm of the solubility product of various compounds listed in the table.

Source: Adapted from Sample et al., 1980.

saturated with one or more of the reaction products of widely varying solubilities, thus causing their precipitation. The nature of the compound precipitating and, hence, the plant-availability of the concentrated P solution diffusing out of the application site would thus depend on the nature of the soil environment surrounding the fertilizer application site. In acid soils, where Fe and Al concentrations are high, Fe and Al phosphates are the primary reaction products, whereas in neutral and alkaline soils, calcium and magnesium phosphates precipitate out (Hagin, 1985), and with time, probably the product would be octacalcium phosphate (Sample et al., 1980). The relative solubilities of some of these P reaction products in soil are given in Table 9.

Ammonium ortho- and polyphosphate fertilizers, on the other hand, dissolve completely in soils without leaving any residues at the application site (Khasawneh et al., 1974). However, when other fertilizer sources are also incorporated, for example, micronutrient sources, a number of reactions may take place involving the initial reaction products of ammonium phosphates in soil, leading to soluble P immobilization. However, the amounts immobilized are generally small because only small quantities of micronutrients are usually incorporated (Sample et al., 1980). The following two immobilization reactions were proposed in this respect (Lehr, 1972):

$$M^{II}SO_4 + (NH_4)_2HPO_4 \rightarrow M^{II}NH_4PO_4 + NH_4HSO_4 \quad \text{(c)}$$

$$M^{II}SO_4 + (NH_4)_4P_2O_7 + H_2O \rightarrow M^{II}(NH_4)_2P_2O_7 \cdot H_2O + (NH_4)_2SO_4, \quad \text{(d)}$$

where M^{II} = Cu(II), Fe(II), Mn(II), or Zn.

B. Phosphate Rocks as Fertilizers

Finely ground indigenous phosphate rock (PR) has been directly applied to the soil in several countries because of low input of capital investment and energy required to prepare the product (Chien and Hammond, 1989). In general, these materials were found most effective in phosphorus-deficient acid soils. Several aspects of chemical reactivity of these materials as direct application P fertilizers are discussed.

1. Mineralogy of Phosphate Rocks

Phosphate rock deposits may be of igneous, metamorphic, or sedimentary type. Although most commercial deposits are of sedimentary marine PR, but a significant amount of phosphorus (P) is also obtained from alkaline igneous complexes and residual deposits obtained on weathering of sedimentary phosphatic limestones and igneous carbonatite complexes (Notholt, 1975). According to McClellan and Gremillion (1980), PRs may be classified into three broad classes according to their mineralogical composition:

1. Fe–Al phosphates, for example, wavellite, $Al_3(PO_4)_2(OH)_3 \cdot 5H_2O$; variscite, $AlPO_4 \cdot 2H_2O$, and strengite, $FePO_4 \cdot 2H_2O$; and some other Fe–Al phosphates of less common occurrence.
2. Ca–Fe–Al phosphates, for example, crandallite, $CaAl_3(PO_4)_2(OH)_5 \cdot H_2O$ and millisite, $(Na, K)CaAl_6(PO_4)_4(OH)_9 \cdot 3H_2O$.
3. Apatitic group of minerals having a common crystalline structure, for example, fluorapatite, $Ca_{10}(PO_4)_6F_2$. The chemical composition of apatites varies widely among the members, resulting primarily from isomorphous substitution in the apatite crystal lattice. Accordingly, the chemical properties (for example, reactivity in soil) of these PRs also vary considerably, a fact that has been exploited commercially for the use of these rock deposits as source materials to produce various P fertilizers and for direct application as PR fertilizer for crop production.

In terms of weathering sequence, Fe and Al phosphates are often the most weathered stable end products in a sequence, wherein apatites are the least weathered (Khasawneh and Doll, 1978).

The apatites in igneous and metamorphic PR deposits are relatively inert, being coarse-grained with little internal surfaces. Their chemical composition is close to the theoretically required stoichiometric composition of fluorapatite, having less of impurities as accessory minerals, for example, cristobalite (silica), feldspars, amphiboles, pyroxenes (silicates), calcite, dolomite (carbonates), goethite, gibbsite, and boehmite (Fe and Al oxides and hydroxides). Thus, the P content of these deposits is relatively high. This factor is important in the production of soluble and partially soluble P fertilizers from PR, but is of no concern when determining the reactivity of the PR to be applied directly to soil as fertilizer.

The sedimentary rocks, on the other hand, contain apatite minerals that are microcrystalline. They consist of fairly open, loosely consolidated aggregates of microcrystals with relatively large specific surface. Chemically, these are usually carbonate fluorapatites (FA), known as francolites (Khasawneh and Doll, 1978) with varying degrees of substitution of PO_4^{3-} by CO_3^{2-} and F^- (fluoride) in the lattice. They also contain various amounts of several accessory minerals such as calcite, dolomite, gypsum, Fe, and Al sesquioxides and hydroxides, silicate clays, organic matter. McClellan and Gremillion (1980), from petrographic examination of different sedimentary PR, reported homogeneity of the apatites in most sedimentary PR. The physical and chemical charactertistics of these apatites are relatively uniform.

The observed effectiveness of PR relative to soluble P fertilizers is strongly dependent on the mineralogy and chemistry of the source material (Hammond et al., 1986). In addition, soil and plant factors, and associated parameters should also be given importance.

2. Chemical Characteristics of Phosphate Rocks and Crystal Chemical Structure

The chemical composition of PR from several deposits of varying geographic and geological origin has been established (Chien and Hammond, 1978a; McClellan and Gremillion, 1980; Hammond et al., 1986). Compositions of apatites in sedimentary PR can be adequately described in terms of the contents of six major constituents, namely, Ca, Na, Mg, P, CO_2, and F. Compositions of some representative apatites are given in Table 10.

Pure fluorapatite rarely exists in nature. There have been high degrees of substitution of nearly all the ionic constituents in the fluorapatite structure. The main substituting ions (through isomorphous replacement) are shown below (Khasawneh and Doll, 1978).

Constituent ion	Substituting ion
Ca^{2+}	Na^+, K^+, Mg^{2+}, Mn^{2+}, Sr^{2+}, U^{4+}
$PO_4{}^{3-}$	$CO_3{}^{2-}$, F^-
F^-	OH^-, Cl^-.

Table 10. Some typical francolite compositions computed from unit-cell "a" dimensions

	Composition, %					
Source	CaO	MgO	Na_2O	P_2O_5	CO_2	F
Western U.S.A.	55.6	0.13	0.26	40.1	1.59	4.09
Tennessee (U.S.A.)	55.5	0.24	0.47	38.7	2.71	4.31
Florida (U.S.A.)	55.5	0.36	0.72	37.1	3.95	4.56
Morocco (Africa)	55.4	0.43	0.85	36.3	4.53	4.68
North Carolina (U.S.A.)	55.3	0.52	1.04	35.3	5.36	4.85
Tunisia (Africa)	55.2	0.60	1.20	34.7	5.70	4.93

Source: McClellan and Gremillion, 1980.

The CO_3^{2-} substitution for PO_4^{3-} strongly affects the crystal structure and reactivity of the apatites (Hammond et al., 1986). The degree of substitution is expressed as the mole ratio CO_3/PO_4.

The apatites in the sedimentary PR form a series with the end member formulas as $Ca_{10}(PO_4)_6F_2$, fluorapatite, and $Ca_{10-a-b}Na_aMg_b(PO_4)_{6-x}(CO_3)_xF_{2+y}$, carbonate-fluorapatite, or francolite (McClellan and Gremillion, 1980).

Here,

$$\frac{x}{6-x} = 4.90(9.374 - a_0), \quad a = 1.327[x/(6-x)],$$

$$b = 0.515[x/(6-x)], \quad \text{and} \quad y = 0.4x \text{ (generally).}$$

a_0 in these relations refers to the unit cell dimension along the a-axis of the hexagonal apatite crystals. The theoretical and observed limit on the values of x is derived from a corresponding limit of the value of $x/(6-x)$, which is 0.3 (Khasawneh and Doll, 1978). A higher substitution disrupts the crystal structure.

Isomorphous substitution of $CO_3^{2-} + F^-$ to PO_4^{3-} affects the a and c axes dimensions of apatite crystal structure. It decreases the a value (a_0) from 9.376 Å, which is characteristic of pure fluorapatite, and increases the specific surface area of the apatites. Changes in the a axis dimension are considered more significant to PR reactivity than those in c axis dimensions. A more detailed discussion of this aspect is given later.

Once the value of a_0 is obtained from X-ray diffraction analysis, the above relationships may be used to compute for x, a, and b, hence, the empirical formula of the given apatite.

3. Dissolution of Phosphate Rocks in Soil

The extent and rate of PR dissolution is primarily important in determining its effectiveness when directly applied as fertilizer. A better understanding of the reactions of PR in soils will be useful in assessing on which soils PR sources could be best used. Several factors, some of which are interrelated, influence PR dissolution in soil.

a. Influence of Soil Factors on Phosphate Rock Dissolution

In addition to chemical composition and particle size of the PR, soil properties, such as pH, moisture content, and Ca and P concentration in soil solution are likely to affect the rate and amount of dissolution (Khasawneh and Doll, 1978; Chien et al., 1980a; Smyth and Sanchez, 1982; Wilson and Ellis, 1984; MacKay et al., 1986; Kanabo and Gilkes, 1988; Chien and Hammond, 1989). The PR dissolution in soil solution (acidic) may be represented as follows (Hammond et al., 1986):

$$Ca_{10}(PO_4)_6F_2 + 12H^+ = 10Ca^{2+} + 6H_2PO_4^- + 2F^-. \qquad \text{(e)}$$

It follows from reaction (e) that the law of mass action favors PR dissolution in soil solution under conditions of low (1) soil pH, (2) soil exchangeable Ca^{2+}, and (3) P in soil solution.

Cook (1935) and Graham (1955), while investigating the solubility and plant growth in hydrogen (H-) and Ca-saturated systems, concluded that the H-ion-solubilized PR colloidal systems caused greater PR dissolution by providing a sink for the released Ca. Graham (1955) and Howe and Graham (1957) also showed that bonding energy for Ca in clay suspensions, acid soils, and in cation-exchange resin systems were important criteria for PR dissolution in the soil. Furthermore, Khasawneh (1977, cited in Khasawneh and Doll, 1978) demonstrated that the affinity of a soil for Ca is a significant factor influencing PR dissolution in soils. However, he examined the effect of increasing Ca by adding either $CaCO_3$ or $SrCO_3$ at three liming rates in a greenhouse experiment. The pH of the system was thus increased by the CO_3^{2-} rendering it difficult to examine the effect of increasing Ca at a constant pH. Chien (1979a) reported that increasing $CaCl_2$ concentration depressed P solubility from two PR sources.

As the dissolved P (from PR) is sorbed on surfaces, or reacts with Fe and Al to form compounds less soluble than the original apatite, Ca remains in the soil solution predominantly as an exchangeable cation. Changes in exchangeable soil Ca are therefore often taken as providing an indirect estimate of PR dissolution. This is particularly useful for soils, having high P sorption capacities, in which soil solution P levels remain low.

MacKay et al. (1986), using 30 contrasting soils, found that PR dissolution increased as exchangeable Ca decreased and as P sorption capacity of the soils increased. Chien et al. (1980b), Smyth and Sanchez (1982) and Kanabo and Gilkes (1987a) also reported that PR dissolution increased with P-sorption capacity of soil, thus maintaining low P level in soil solution and favoring the reaction (e).

Kumar and Mishra (1986) suggested that P fixation capacity of soils strongly influence PR dissolution in soils. Hughes and Gilkes (1986a) noted a strong positive correlation between the extent of PR dissolution in soils and oxalate and pyrophosphate-extractable iron and aluminum contents of the soils. Organic carbon, exchangeable Ca, and soil pH (negatively correlated) were found to be subsidiary factors.

Kanabo and Gilkes (1987b) reported PR dissolution in soil to increase linearly with decreasing pH. Khasawneh and Doll (1978) listed soil factors, such as soil pH, soil Ca, soil P, soil texture, and soil organic matter content, as factors affecting the PR dissolution in soils.

The influence of soil exchangeable Ca on PR dissolution in soil, however, needs careful attention because many studies have evaluated Ca in relation to liming at various soil pH (Khasawneh, 1977, cited in Khasawneh and Doll, 1978). Wilson and Ellis (1984), on the other hand, investigated the solubility of several PRs in neutral normal ammonium citrate solutions with varying Ca^{2+} ion activity. They used the solubility product principle to interrelate P solubilized from PR and Ca^{2+} activity in solution from the consideration of the solubility equilibria of the end members, namely hydroxyapatite (HA), $Ca_{10}(PO_4)_6(OH)_2$, and fluorapatite (FA), $Ca_{10}(PO_4)_6F_2$. For the former, the solubility product is given as

$$K_{sp_{HA}} = (Ca^{2+})^{10}(PO_4^{3-})^6(OH^-)^2, \tag{15}$$

where (a_i) represents the activity of a_i in a saturated solution.

The total P measured in the solution is obtained as

$$\text{Total P} = \frac{(H_2PO_4^-)}{f_1} + \frac{(HPO_4^{2-})}{f_2} + \frac{(PO_4^{3-})}{f_3}, \tag{16}$$

where the various f-terms are the activity coefficients of $H_2PO_4^-$, HPO_4^{2-}, and PO_4^{3-} ions. The ion pair H_3PO_4 was neglected at pH 7. On combining Eqs. (15) and (16), and remembering that pH and ionic strength were constant, Wilson and Ellis (1984) obtained Eq. (17) for total P.

$$\text{Total P} = K'(Ca^{2+})^{-5/3}(OH)^{-1/3} \tag{17}$$

where K' is a constant involving $K_{sp_{HA}}$ and the f-terms.

Analogous equations were derived for the solubility equilibrium of FA. Thus,

$$\text{Total P} = K''(Ca^{2+})^{-5/3} \cdot (F^-)^{-1/3} \tag{18}$$

Unlike the case of HA (Eq. 17), where pH is controlled, F^- activity for FA is unknown and variable. However, the F^- released is related to P released, but would only be one-third as great (Eq. 18). However, if the solution activity of the Ca^{2+} ion is high enough to cause CaF_2 precipitation, then the otherwise linear relationship between log P and $-$log Ca (as expected from, for instance, Eq. 17) may become nonlinear tending to show a larger negative slope than for the HA plot.

Figure 9 demonstrates the effect of Ca^{2+} ion activity on the solubility of six PR samples of widely varying compositions. The plots between log P and $-\log(Ca^{2+})$ were nearly linear for all cases, with Indian PR showing (numerically) the lowest slope and the Missouri sample the largest. Substitutions by Na^+ and Mg^{2+} of Ca^{2+} ion in the PR make the slope more negative, whereas that by CO_3^{2-} and F^- for PO_4^{3-} render it less negative.

The role of exchangeable Ca in soil in PR dissolution was also highlighted by Smillie et al. (1987), who suggested that P (from PR dissolution) sorbed in soil may enter into chemical association with Ca, possibly at metal oxide surfaces. This surface complex of Ca–P is expected to behave similarly to that of Ca–phosphate, hence, the qualitative similarity in behavior usually shown by all forms of added and native P.

The positive effect of soil organic matter on PR dissolution in soil was also attributed to a concomitant change in Ca activity in soil solution as induced by chelation of Ca^{2+} ion in soil solution by organic anions (e.g., tartrate, citrate, oxalate) (Chien, 1979b). Solubilization of PR when incorporated dur-

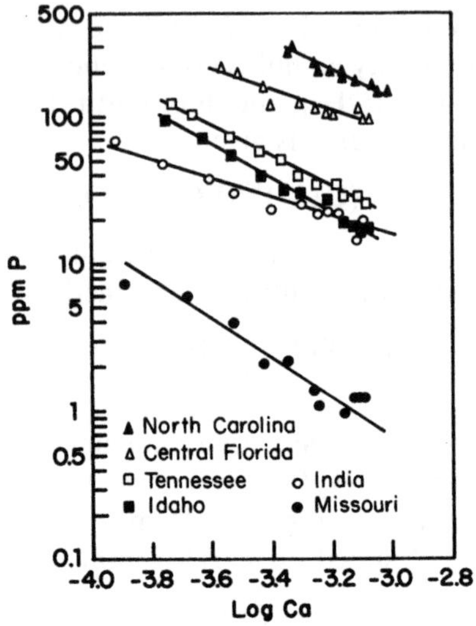

Figure 9. The effect of Ca^{2+} ion activity (mol/l) on the solubility (ppm P) of six phosphate rocks. (*Source*: Wilson, N.A. and B.G. Ellis. Influence of calcium solution activity and surface area on the solubility of selected rock phosphates. Soil Science 138, 354–359. © by Williams & Wilkens, 1984).

ing composting of organic wastes (Mishra et al., 1982; Bangor et al., 1985), or the beneficial effect of farmyard manures on enhancing P availability from PR (Hammond et al., 1986) may also be attributed to the action of organic acids produced in complexing Ca^{2+} ions in soil solution. Organic matter also helps incorporate inorganic P released from PR into the organic pool of soil P, thereby offering an additional route of releasing P from PR for eventual plant uptake through mineralization (Khasawneh and Doll, 1978).

Soil moisture content is another important soil parameter that affects PR dissolution in soils (Debnath and Basak, 1986; Kanabo and Gilkes, 1988). Thus, PR fertilizers may be poorly active in soils that remain dry for much of the year, such as that occurs under mediterranean climate (Bolland et al., 1986) because of insufficient soil solution to effect PR dissolution. In contrast, performance of PR fertilizers in moist soils, such as the pasture land in New Zealand, is usually better (Quin, 1981; Gregg et al., 1981).

A higher level of biological activity has also been suggested by Kanabo and Gilkes (1988) to be responsible for enhanced PR dissolution in moist soil. Also, they found that the water retained in the soils at field capacity is sufficient to support near potential maximum dissolution, although some differences were noted between soils kept continuously wet and those receiving wetting–drying treatments. They concluded that a short period of drying does not substantially affect PR dissolution in soil, especially if the drying period is imposed after much of the potential dissolution has already taken place. However, for a longer period of the dry spell where the near-surface soil is affected during the growing season, the effect of soil moisture content on PR dissolution may be of more important consideration (Bolland et al., 1986). Furthermore, an increase in concentration of Ca^{2+} and $H_2PO_4^-$ ions in soil solution accompanying a decrease in soil moisture content, adversely affects PR dissolution (MacKay et al., 1986) according to reaction (e).

Note, however, that an increased PR dissolution in soils does not guarantee an increase in the amount of plant-available P. Thus, dissolved P from PR may be taken up by the plant or it may be fixed by soil solids, though sometimes temporarily. Hence, the factors that promote PR dissolution should be distinguished from those that control the plant-available P pool in the soil to assess truly the agronomic effectiveness of PR (Syers and MacKay, 1986; Kanabo and Gilkes, 1987c).

Thus, in a given soil–crop situation where the P released from a water-soluble P fertilizer and a PR is subjected to strong retention by soil sesquioxides, agronomic effectiveness of the two may not differ much (Kanabo and Gilkes, 1987a, c). In contrast, in certain acid soils, characterized by low P sorption capacity, which may lead to considerable leaching losses of P, a poorly soluble fertilizer, such as PR, may be more effective despite its low dissolution rate (Hammond and Leon, 1983; Kanabo and Gilkes, 1987a). On the other hand, Bolland et al. (1986) found the water-soluble P fertilizers to be superior to PR in soils with a moderate P sorption capacity and poor ability to encourage PR dissolution.

b. Measurement of Dissolution of Phosphate Rock in Soil

Direct measurement of the extent of dissolution of PR applied to soil is beset with difficulties. Thus, most methods used are indirect. Various chemical extractants have been used to measure increases in different forms of solid and solution P and Ca (Chien, 1977a; Rajan, 1983a; Hughes and Gilkes, 1984; Bolland et al., 1984; Syers and MacKay, 1986; MacKay et al., 1986; MacKay and Syers, 1986; Kirk and Nye, 1986; Apthorp et al., 1987; Bolan and Hedley, 1989). Two important prerequisites for such extractants are that they should (1) not promote any dissolution of PR during extraction, and (2) extract most of P or Ca released during dissolution. The following extraction methods are more commonly used (Khasawneh and Doll, 1978):

1. Inorganic P fractionation procedure of Chang and Jackson (1957), or one of its many modifications. Thus, increases in Fe–P and Al–P fractions in a soil treated with PR (over non-PR treatment) are taken to provide an estimate of P that has dissolved from PR, whereas increases in Ca–P are considered to be associated with unreacted PR.
2. Dilute acid-NH_4F extraction (MacKay et al., 1984).
3. Alkaline extraction with 0.5 *M* $NaHCO_3$, 1.5% Na_2CO_3 (Joos and Black, 1950), and 0.5 *M* NaOH (MacKay et al., 1986). MacKay et al. (1986) claimed that their single extraction procedure involving 0.5 *M* NaOH is a much superior direct method to the more involved and time-consuming fractionation scheme. However, Bolan and Hedley (1989) pointed out that

this method is liable to give erroneous results in cases where there is an active net mineralization or immobilization of soil P.

4. Extraction with anion-exchange resin.
5. Labile P as measured by ^{32}p isotopic exchange.
6. Hughes and Gilkes (1984) compared different extractants to measure exchangeable Ca. They observed that the increase in Ca extracted by 2.0 *M* $BaCl_2$ buffered at pH 8.1 with triethanolamine (TEA) gives a good estimate of PR dissolution.

A recent study by Bolan and Hedley (1989) showed that most of the extractants used for measuring PR dissolution tend to remove some of the undissolved PR during the extraction process. The extent of this "induced dissolution" depends on the nature of the extractant and the PR. Both 0.5 *M* NaOH and 0.5 *M* $BaCl_2$/TEA were found to be better extractants of the PR dissolution products from soil than were 0.5 *M* $NaHCO_3$ and 1 *M* NH_4OAc, particularly when soil organic P formation is negligible. Under field conditions where dissolved P or Ca are also subject to plant uptake or leaching loss, a modified HCl or H_2SO_4 method similar to that used by Rajan (1983a), Apthorp et al. (1987), and Chien et al. (1987a) may be more suitable (Bolan and Hedley, 1989).

c. Assessment of Chemical Reactivity of Phosphate Rocks

Conventionally, the probable extent of PR dissolution and its suitability as a P fertilizer are assessed based on its solubility in chemical extractants. Neutral ammonium citrate, 2% formic acid, 2% citric acid, 1% lactic acid, and Na–EDTA are the extractants most commonly used (Chien and Hammond, 1978a; Khasawneh and Doll, 1978; McClellan and Gremillion, 1980; Kanabo and Gilkes, 1988).

Solubility of PR in 2% citric and 2% formic acids is affected by many factors other than the solid/solvent ratio (Braithwaite et al., 1989). The impurities in PR, such as calcium carbonate, Fe, and Al oxides, affect P solubility in PR in these extractants, and as a result, affect the assessment of PR into "reactive" and "unreactive" categories.

The reactivity of sedimentary PRs is also considerably influenced by the amount and nature of accessory minerals present in PR such as calcite, dolomites, and gypsum. PRs containing calcite and dolomite tend to depress the dissolution of apatite-bound P in neutral ammonium citrate due to the partial consumption of the latter by the accessory minerals (Hammond et al., 1986). Hence, it is suggested to discard the first extract with citrate and use the second successive extract in evaluating apatite-bound P in the rock (Chien and Hammond, 1978a; Khasawneh and Doll, 1978). Also, sequential extractions used by MacKay et al. (1984) can be used to estimate the agronomic potential of the PR.

The influence of other accessory minerals (for example, siliceous minerals) on the results of chemical solubility tests of PR materials has been discussed by Hammond et al. (1986). They gave a detailed comparison of various methods of evaluating the agronomic potential of PRs, involving the use of several chemical extractants. Generally, neutral ammonium citrate (second extraction) (Mishra et al., 1985; Hammond et al., 1986) and 2% formic acid extractable P correlated better with the agronomic data (Hammond et al., 1986). Chien and Hammond (1978b) suggested the use of H-resin in place of citrate extraction to evaluate the reactivity of granulated PR.

The chemical and physical characteristics of PR materials in relation to their chemical reactivity were discussed by Syers et al. (1986). The reactivity of PRs as measured by ammonium citrate extraction method increased with increasing $CO_3^{2-} + F^-$ substitution of PO_4^{3-} in the apatite crystal. A linear correlation between citrate-soluble P and the mole ratio of CO_3/PO_4 or weight ratio of F/P_2O_5 was observed. Furthermore, a negative correlation was also found between the a_0 dimension of the apatite and its reactivity (Khasawneh and Doll, 1978; McClellan and Gremillion, 1980; Dash et al., 1982a; Marwaha et al., 1983; Bhujbal and Mistry, 1985; Hughes and Gilkes, 1986b).

The CaO/P_2O_5 ratio in the PR and the unit–cell crystallographic dimension, a_0, of the apatite are two properties recognized as guiding the reactivity of PR as directly applied fertilizer. The latter provides an indirect measure of the amount of CO_3^{2-} substitution for PO_4^{3-} in the apatite structure for members of the francolite series, and is inversely related to the mole ratio, CO_3/PO_4, for several apatites. This relationship is illustrated in Figure 10.

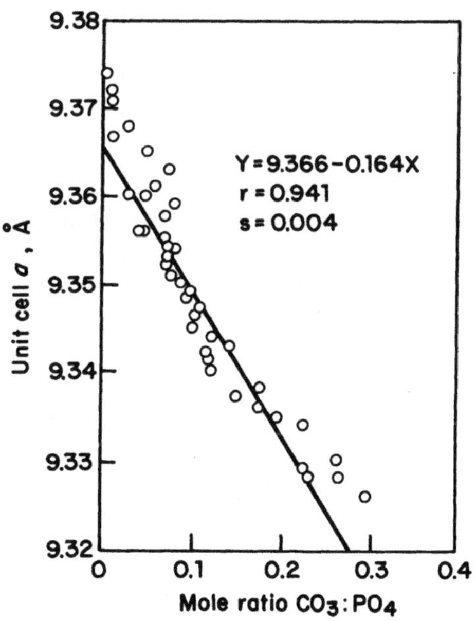

Figure 10. Relationship between unit–cell *a* dimensions and molar ratio CO_3/PO_4 in apatite (*Source*: McClellan and Gremillion, 1980).

The grade of PR refers to the apatite concentration in the given deposit, that varies with location in the deposit. Accordingly, PR reactivity from a given deposit cannot be related to the grade of PR (McClellan and Gremillion, 1980). To avoid the error for PR concentrates of different degrees of beneficiation, Lehr and McClellan (1972) defined a reactivity parameter for PR samples, namely the *absolute citrate solubility index* (ACSI), in preference to the conventional citrate solubility parameter (Khasawneh and Doll, 1978). In ACSI, the quantity of citrate-soluble P is expressed as a fraction of P concentration of apatite phase in the PR material, rather than of P concentration in the entire sample unit. Thus,

$$\text{Conventional citrate solubility} = \frac{\text{citrate-soluble P}}{\text{total P in the PR sample}} \times 100$$

$$\text{and} \quad \text{ACSI} = \frac{\text{citrate-soluble } P_2O_5(\%)}{\text{theoretical } P_2O_5(\%) \text{ of apatite}} \times 100$$

For 2% citric and 2% formic acid extractants, the absolute solubility index (ASI) (CtA) and ASI (F_mA) have been similarly defined. These three indexes are interrelated in a large number of PR samples:

$$\text{ASI}(CtA) = 1.26\ \text{ACSI} + 7.29 \quad \text{and}$$

$$\text{ASI}(F_mA) = 2.42\ \text{ACSI}.$$

The a_0 dimension of apatite crystal has also been shown to be inversely related to PR reactivity and, hence to their solubility in various extractants. The following interrelationships were proposed (Khasawneh and Doll, 1978):

$$\text{ACSI} = 341(9.376 - a_0)$$

$$\text{ASI}(CtA) = 429(9.393 - a_0)$$

$$\text{ASI}(F_mA) = 823(9.376 - a_0).$$

An estimate of the citrate solubility of a PR material may thus be obtained from the length (a_0) of "a" axis of the unit apatite crystal as obtained by X-ray diffraction analysis. The ACSI was found to be highly correlated with the plant-available P from the PR directly applied in soils (Hammond et al., 1986).

Dash et al. (1982a) reported a relatively simple and indirect technique, based on linear regressions, to determine a_0, CO_3/PO_4, F/P_2O_5, and ACSI values of several Indian PRs, without using the X-ray diffraction technique. For this purpose, the linear regression equations

$$y = a + bx$$

were worked out separately between the cumulative citrate-soluble P (x), expressed as a percent of rock, or as percent of total P in the rock in the first, second, 1 + 2, 1–5 or 1–8 extracts obtained in the given experiment, and

Table 11. Computed or reported x-ray parameters of different rock phosphates

Soil no.	Rock phosphate	$a_0 Å$	CO_3/PO_4	F/P_2O_5	ACSI[b] % extract first	ACSI[b] % extract second
1.	Mussoorie (India)*	9.3591 ± 0.0022	0.0629 ± 0.0108	0.1018 ± 0.0015	2.6	2.1
2.	Jhamarkotra (India)*	9.3626 ± 0.0021	0.0415 ± 0.0103	0.0991 ± 0.0016	0.8	1.1
3.	Kasipatnam (India)*	9.3633 ± 0.0023	0.0402 ± 0.0113	0.0989 ± 0.0017	0.7	1.3
4.	Purulia (India)*	9.3612 ± 0.0025	0.0513 ± 0.0124	0.1003 ± 0.0018	2.3	1.5
5.	Meghnagar I (India)*	9.3634 ± 0.0020	0.0396 ± 0.0101	0.0988 ± 0.0016	0.7	1.3
6.	Meghnagar II (India)*	9.3650 ± 0.0022	0.0307 ± 0.0101	0.0978 ± 0.0016	0.3	0.3
7.	Meghnagar III (India)*	9.3639 ± 0.0021	0.0368 ± 0.0101	0.0985 ± 0.0016	0.4	0.9
8.	Udaipur (India)**	9.365	0.028	0.090	2.5	2.0
9.	North Carolina (U.S.A.)**	9.323	0.262	0.119	26.3	23.0
10.	Morocco (Africa)**	9.347	0.098	0.114	9.4	9.4
11.	Gafsa (Africa)**	9.326	0.285	0.131	22.9	16.9
12.	Jordan (Middle East)**	9.332	0.197	0.121	15.1	14.9
13.	North Florida (U.S.A.)**	9.330	0.178	0.118	19.1	16.0
14.	Central Florida (U.S.A.)**	9.345	0.164	0.118	10.6	8.0
15.	Missouri (U.S.A.)**	9.373	0.008	0.096	0.4	0.4
16.	Idaho (U.S.A.)**	9.356	0.089	0.104	3.0	2.0
17.	Tennessee (U.S.A.)**	9.351	0.089	0.104	1.2	2.1

[b] Calculated from the equation ASCI = 100 (citrate-soluble P_2O_5/P_2O_5 content of apatite determined from a_0).

*$n = 10$ for a_0, CO_3/PO_4, F/P_2O_5.

**Data obtained from Lehr and McClellan (1972).

Source: Adapted from Dash et al., 1982a.

a_0, CO_3/PO_4 or F/P_2O_5 (y), for 10 PR samples for which data are available in literature (Lehr and McClellan, 1972). Correlation coefficients for the above linear regressions were highly significant (Dash et al., 1982a). From the 10 regression equations, the a_0, CO_3/PO_4, and F/P_2O_5 values for seven unknown Indian PRs were calculated. Table 11 shows the mean values of these parameters along with their standard deviations, ACSI values for the samples, and the values of these parameters for the 10 PR samples used to derive the values of regression coefficients (a and b).

The degree of CO_3^{2-} and F^- substitution to PO_4^{3-} in the apatite structures of Indian PRs were much lower than that for North Carolina, Gafsa, or

Jordan rocks. The corresponding ACSI values were also lower. The a_0 values of the Indian PR samples (computed) were close to that for unsubstituted fluorapatite (9.376 Å) and, hence have lower reactivity. These and other Indian PR samples were thus assessed to be unavailable as fertilizers for direct application to the soil or for P fertilizer production on acidulation (Dash et al., 1981a, 1982a).

Hughes and Gilkes (1986b) also assessed the reactivity of several apatite PRs vis-a-vis calcined PRs and mono- and dicalcium phosphate fertilizers on acid soils. They found that percent dissolution of noncalcined PR followed the expected order, such that the sedimentary high-carbonate apatites were most reactive and the igneous carbonate apatites, the least soluble (Lehr, 1980). In agreement with earlier observations in pot and field trials (Palmer et al., 1979; Bolland and Bowden, 1982; Werner and Solle, 1983), they found that the performance of calcined Fe–Al phosphates was poor. Furthermore, the order of PR solubility in the soils did not vary greatly, but the extent of dissolution varied markedly, thereby emphasizing that the inherent reactivity of a PR sample simply provided an indication of the position of the sample as a fertilizer on a relative reactivity scale. This obviously could not assess fully the agronomic effectiveness of PRs in terms of actual dissolutions in a given soil–crop situation.

d. Thermodynamics of Solubility of Phosphate Rocks

Khasawneh and Doll (1978) discussed the driving force for PR dissolution in soils in terms of solubility isotherms of fluorapatite, hydroxyapatite, and substituted (carbonate) apatites, that were constructed based on the free energies of formation of these products (Chien and Black, 1976). Chaverri and Black (1966) also considered the solubility of FA and HA in terms of the corresponding solubility isotherms.

Chien (1977b) who considered the free energy of acid dissolution of carbonate apatites CA and HA, gave the standard free energy of reaction of these products in the acid environment as,

$$(\Delta G_R^\circ)_{CA} \simeq -(2.5 + 5.1x), \tag{19}$$

where x is the number of moles of CO_3^{2-} substituted for PO_4^{3-} in the apatite structure represented by the average stoichiometric formula,

$$Ca_{10-0.42x}Na_{0.3x}Mg_{0.12x}(PO_4)_{6-x}(CO_3)_xF_{2+0.4x}.$$

The standard free energy of hydroxy fluorapatite, $Ca_{10}(PO_4)_6(OH)xF_{2-}x$, is given as (Chien, 1977b)

$$(\Delta G^\circ)_R \simeq -(2.5 + 14.8x). \tag{20}$$

Thus, the free energy of reaction (dissolution) of both the substituted products CA and HA becomes more negative with increasing substitution (Eqs. 19 and 20). This indicates a higher thermodynamic tendency to react and

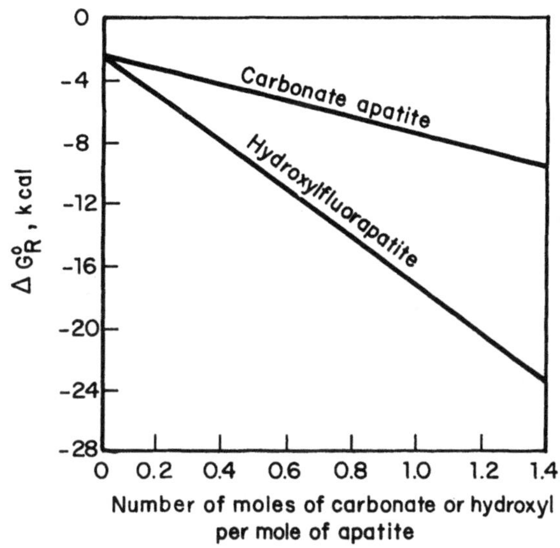

Figure 11. Standard free energy of reaction, ΔG_R°, of apatite mineral in phosphate rock as a function of isomorphic substitution in apatite structure (*Source*: Chien, S.H. Thermodynamic Considerations on the solubility of phosphate rock. Soil Science 123, 117–121. © by Williams & Wilkens, 1977).

dissolve at equilibrium with increasing carbonate and hydroxyl substitution, respectively, in acid soils. Figure 11 shows a plot of Eqs. (19) and (20).

Chien (1977b) further established that the net standard free energy of reaction, ΔG_R°, of a PR material in acid solution is a sum of free energy of dissolution of the material and of neutralization of anions of the dissolved apatite with H^+ ion. The standard free energy of dissolution of the apatite minerals is positive, and is relatively constant, whereas the free energy of neutralization is strongly negative and depends on the extent of isomorphism. The free energy of neutralization thus provides the thermodynamic driving force for the PR materials to dissolve in acidic environments.

It is important to note from Figure 11 that the ΔG_R° is more negative for substituted hydroxyfluorapatites than for CAs at a given value of x. Chien (1977b) emphasized this fact to examine the possibility of using the PRs containing the former materials as direct application fertilizer in acid soils.

Khasawneh and Doll (1978) observed that PR reacts only in acid soils primarily because the lowest free energy form of soil P in these soils is Al and Fe phosphates and not fluorapatite or hydroxyapatite, which are the usual end reaction products of soluble fertilizer P applied to neutral and alkaline soils, or previously limed acid soils. Also, they identified from the solubility product isotherms for various carbonate-substituted fluorapatites and hydroxyapatites the following specific driving forces for PR dissolution.

1. Gradients in pH where solubility (of PR) increases with decreasing pH.

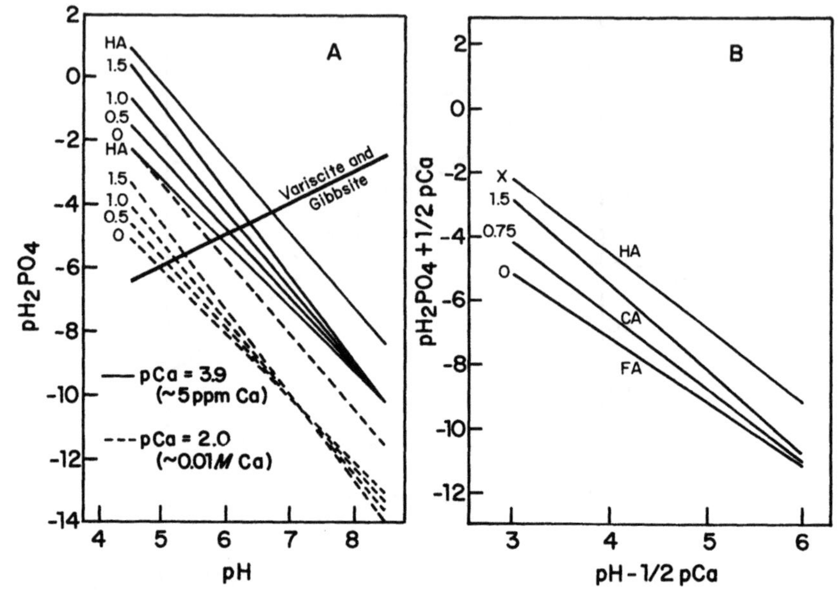

Figure 12. The effect of isomorphous substitution of CO_3^{2-} in the apatite structure and of pCa on solubility isotherms of fluorapatite and carbonate apatite in equilibrium with fluorite. Numbers to the left of the isotherms indicate the number of moles of CO_3 per formula weight of apatite. Isotherms for hydroxyapatite and for variscite in equilibrium with gibbsite are shown for comparison (*Source*: Khasawneh and Doll, 1978).

2. Gradients in pCa where solubility increases with decreasing activity of Ca^{2+} ion in the soil solution.
3. Gradients in $H_2PO_4^-$ activity in the soil solution where solubility is enhanced if there is a sink for $H_2PO_4^-$, and hampered if the activity of $H_2PO_4^-$ is at an elevated level.
4. At pH values below certain levels (see below), solubility increases with increasing CO_3^{2-} substitution for PO_4^{3-}.

To appreciate the significance of factor 4, which is a PR factor (whereas 1, 2, and 3 are soil factors), the solubility isotherms of fluorapatite ($x = 0$) and several carbonate apatites (CA) ($x = 0.5$–1.5), in equilibrium with fluorite, are shown in Figure 12. Thus, the solubility isotherms of CAs for two pCa (where pCa is the negative logarithm of Ca^{2+} ion activity) levels in solution intersect at different pH values (Fig. 12A). At the pH above these intersection points, FA becomes more soluble than CA does with solubility decreasing with increasing substitution. This implies that above certain critical pH levels, CAs are thermodynamically more stable than FA is (Khasawneh and Doll, 1978). However, as these pH values for the given CAs are higher than the pH range (3.5–6.5), considered by Chien (1977b), Figures 11 and 12 are not in contradiction. Khasawneh and Doll (1978) further emphasized that the increasing degree of carbonate substitution tended to bridge the gap between the isotherms of pure FA and HA, thus, indicating that the naturally occurring apatites are neither pure FA nor pure HA, but a series of isomorphically substituted products between these two products.

e. Kinetics of Dissolution of Phosphate Rocks

There have not been many kinetic studies on PR dissolution in soils. Chien (1977a) described the PR dissolution rate in 1 *N* NH_4OAc (pH 4.8), whereas Olsen (1975) described the dissolution rates in solutions saturated with EDTA. In both cases, the kinetic equations were similar and of the form,

$$c = kt^{1/n}, \tag{21}$$

where c is the concentration in solution at any time t, and n is a coefficient.

Plots of log c versus log t were constructed to find the kinetic coefficient k assumed to be related to PR reactivity. Khasawneh and Doll (1978) pointed out serious limitations of both methodologies and of correlating the resulting reactivity of PR source to the actual field situation. They further emphasized the importance of the intraparticle surfaces of sedimentary PR particles, having porous aggregate structures (Wilson and Ellis, 1984) in PR dissolution kinetics. They proposed that the limiting factor for PR dissolution in soil could well be the diffusion of dissolved Ca and P across the stagnant, saturated solution layers, both within and outside the particles. The total surface area, which varies with PR, was considered to cause the differences in their reactivities. However, the inherent chemical reactivity of PR to dissolve in acid solution will obviously depend also on the chemical and structural factors associated with the given PR, which do not need to be fully correlated with the surface area and should be considered in a discussion on the reactivity of materials in soil.

Chien et al. (1980a, b) fitted their data on the kinetics of dissolution of some PRs in soils to the modified Elovich equation (Eq. 14) to obtain values of the coefficients α and β. The reaction that was continued from 0.5 h to 1 week indicated that the water-soluble P concentrations in the treated soils decreased as the reaction progressed (owing to refixation of the solubilized P by soil) until a steady state was attained in which the dissolution rate of PR matched the P sorption rate by the soil. In this context, the term C_0 in Eq. (14), which was the value of C at $t = 0$, was taken to give the maximum P concentration that a PR can provide in a soil. Hence, the higher the value of C_0, the more likely the greater initial P uptake by plants from the given PR sample is (Chien et al., 1980a).

Significance of the coefficients α and β of the Elovich equation was discussed earlier in section II. The values of β were found to be linearly correlated with reactive Al, which was also correlated with α values on a log–log plot for several acid soils (Chien et al., 1980b). The P sorption capacity of these soils was further correlated with the amounts of reactive Al or active Al, but not with the amounts of free Fe_2O_3 and exc! ¬geable Al in the soils. It was further demonstrated that the relative amounts of water-extractable P of the PR with respect to concentrated superphosphate increased as the P sorption capacity of the soils increased (Chien et al., 1980b). This finding may have an important bearing in deciding the agronomic effectiveness of PR compared with CSP for use in strongly P-fixing soils.

The effect of temperature on the PR dissolution rate in tropical soils was found to be small (Chien et al., 1980a), implying that the effect of temperature on efficiency of PR in these soils is much less than that on water-soluble P fertilizers.

The extent and rate of PR dissolution in 30 acid soils was investigated by Hughes and Gilkes (1986a). An initial rapid dissolution within 1 day was followed by a continued slower dissolution up to 124 days. An increase in PR level caused a smaller proportion of the PR dissolving. Pyrophosphate- and oxalate-extractable Fe and Al were the main soil properties that control the amount dissolved. Soil pH, organic carbon, silt content (but not clay), and exchangeable Ca were found to be subsidiary predictive properties for some of the soils studied.

Syers and MacKay (1986) noted that the rate and content of PR dissolution increased with P sorption capacity of soils. In the same study, however, P uptake by ryegrass was poorly correlated with the extent of PR dissolution in 90 days, although a very good correlation of P uptake was obtained with Bray-extractable P from the PR-treated soil. It is thus imperative that the soil properties promoting PR dissolution be identified from those that control the subsequent amount of plant-available P in soil.

4. Agronomic Effectiveness of Phosphate Rocks

As stated earlier, the efficiency of PR directly applied to soil as P fertilizer depends on the properties of PR and soil. PO_4^{3-} substitution in the apatite structure by CO_3^{2-} and F^-, which causes a fall in the *a* axis dimension of the unit cell of apatite crystal, improves the reactivity of PR as measured by its solubility in neutral ammonium citrate and by crop responses. However, whether or not a particular PR is more effective in a given soil–crop situation often depends largely on the soil properties. Thus, a careful assessment of the situation is necessary in evaluating the agronomic effectiveness of the PR.

Particle size is one of the properties of PR that affect PR agronomic effectiveness. This is discussed below.

a. Particle Size of PR and its Efficiency

Evidently, the extent of reaction between PR and soil increases with an increase in surface area of contact between the two. Grinding the PR should therefore favor its dissolution in soil. However, the increase in solubility on grinding was not significant. Lehr and McClellan (1972) attributed the solubility differences to the chemical composition, primarily due to PO_4^{3-} substitution by CO_3^{2-} in the apatite structure. Reviews on experiments conducted using finely ground PR (Cooke, 1956; Khasawneh and Doll, 1978; Hammond et al., 1986) suggest that no additional benefit can be derived by having PR particle size < 100 mesh (150 μm). For optimum particle size of PR, it is now recommended that grinding may be done to ensure at least 80% of the material to pass through a 100-mesh sieve. The possibility of a plant root intercepting a PR dissolving zone in soil has also been considered, and this increases with a decrease in the particle size of PR (Hammond et al., 1986). The latter may guide, in part, the plant availability of P dissolving from PR. The possibility of root interception also increases with P fertilizer application rate, and with a more intimate PR incorporation into soil.

Measurement of surface area of the −150 + 270-mesh fraction of several PR sources (Wilson and Ellis, 1984) indicated that PR dissolution is partially governed by the amount of reactive surface that is determined by the particle size as well as the inherent porosity of the sample.

b. Results from Field Trials

Several experiments (Khasawneh and Doll, 1978; Engelstad and Terman, 1980; Hammond et al., 1986) have examined the agronomic effectiveness of PRs directly applied to the soil as P fertilizers and their residual effects. Many of these studies have also looked into the relative efficiency of PR materials with respect to other P fertilizer sources, notably the water-soluble fertilizers, for example, superphosphates. This section reviews the more recent studies and pertinent findings.

In a long-term field experiment conducted on a tropical Oxisol, Chien et al. (1987a) found that the decomposition rate of PRs was faster than it is usually believed to be in temperate soils, and the reaction products, rather than the untreated PRs, provided the residual available P to the plant. They further noted that the reaction products formed from PRs, even though of the same forms (Al–P and Fe–P), may be less crystalline than are those from triple superphosphate (TSP) which explains the higher residual available P value of PRs than that of TSP. These authors (1987b) also found that a 50:50 mixture of TSP and a reactive PR can be as effective as pure TSP in increasing plant yield in a limed soil. This is in contrast with the findings of Terman and Allen (1967) and Hammond et al. (1980) in which less reactive PR was used.

From a study of the agronomic potential of 11 Latin American PRs applied to acid soils, Leon et al. (1986) confirmed from the solubility in chemical extractants and crop response data that the extent of dissolution of these PRs was strongly correlated with both yield and plant uptake of P. The sources used were further grouped into four classes representing high, medium, low, and very low reactivity. Standard sources included for comparison were TSP,

Table 12. Average effectiveness and solubility measurements for sources within relative potential groupings

	Relative agronomic potential (RAP)			
	High	Medium	Low	Very low
Numer of samples in the group	4	4	5	3
Relative agronomic effectiveness (%)				
Mean	94	76	55	21
Range	85–99	74–79	42–67	12–28
Neutral ammonium-citrate-soluble P (%)				
Mean	2.6	1.2	0.5	0.2
Range	2.4–2.9	0.8–1.5	0.2–0.8	0.1–0.3
2% Citric-acid-soluble P (%)				
Mean	5.8	2.9	2.5	1.3
Range	5.0–6.7	2.1–3.8	2.1–3.1	1.2–1.3
2% Formic-acid-soluble P (%)				
Mean	9.1	3.1	2.2	1.4
Range	7.6–10.4	2.2–3.6	1.0–3.0	1.4–1.5
Ammonium citrate, pH 3, soluble P (%)				
Mean	10.9	4.5	2.0	0.6
Range	6.1–13.8	2.8–6.0	1.0–4.4	0.1–1.2

Source: Leon et al., 1986.

and five PR from the U.S.A., Tunisia, and Israel. The average values by group for relative agronomic effectiveness (RAE) based on dry matter yield and solubility of the sources are given in Table 12.

In a study of several P sources for a subterranean clover pasture (Bolland et al. 1984), the residual value of superphosphates, as measured by bicarbonate-extractable P, decreased from year 2 to 7, whereas that of calcined Christmas Island C-grade ore (calciphos) was very low for year 2, followed by an increase up to year 4, and finally a decline by year 7. The residual value of C-grade ore was low throughout, thereby rendering both the C-grade ore and calciphos unsuitable replacements for superphosphate fertilizers.

In another study, Bolland and Bowden (1984) confirmed that the effectiveness of all PR fertilizers used for subterranean clover remained approximately constant throughout the 6-year field experiment with successive crop seasons. Furthermore, the proportion of total P present in the PR, which was initially soluble in neutral ammonium citrate, was a poor predictor of the agronomic effectiveness of PR. Evidently, the soil properties largely influenced the effectiveness of the PR sources.

Phosphorus availability of plants results when PR is granulated with *S* before application to soil (Rajan, 1982a,b, 1983a,b). Increased P availability has been attributed to enhanced PR dissolution by sulphuric acid produced on S oxidation by *Thiobascillus* spp. bacteria. PR/S granules (Rajan, 1983a,b) and "biosuper" (Rajan, 1982a,b), which is a granulated PR-S mixture that has been inoculated with *Thiobascillus* spp., were used to assess the plant-availability of P to perennial ryegrass grown in volcanic ash soils of acidic pH. In both cases, association of S improved PR effectiveness. A greater residual effect of PR/S granules was suggested relative to superphosphate from the fractionation of the inorganic soil P, which indicated no Fe– and Al–P accumulation (Rajan, 1983a).

In more recent studies, the beneficial effect of S and *Thiobascillus* application along with apatite in increasing PR dissolution in soil has been observed leading to an increased crop yield and P uptake (Pathiratna et al., 1989) and available P-levels in an Oxisol (Muchovej et al., 1989).

Salih et al. (1989) highlighted the role of some P-dissolving fungi in increasing the plant-availability of P for sorghum grown on a calcareous soil from PR and also from superphosphate. The fungi, which proved particularly useful for the purpose, were one *Penicillium* spp. and two *Aspergillus foetidus* isolates. Tiwari et al. (1988) reported that composting rice straw with PR increased both citrate and water-soluble P, which was further increased by inoculation with *Aspergillus awamori.*

Hedley et al. (1982c), in another study, showed that the plant roots can use calcium phosphates in soil such as apatites under P-deficient conditions as a result of excess cation uptake by the plant. The latter causes the plant roots to excrete excess H^+ ions to restore the cation–anion balance within the plant, and consequently, PR dissolution in the rhizosphere zone is facilitated (Hedley et al., 1982a).

The agronomic value of six PR as a Ca source was assessed by Hellums et al. (1989) in an acid soil in the presence of 400 mg of P/kg as KH_2PO_4 to ensure that P was not a limiting factor for maize plant growth. Results showed that PRs with medium or high reactivity have potential Ca value, in addition to their use as a P source, when directly applied to acid soils also low in exchangeable Ca.

C. Partially Acidulated Phosphate Rocks

Partially acidulated phosphate rock (PAPR) refers to P fertilizers obtained upon PR acidulation with H_2SO_4 or H_3PO_4 at a smaller amount than that required for complete PR acidulation to produce single superphosphate (SSP) or TSP, respectively, at the expense of tricalcium phosphate. The product consists of water-soluble monocalcium phosphate and the unreacted PR. The presence of small amounts of dicalcium phosphates has been reported (McSweeney and Charleston, 1985). The nomenclature *% PAPR* generally denotes the proportion of acid used to prepare the PAPR relative to the quantity of acid necessary to produce the completely acidulated product from the same PR, that is, single superphosphate (SSP) or TSP. Hagin (1985) reported that the X-ray diffraction patterns of a PAPR is essentially similar to those of a mechanical mixture of PAPR components superphosphate and PR. Both patterns show clear peaks for apatite and several peaks for MCP,

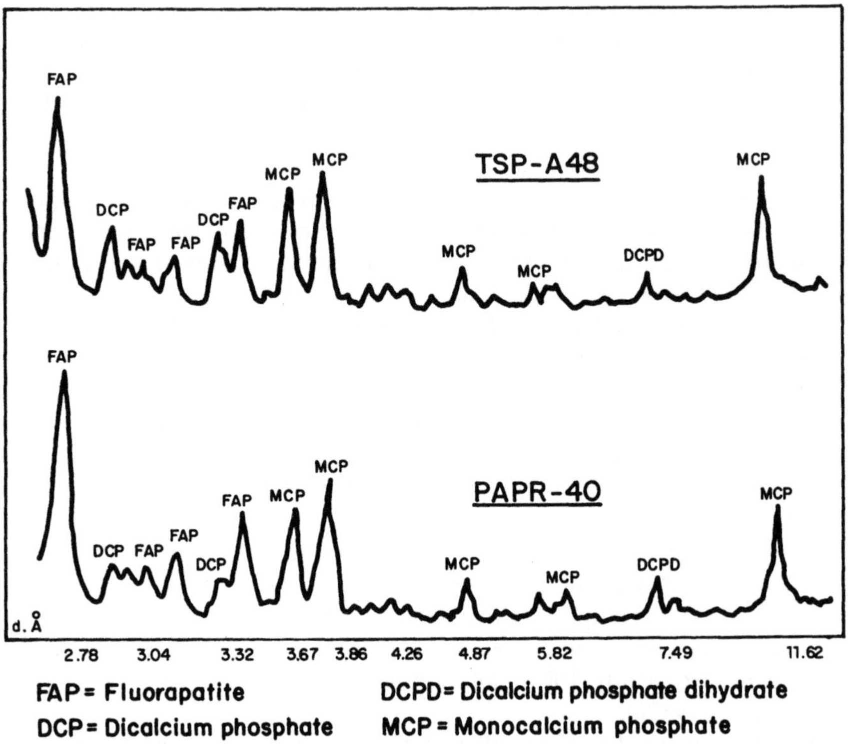

Figure 13. X-ray diffractions of a triple superphosphate mixture with phosphate rock (TSP-A48) and a partially acidulated phosphate rock (PAPR-40) (*Source*: Hagin, 1985).

apart from indicating the presence of some DCP and DCPD. Figure 13, taken from Hagin (1985), clearly illustrates this point.

1. Manufacturing Reactions

The chemical reaction for the production of PAPR from a PR and H_2SO_4 may be represented as (Hammond et al., 1986)

$$Ca_{10}(PO_4)_6F_2 + 7YH_2SO_4 + 3YH_2O \rightarrow 3YCa(H_2PO_4)_2 \cdot H_2O + 7YCaSO_4 + 7YHF + (1 - Y)Ca_{10}(PO_4)_6F_2, \quad \text{(f)}$$

where Y represents the degree of acidulation. For 100% acidulation, $Y = 1.0$, and $Y = 0.5$ for 50% PAPR.

When H_3PO_4 was used for treating PR, the P content of the product increased with increasing degree of acidulation, thus yielding a Mitscherlich curve (Rajan, 1985).

Although Sanchez and Salinas (1981) reported that the cost of PAPR production may be similar to that of imported TSP, Hammond et al. (1986) pointed out that in situations where use of local low-reactivity PR materials for PAPR production is considered important to substitute the imported products, the former may be considered advantageous for the following reasons:

1. In agronomic terms, PAPR can provide a portion of the P in a readily plant-available form and the remainder in a form that enhances the residual value.
2. When H_3PO_4 is used, the PAPR increases the soluble P concentration above that of the unacidulated PR.
3. When H_2SO_4 is used, sulfur is added as an additional nutrient.
4. The amount of acid requirement is reduced.
5. PRs that are chemically unsuitable for superphosphate production, or have too low a reactivity for use as a directly applied material in soil, can be used in the form of PAPR.

2. Reactions of PAPR in Soil

PAPR, being essentially a mixture of MCP and unreacted PR, undergoes reactions in soil in the same manner as do superphosphates, producing phosphoric acid and DCP upon initial wetting in soil (Nordengren, 1957; Garbouchev, 1981; Marwaha et al., 1983). However, unlike in TSP or SSP, the H_3PO_4 produced on MCP and DCP hydrolysis in PAPR tends to react with the unreacted PR within the fertilizer granule, forming more of water-soluble MCP, rather than diffusing out of the granule and reacting with soil components producing less-soluble Fe and Al phosphates. Accordingly, the reaction product of H_3PO_4 and apatite in the granule will be appreciably more soluble than the apatite fraction itself (Hagin, 1985; Rajan, 1985; Hammond et al., 1986). Indirect evidence of interaction between the H_3PO_4 produced and the initially unreacted PR was obtained by Logan and McLean (1977) and Mokwunye and Chien (1980).

a. Factors Affecting PAPR Reactivity

Several factors affect PAPR efficiency as a P fertilizer (Marwaha, 1983; Hammond et al., 1986). The degree of acidulation and the reactivity of the original PR obviously are important considerations (Khasawneh and Doll, 1978; Chatterjee et al., 1983; Stephen, 1985; Hammond et al., 1986; Stephen and Condron, 1986).

Factors Related to PR and the Acid Used for PAPR Production. The inherent PR reactivity to dissolve in acid solution (Section IIIB) depends on CO_3^{2-} + F^- substitution for PO_4^{3-} in the apatite structure. The particle size (applied granular or as powder), and the mode of placement of the fertilizer in soil also determines PAPR efficiency. Conflicting reports on the effect of particle size and mode of placement in soil on PAPR efficiency have been existent. Much of this controversy, however, may be reconciled (Stephen and Condron, 1986) by considering several aspects of the procedures employed by the various workers. Braithwaite (1987) found that although the fineness of PR and the quality of acid used directly affect the rate of reaction between PR and

H_3PO_4, none of these had any major influence on the total percentage of soluble P in the matured PAPR product.

Increase in the degree of H_3PO_4-acidulation of PR increased the cumulative total P uptake by ryegrass in a glasshouse experiment (Stephen, 1985), and improved the water and citrate solubility of PAPR (Rajan, 1985). However, Garbouchev (1981) found that PR acidulation with H_3PO_4 to 26%–30% gave a product with a 60:40 MCP/PR ratio. It was the most efficient product, requiring 68%–70% less of H_3PO_4 for its production. For certain Indian and Sri Lankan PRs, 20%–40% of acidulation was found to be most suitable for the purpose (Chatterjee et al., 1983). On increasing the extent of acidulation, the product quality deteriorated.

Another experiment demonstrated the effect of the acid generated in PAPR upon MCP hydrolysis on the untreated PR. Mokwunye and Chien (1980) found that water-soluble P in PAPR (20% by H_3PO_4) was higher after incubation than that after continuous shaking. Hagin (1985) explained this observation in terms of the more effective dissipation of acidity produced on MCP hydrolysis during shaking than during incubation. The latter was more effective in conserving the generated acidity for reaction with unreacted PR. Mokwunye and Chien (1980) further confirmed that the presence of PR in PAPR or in the mixture with concentrated superphosphate slowed down the immobilization of water-soluble P by reacting with some of the acidity produced during MCP hydrolysis, thus reducing the amount of acid available to solubilize soil Al and Fe.

Logan and McLean (1977) also found that more phosphate diffused out of ^{32}P-labeled 20% PAPR than from a 100% acidulated PR in an acid soil. Apparently, this can be attributed to a higher soluble P concentration maintained in 20% PAPR (Fig. 14).

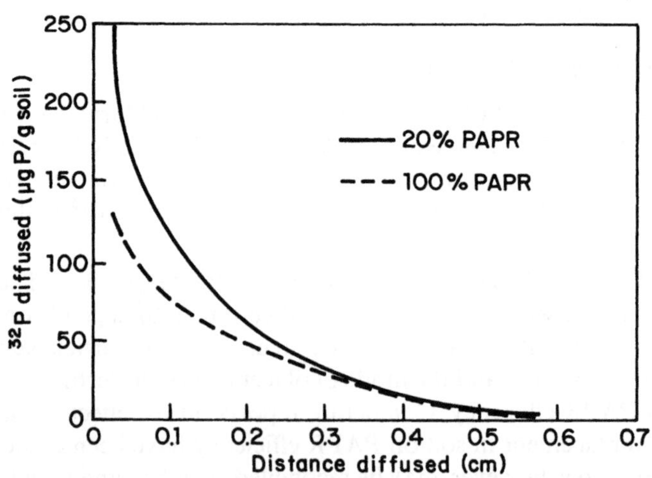

Figure 14. Distribution of ^{32}P in Venango soil from 20% and 100% acidulated rock phosphate (*Source*: Logan, T.J. and E.O. McLean. Diffusion of ^{32}P from partially acidulated rock phosphate. Soil Science 123, 203–206. © Williams & Wilkens, 1977).

Such interaction between the MCP component and the untreated PR was also noted by Rajan (1985). Stephen (1985) and McSweeney and Charleston (1985), however, found no evidence of soluble-P contribution from the unreacted PR. Similar conclusions were reported by Terman and Allen (1967) using Florida PR mixed with either SSP or TSP. On the other hand, a study examining the effect of addition of a reactive PR to immature SSP (being produced on acidulation of unreactive PR with H_2SO_4) showed a preferential consumption of the residual free acid by the reactive PR (RPR), leading to underacidulation of the unreactive PR and partial acidulation of RPR. Essentially, as a result, the SSP–RPR product consisted of an underacidulated SSP component and a PAPR (Bolan et al., 1987a).

The reactivity and plant-availability of the nonacidulated PAPR fraction was found to be markedly lower than the "soft" reactive PR when measured under greenhouse conditions (Resseler and Werner, 1989). Bolan et al. (1987a), Braithwaite (1987), and Timmermann (1972) suggested a decrease in reactivity of the residual PR component. Such a reduction in efficiency was assessed by Junge and Werner (1989) by means of the solubility data, P transformation, as well as the plant response data in acid soils. These authors further demonstrated that transformation and plant response to PAPR products were due to their acidulated P content, rather than due to any MCP hydrolysis-induced acidulation of the unreacted PR.

Electron-beam microanalysis of the nonacidulated P fraction of PAPR showed a surface "coating" with highly increased fluorine content surrounding the unreacted PR particles (Resseler and Werner, 1989). The latter was postulated to be responsible for low reactivity of the residual PR and its inferior agronomic effectiveness to that of the original mother PAPR. However, the surface characteristics of North Carolina PR, subjected to partial acidulation, were shown by scanning electron microscopy to have remained unaffected (reported by Stephen and Condron, 1986).

Hammond et al. (1989) also observed that the agronomic effectiveness of PAPRs with respect to SSP (in terms of dry-matter yield of maize) decreased with increasing $Fe_2O_3 + Al_2O_3$ content in PR. These authors suggested that the detrimental effect of $Fe_2O_3 + Al_2O_3$ content on the solubility and P availability of PAPR should be considered while selecting a PR for PAPR production.

With the conflicting reports available, there is not enough convincing evidence to support the earlier supposition (Nordengren, 1957) that the beneficial effect of using PAPR as a P fertilizer is due to the reaction between the dissolution products of the MCP component of a PAPR and the PR component.

The acid used for partial acidulation and the temperature used for drying the product also influence the efficiency of PAPR fertilizer. Thus, Hammond et al. (1980) found that 20% acidulation of Pesca PR with H_2SO_4 resulted in a product having water-soluble P considerably lower than that obtained by 20% acidulation with H_3PO_4. Furthermore, such partial acidulation with H_2SO_4 may lead to nonreversible agglomeration of the fertilizer granule by the cementing action of the calcium sulfate formed (Lehr, 1980). Hammond et al. (1980) also noted that partial acidulation of PR followed by drying the products at 110°C reduced the water solubility and agronomic effectiveness of the product. The latter might well have arisen from the dissolution rate of the residual PR in the PAPR falling due to the presence of $CaSO_4$, which also exerts a chemical retarding effect (Stephen and Condron, 1986). However, no reduction in the efficiency of such a PAPR product was observed when the partially acidulated product was dried at 25°–60°C instead of 110°C (Hammond et al., 1980).

Hydrochloric acid was also used to obtain PAPR from PR. The products were found to be comparable to those obtained on acidulation with H_2SO_4 (Shinde et al., 1978; Dash et al., 1981b, 1982b). One of the studies indicated that a 50%–75% HCl or H_2SO_4-acidulated PAPR may be used as a single application to an upland crop in an upland crop–rice rotation, especially on acid soils, where the water-soluble fractions of the product are used by the upland crop (wheat). During upland crop growth under aerobic soil conditions, the citrate-soluble and insoluble fractions undergo such transformations that make it possible for the following rice crop to utilize them under waterlogged conditions (Dash et al., 1981b).

Soil Factors. Soils of high P-fixing capacities, especially if they have low pH, tend to have high Fe and Al concentrations in soil solution. Addition of SSP or TSP encourages further dissolution of Fe and Al compounds through the acidity generated on MCP hydrolysis. This would cause rapid soluble-P fixation as insoluble Fe and Al phosphates. On such soils, several studies have demonstrated that finely ground PAPR (10%–20% acidulation with H_3PO_4) performed as good or even better than did TSP (McLean and Wheeler, 1964; McLean et al., 1965; McLean and Balam, 1967; McLean and Logan, 1970).

In soils with relatively low P-fixing capacity, on the other hand, granulated PAPR was reported to be less efficient than TSP, in greenhouse studies (Terman et al., 1964, 1970; Terman and Allen, 1967; Hammond et al., 1980). Recently, Chien and Hammond (1989) showed that PAPR effectiveness in increasing dry-matter yield and P uptake by maize (obtained with SSP) increased linearly as the soil P-fixing capacity of six soils increased. PAPR and SSP were equally effective at P-fixing capacities of 28%–36%, whereas PAPR was found to be superior to SSP in soils having higher P-fixing capacity. Indeed, McLean and Logan (1970) observed that the crop yield was best with 100% acidulated PAPR (i.e., superphosphate) in soils with lowest P-fixing capacity, whereas it was best with 20% acidulated PAPR product in soils with moderately high P-fixing capacity. For soils having intermediate P-fixing capacities, materials ranging from 20% to 100% acidulation were considered to be equally effective (McLean and Logan, 1970).

Stephen and Condron (1986) discussed the possible effects of soil pH on the effectiveness of PAPR materials, and observed that the relationship between P retention capacity of a soil and its pH, *vis-a-vis* the response to a PAPR, requires closer examination to generate important knowledge.

IV. Chemistry of Phosphorus Transformations in Submerged Soil

A. Physicochemical Changes on Flooding that Affect Phosphorus Availability

The behavior of phosphorus (P) in flooded lowland soils remarkably differs from that in upland soils. Moreover, the chemistry of P transformations in flooded soils has received rather scant attention compared with those in nonflooded soils. Flooding the soil increases the availability of native and added P (Patrick and Mahapatra, 1968; Ponnamperuma, 1965, 1972, 1985). Consequently, yield responses of lowland rice to fertilizer P are generally lower than those to *N*, or even to P for upland crops grown on the same soil (De Datta and Gomez, 1982). Lowland rice therefore has access to soil P sources ordinarily unavailable to other crops (Patrick and Mahapatra, 1968; Mahapatra and Patrick, 1969).

Several investigators have looked into the physicochemical changes that accompany flooding, and that are distinctly different from those in upland soils (Goswami and Banerjee, 1978; Jones et al., 1982; Patrick et al., 1985; Ponnamperuma, 1972, 1985; Tian-ren, 1985; Willett, 1986, 1989; De Datta et al., 1989). These changes indirectly affect the behavior of soil P that by itself does not participate in these redox processes.

Rice crop easily adapts to the environment. It can grow in various types of soils under a wide range of climatic and soil moisture conditions. Rice can be grown with a thin film of moisture on the soil surface, to about 10–50 cm of standing water (Mandal, 1984). It is mostly grown, however, in submerged soils with 10–30 cm standing water during most of its growth period.

Rice can also be grown under continuous, flooded soil, or under alternate wetting and drying conditions. These changes in soil moisture conditions in the rice fields affect the changes in soil, which in turn influence the transformation of native and applied P, P availability, and consequently, rice nutrition and growth. Changes in P availability in alternately flooded and drained soils are also important with regards to the growth of subsequent crops in rotation with rice.

1. Phosphorus Transformations Under Continuous Flooding

a. Causes of Changes in Extractable Phosphorus

Phosphorus availability in soil increases upon submergence due to the following changes (Goswami and Banerjee, 1978; Ru-kun et al., 1982; Ponnamperuma, 1985; Willett, 1986, 1989).

1. Reduction of ferric compounds. The reduction of free hydrous Fe oxides during flooding, and the liberation of sorbed and coprecipitated P as a result increased the levels of solution or extractable P in flooded acidic soil (Willett, 1986). The subsequent release of occluded P from within the

structure of amorphous Fe oxides has also been proposed (De Datta et al., 1989).

The chemical equilibria equations of the following types have been used to describe the activity of Fe^{2+} in solutions of flooded soils (Ponnamperuma et al., 1967).

$$Fe(OH)_3 + 3H^+ + e^- = Fe^{2+} + 3H_2O \qquad (g)$$

for reduction in the early stages of flooding, and

$$Fe_3(OH)_8 + 8H^+ + 2e^- = 3Fe^{2+} + 8H_2O \qquad (h)$$

in soils after prolonged flooding.

There is thus an increase of exchangeable Fe^{2+} ions in soil with a concomitant rise in soil pH and a decline in E_h.

Among the ferric hydrous oxides, ferrihydrite (of standard free energy of formation, $\Delta G_f^\circ = -677$ kJ/mol), the least stable oxide, has been postulated to undergo reductive dissolution (e.g., reaction g) first, releasing the sorbed P, in advance of the more stable oxides such as goethite ($\Delta G_f^\circ = -742$ kJ/mol) (Munch et al., 1978; Munch and Ottow, 1980). Recently, Willett (1989) showed that ferric oxide reduction was the dominant source of P released during flooding. However, the amount of P released was strongly inhibited by resorption. It was suggested that direct measurement of the amount of ferric iron reduced during flooding, and of P sorption are required to predict the net amount of P released during flooding (Willett, 1989).

Reduction of $FePO_4 \cdot 8H_2O$ to more soluble $Fe_3(PO_4)_2 \cdot 3H_2O$ or $Fe_3(PO_4)_2 \cdot 8H_2O$ has also been proposed under submerged conditions in soil (Ru-kun et al., 1982; De Datta et al., 1989). Fischer (1983), from a theoretical study of simple chemical systems, suggested that reduction of ferric hydrous oxides in the presence of P in solution takes place more easily than in similar solutions without P, because of the precipitation of vivianite $Fe_3(PO_4)_2 \cdot 8H_2O$. However, it was also shown (Willett and Cunningham, 1983; Willett, 1985) that if P, sorbed onto ferric hydrous oxide, is a significant source of P in soils, then it is the rise in pH (associated with the reduction of Fe^{3+} compound) rather than the fall in redox potential, E_h (favoring the reductive dissolution of ferric hydrous oxides) that is responsible for the relatively high P concentration in waterlogged soil solutions. An increase in pH would, in fact, favor P desorption from clay, aluminum oxides, and excess (not yet reduced) ferric oxide surfaces through a decreased surface positive charge. Such pH changes were, however, found to favor desorption of freshly applied P only, but did not affect P release in untreated soils (Willett, 1989).

Reductive dissolution of Mn (III) and Mn (IV) has not been found to affect P release during flooding (Willett, 1986). Thus, P release during flooding follows the reduction of ferric compounds, which, in turn, occurs after the reduction of manganese oxides (Willett, 1986).

2. Higher solubilities of $FePO_4 \cdot 2H_2O$ and $AlPO_4 \cdot 2H_2O$ resulted from hydrolysis due to increased soil pH in acid and strongly acid soils.
3. Organic transformations influencing P release. Organic acids released during anaerobic decomposition of organic matter under flooded soil conditions (Tsutsuki and Ponnamperuma, 1987) can increase the solubilities of Ca–P compounds by complexing Ca^{2+} ions, and thereby disturbing the solubility equilibria of Ca–P (Willett, 1986). Mandal and Mandal (1973) also attributed the observed lowering of fixation of applied P in the presence of added organic matter in flooded acidic lowland rice soils to the complexation of soil Fe and soil Al by the decomposition products of organic matter (Debnath et al., 1974; Mandal, 1979). Moreover, Welp et al. (1983) reported that up to 70% of the total soluble P in flooded soil solution is in organic combination.

   Organic matter in soil may also have an important effect on ferric iron reduction through its promoting influence on the bacterial activity in flooded soil. Willett (1986) reported that the level of organic matter governed the amount of P released in several soils, due to its effects on ferric reduction.

   Mineralization of organic P is generally too slow to be significant in plant nutrition (Tate, 1984), although mineralization rates increase under flooding (Islam and Mandal, 1977). Mineralization of organic P has been considered as a minor source of P in flooded soils (Patrick and Mahapatra, 1968; Uwasawa et al. 1988a) except in flooded organic soils (Racz, 1979; Reddy and Rao, 1983; Uwasawa et al., 1988b). On the other hand, Goswami and Banerjee (1978) found an increased mineralization of organic P upon flooding, particularly Fe-phytates in acid soil.

   Uwasawa et al. (1988a) and Willett (1989) also reported that the contribution of organic matter to P release during flooding appears to be mainly through organic P mineralization rather than by accelerating reduction of ferric compounds. Further, Kuo and Jellum (1987) suggested that the biological mineralization of organic N and the transformation of NH_4^+–N to NO_3^-–N is likely to play an active role in the seasonal pattern of water-soluble P in soils, which increased during the high rainfall periods compared with that in the dry months.
4. Release of phosphate ions from the exchange between organic anions and phosphate ions in Fe–P and Al–P compounds.
5. Increased solubility of Ca–P in calcareous soils as a result of pH depression due to CO_2 accumulation by organic matter decomposition.

   Thus the solubility of several Ca–P compounds, such as octacalcium phosphate, β-tricalcium phosphate, hydroxypatite, and fluorapatite, has been suggested to increase following a fall in pH after flooding a calcareous soil (Ponnamperuma, 1985).

   In acidic soils, such an accumulation of CO_2 under anaerobic conditions, would tend to bring down the pH, opposing thereby a pH rise due to reaction (g). This will also cause an increase in HCO_3^- concentration in the solution phase through the solvent action of CO_2 on carbonates, and would cause desorption of several exchangeable cations (e.g., Fe^{2+}, Ca^{2+}, Mg^{2+}, and NH_4^+) to maintain the electroneutrality in solution. These effects, coupled with that accompanying reduction of soil Fe^{3+} compounds, would increase the specific conductance and the ionic strength of the soil solution (Ponnamperuma, 1985).

6. Increased P diffusion under submerged conditions. Tian-ren et al. (1989) showed that flooding increases the buffer capacities for soil P. Ren (1987, quoted by Tian-ren et al., 1989) and Roy and De Datta (1985) showed that the buffer capacities for P sorption at low solution-P concentration (<1 μg/ml) increased up to fivefold. Thus, although the increased soil moisture content tends to bring down the soil impedance factor, and increase the P diffusion coefficient, a simultaneous steep rise in the buffer capacity may more than offset such increases. As a result, the P supply by diffusion in the soil to the rhizosphere zone may become the controlling factor for P uptake by lowland rice (Tian-ren et al., 1989).

To illustrate the complex nature of such effects, Tian-ren et al. (1989) considered the idealized buffer curves using the salient features of these curves given by Ren (1987 cited in Tian-ren et al., 1989) and Roy and De Datta (1985) (Fig. 15). Although the buffer capacity of the reduced soils was uniformly higher than that of the nonreduced soils, the amounts of added

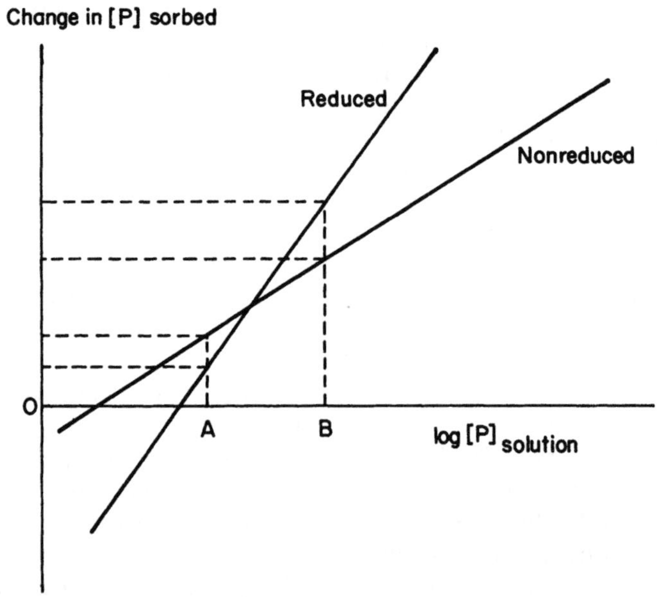

Figure 15. Idealized linear–log P buffer curves under reduced and nonreduced conditions. The true, unamended soil $[P]_{solution}$ values are those for which P addition has resulted in no change in $[P]_{sorbed}$; the ratio of the buffer capacities at a particular $[P]_{solution}$ is given by the ratio of the slopes. (*Source*: Tian-ren et al., 1989).

P required to maintain a given equilibrium P concentration for the reduced and nonreduced soils do not follow a uniform trend. Thus, at a lower solution concentration (A), the reduced soils require less of added P, but to maintain a higher equilibrium concentration (B), the amount of P added is more for the reduced soil.

Increased buffer capacity due to flooding has been attributed to P adsorption from soil solution by the reprecipitated poorly crystalline ferrous hydroxides or carbonates from Fe^{2+} ions formed by soil reduction (Patrick et al., 1985; Ponnamperuma, 1985; Tian-ren et al., 1989). An increase in pH opposed further P adsorption due to an increase in negative charge of the variable-charge P-adsorbing surfaces in the flooded soil. At the same time, an increase in ionic strength of the solution depresses the activity coefficients of the ionic species in the solution phase. The latter would tend to raise the concentration of phosphate ions and, hence, affects the ionic equilibria between solid-phase P and soil-solution P, opposing P desorption due to pH rise. Such a moderating effect of ionic strength on P sorption reduction at elevated pH was clearly demonstrated in Figure 16. In this case, increasing the salt concentration increases the P sorption at high pH, but decreases it at a lower pH. There is thus a point where the P sorption is independent of salt concentration, that is a point of zero salt effect on P sorption. The latter decreases with increasing levels of P sorption (Fig. 16), because the latter increased the negative charge of the soil. A similar behavior was noted by Bolan et al. (1986).

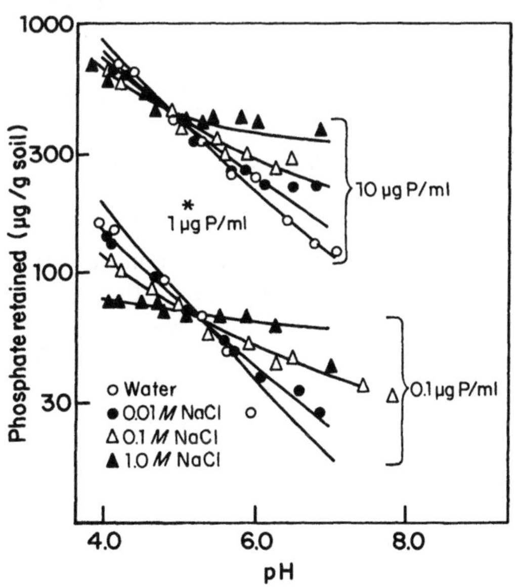

Figure 16. Effect of concentration of sodium chloride on phosphate sorption at the indicated concentration in solution. (*Source*: Barrow and Ellis, 1986).

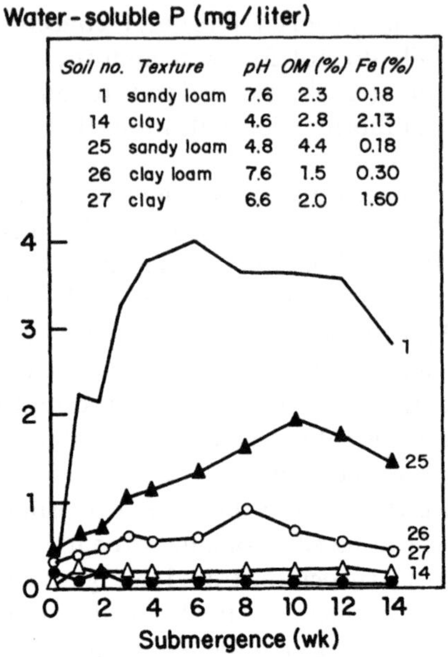

Figure 17. Kinetics of water-soluble *P* following flooding of a range of soils. (*Source*: Ponnamperuma, 1985).

7. Phosphorus mobilization resulted from an increased microbial activity in the presence of physiologically active rice roots and from the capacity of rice plants to reoxidize the rhizosphere during the later phase of the growing period (Alva et al., 1980).
8. In soils poor in free iron oxides, under highly reduced conditions, another process, shown in the following conversion increases the availability of P in flooded soils (Patrick and Mahapatra, 1968; Mandal, 1979):

$$Fe_3(PO_4)_2 + 3H_2S = 3FeS + 2H_3PO_4 \quad \text{(i)}$$

Flooding a soil increases the soluble P concentration in the soil and reaches a maximum before falling. Figure 17 illustrates this behavior. The subsequent fall in P concentration after reaching the peak has been attributed to readsorption of P on clays and Al hydroxides, precipitation, or microbial degradation of organic anions at the exchange sites, causing P resorption from soil solution (Patrick et al., 1985; Ponnamperuma, 1972, 1985; Tian-ren et al., 1989).

During prolonged flooding, the level of Fe^{2+} iron in solution stabilizes but the level of acid-extractable Fe^{2+} iron continues to increase (Willett, 1986). Precipitation of ferrosic (or ferrosoferric) hydroxide, $Fe_3(OH)_8$, on prolonged flooding was proposed by Ponnamperuma et al. (1967), and this compound was stated to have a large surface area with a high P-sorption capacity. The latter could well contribute to the decline of P concentration in soil solution on continued submergence (Willett and Higgins, 1978). However, P thus sorbed may still remain acid-extractable, and contribute to the labile pool of P in soil, and thus remain available to plants (Holford and Patrick, 1979).

However, ferrosic hydroxide has not been isolated or synthesized so far, and its existence has been speculated on the basis of conformation of the E_h, pH, and Fe^{2+} activity in soil solutions to Eq. (h) (Schwab and Lindsay, 1983; Willett, 1986).

The increase in water-soluble P concentration after flooding a soil is, however, less than that in stagnant lake waters, and is also strongly dependent on soil properties (Fig. 17). Maximum P concentration was found highest in sandy calcareous soils, low in Fe, moderate in acid sandy soils, and small in nearly neutral clay. Values were the lowest in acid ferrallitic clays (Ponnamperuma, 1972).

b. Forms of Phosphorus in Flooded Soils

Extractable P content in flooded soils increases, depending primarily on the distribution of different inorganic P fractions and the intensity of soil reduction.

Broeshart et al. (1965) observed a substantial increase in the available P in submerged rice soils free of $CaCO_3$. In general, higher P availability in flooded soils is attributed mainly to Fe–P; the role played by Al–P or Ca–P is usually secondary (Mahapatra and Patrick, 1969; Goswami and Banerjee, 1978; Mandal, 1979; Kuo and Mikkelsen, 1979; Lal and Mahapatra, 1979; Verma and Tripathi, 1982; Sah and Mikkelsen, 1986a). Willett and Higgins (1978) found that acetate and oxalate-extractable Fe and P sorptivity of soil largely increased upon flooding. On prolonged waterlogging, oxalate-Fe and P sorptivity levels reached values dependent on the free iron oxide content of the soils.

Mandal (1964) observed that in the presence of starch, 0.5 *N* acetic acid-extractable P considerably increased with a decrease in the Ca–P fraction. The large amounts of CO_2 formed by starch decomposition were said to have converted some insoluble tricalcium phosphates to MCP and DCP (Mandal, 1979).

Sah and Mikkelsen (1986b) also found that the anaerobic decomposition of the added cellulose in flooded soils decreased Al–P and increased Fe–P and reductant soluble (RS)-P. They attributed this to an increase, mediated by the anaerobic decomposition of the organic matter, of crystalline Fe transformation into amorphous forms, which, in turn, increased Fe–P and severe occlusion of P, resulting in the increase of RS–P fraction upon subsequent soil drainage.

Sarkar et al. (1986), while studying the thermodynamics of P equilibria in flooded acid soil with water-soluble and insoluble P fertilizers, suggested the formation of strengite as a fertilizer reaction product in soil. Singhania and Goswami (1978) also found a general trend of increases in Fe–P, Al–P, and RS–P in several flooded soils of India, with Ca–P registering an increase in black and laterite soils only. In a recent study in Thailand, Uwasawa et al. (1988b) reported a mixed trend of dominance of Fe–P, Al–P, and Ca–P in soils, depending on the soil characteristics.

2. Phosphorus Transformations Under Alternate Wetting and Drying

When flooded rice fields become dry, the reduced soil constituents are reoxidized with concomitant changes in E_h, pH, and ferrous iron concentrations. Drying a soil subsequent to flooding generally decreases the solubility of both native and applied P. Phosphorus applied before flooding was found to be immobilized to a greater degree than when P was applied after draining a soil rich in organic carbon and reducible Fe. In soils low in these, however, P applied before flooding was immobilized but P applied after drying was not (Willett, 1982).

Indeed, nonflooded crops grown in rotation with flooded rice have often been reported to develop P deficiency, but not rice (Willett et al., 1978; Willett, 1979; Brandon and Mikkelsen, 1979; Sah and Mikkelsen, 1986a,c). The P deficiency is more acute than in similar soil that has not been recently flooded (Willett, 1986). Rotation crops are thus expected to respond to P fertilizer under similar conditions.

However, some investigators also noted an increased native P-availability to rice upon submergence followed by soil drying (Savant et al., 1970; Shi et al., 1979, cited in Ru-kun et al., 1982). Ru-kun et al. (1982) suggested that this may have resulted from organic P mineralization in soil, whereas, Fe–P- and Al–P-availability may have actually decreased. Satyanarayana and Ghildyal (1970), on the other hand, observed that rice crop grown under flooded conditions produced better shoot growth and higher grain yield than when grown under 60 cm of soil moisture tension and saturated conditions.

Soil P sorption capacity and bonding energy for P were found to increase upon flooding, and then upon drying conditions (Willett, 1979, 1982; Sah and Mikkelsen, 1986c,d). Also, drying increased the amount of acid ammonium oxalate-extractable Fe (Willett, 1979; Sah and Mikkelsen, 1986e; Sah et al., 1989a) and the Fe-bound P at the expense of Al-bound P (Mahapatra and Patrick, 1969). Consequently, flooding and drying were suggested to increase the activity of ferric hydrous oxides in sorbing P (by way of decreasing their crystallinity) that resulted in added P immobilization after draining the rice soils (Mandal and Khan, 1975; Willett, 1982, 1986; Bradley et al., 1984). This was associated with the decreased plant-availability of P.

Much earlier, Patrick and Mahapatra (1968) suggested that the biological reduction of Fe during flooding, followed by reoxidation during drying, enhanced reactivity of the sesquioxide fraction of the soil, consequently increasing the P-fixing capacity, and hence, decreased P solubility. In agreement with this observation, a finding suggests that the induced P deficiency in soils subjected to flooded–drained conditions was due to high P sorptivity and low P desorption as a consequence of Fe transformations in soil (Sah et al., 1989b). Olsen and Court (1982), on the other hand, proposed that the alternate wetting and drying effects on P adsorption and desorption in soils are associated with changes in soil structure caused by the rewetting of dry soil samples.

The P sorption capacity as well as bonding energy for P sorption of soils from the flooded–drained systems increased with temperature and duration of prior flooding (Sah and Mikkelsen, 1986c,d). Willett (1979) also noted the increased Langmuir adsorption maximum and the bonding energy for P sorption by soils subjected to previous flooding. It was further demonstrated (Willett, 1979) that P-availability to maize grown in rotation with flooded rice was more closely related to the bonding energy between soil and P than to the soil's capacity to sorb P. This suggests that the depressed P supply to maize grown in previously flooded soils was due to stronger P sorption by the drained soils, rather than to P immobilization during flooding.

Phosphorus sorptivity and bonding energy of sorption, which increased under flooded–drained soil conditions, after quite some time declined when the previously flooded soil was drained (Sah and Mikkelsen, 1986f), but not to the same levels as that prior to flooding (Willett and Higgins, 1978; Sah and Mikkelsen, 1986f). Therefore, effectiveness of P fertilizer to crops after rice should increase with time after draining the rice soil (Willett and Higgins, 1980). More efficient use of P fertilizer may be achieved by delaying the sowing of the following crop as far as practicable (Willett and Higgins, 1978).

Addition of organic matter (Sah and Mikkelsen, 1986c, 1989) and an elevated temperature (Sah and Mikkelsen, 1989) greatly enhanced P sorption in drained soils from the flooded–drained system, thereby causing a higher P sorption for a relatively shorter period of previous flooding. The effect of organic matter was attributed to an increase of amorphous Fe in soil during the anaerobic decomposition of the organic matter (Sah and Mikkelsen, 1986e) as observed for soils under continuous flooding conditions (Willett and Higgins, 1978; Sah and Mikkelsen, 1986b).

Mandal and Khan (1977a) studied the effect of varying moisture regimes on P availability and transformations. Results showed that a higher amount of applied P in the saloid bound and Bray's available forms (see later) is maintained under saturated moisture conditions than that under alternate submergence and saturation, or under the continuously flooded moisture condition. This was attributed to a higher transformation of the applied P into Fe–P and Al–P under the former moisture regime. Mandal and Khan (1977b) also showed that the fixed Fe–P recorded a decrease in soils of high P-fixing capacity immediately after waterlogging, whereas it remained practically unchanged in low P-fixing soils.

B. Soil Test for Phosphorus in Flooded Soils

A soil test for formulating recommendations for fertilizer P requirement of lowland rice is beset with two problems specific to wetland soils (Chang, 1978; Goswami and Banerjee, 1978; De Datta et al., 1989).

1. A test for available P on an aerobic (air-dried) soil sample may not provide a satisfactory index of P availability after flooding because available P increases significantly due to submergence, the extent of such increase being dependent on the soil characteristics.
2. Wet soils are difficult to sample without altering their physicochemical properties.

Notwithstanding these, several attempts have been made to test for available P in air-dried soil samples from submerged fields, and correlate the findings with crop-response data.

Methods used may be classified into four major groups depending on the nature of the chemical extractants used (Chang, 1978).

1. Solution of a strong acid, such as 0.002 N H_2SO_4 (Truog, Ayres-Hagihara), 0.13 N HCl (Spurway), and 0.05 N HCl + 0.025 N H_2SO_4 (Mehlich).
2. Solution of an organic acid, or an acidified organic salt, such as NaOAc + HOAc (Peech, Morgan), 1% citric acid (Dyer), and CO_2 (McGeorge).
3. Alkaline solutions, such as 0.5 M $NaHCO_3$ (pH 8.5) (Olsen) and 0.1 N NaOH (Saunders).
4. Solution of a strong acid containing complexing radicals for Fe and Al, such as 0.03 N NH_4F + 0.025 N HCl (Bray no. 1) and 0.03 N NH_4F + 0.1 N HCl (Bray no. 2).

There have been several studies that tested the correlation of soil-available P, as determined by the above methods, with rice response to P application. Findings established the superiority of some of the chemical extractants to others for an index of available P in flooded soils (Datta and Datta, 1963; Patrick and Mahapatra, 1968; Mahapatra and Patrick, 1971; Chang, 1978; Goswami and Banerjee, 1978; Verma and Tripathi, 1982; Ponnamperuma, 1985; De Datta et al., 1989). It was particularly suggested that any method that can extract Fe–P and RS–P in aerated samples should provide a reliable measure of available P in lowland rice soils (Mahapatra and Patrick, 1971; Cholitkul and Tyner, 1971). Chang (1978) summarized the findings of several investigators in this regard in terms of the following:

1. When a group of soils is dominant in Fe–P (usually with a pH <5.5), good correlations are often obtained from most of the soil testing methods mentioned.
2. When a group of soils is dominant in Ca–P (usually with pH >6.5), methods employing an alkaline extractant (e.g., Olsen method) are better.
3. When a group of soils has mixed distribution patterns, having either Fe–P or Ca–P or both Fe–P and Ca–P as dominant fractions, an alkaline

extractant (e.g., Olsen method) or a weakly acidic extractant containing a complexing radical for trivalent cations (e.g., Bray no. 1 method) is superior.

Thus in general, Olsen and Bray no. 1 methods, especially the former, have been found to be universally applicable to all soil types.

The suitability of Olsen method to predict the soil P-availability in submerged soils from the values obtained on air-dried soil samples has also been established by correlation studies between aerobic and anaerobic P determinations (by Olsen method) on many soil samples, varying in pH from 4 to 8, and with different soil P distribution patterns (Chang and Maleewan, 1972). A high degree of correlation was found, which was independent of soil pH and only slightly affected by a drop of E_h. Phosphorus determined by the Bray no. 1 or Bray no. 2 method was also moderately correlated only when the soil pH values were >5. However, a decrease in E_h, caused upon submergence and addition of starch, greatly increased the P values as obtained by these two methods.

Figure 18 shows the variation of available P (Olsen P) of a rice soil derived from red earth during a 3-year experiment. It illustrates the point that the Olsen test on air-dried soil samples is a satisfactory measure of P availability. Thus, the variation of Olsen P was closely correlated with the net gains or losses of P, that is, the total P applied to soil minus the P removed by crops.

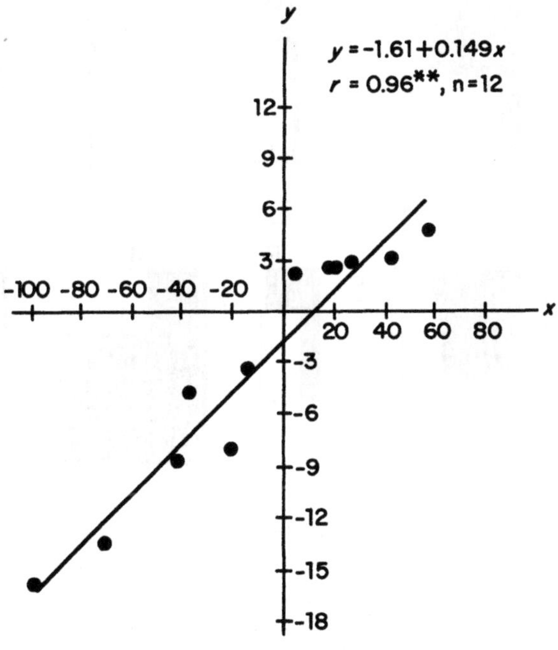

Figure 18. Soil phosphorus balance and Olsen's P change. x, net gain (+) or losses (−) of soil phosphorus (kg P/ha); y, Δ Olsen's P (ppm P). (*Source*: Ru-kun et al., 1982).

The accumulation and the depletion of soil-available P appear to have similar slopes.

V. Soil Organic Phosphorus

A. Chemical Nature of Soil Organic Phosphorus

1. Amount of Soil Organic Phosphorus

The organic P content of soil may vary from traces in arid regions to several hundred ppm in thick forest soils. Often, nearly half of the total P in soil occurs in organic forms, most of which is derived from plant residues and, in part, synthesized by soil organisms from inorganic sources. Figure 19 shows the typical distribution patterns of organic and inorganic P down the profiles of various cultivated and uncultivated soils.

Organic P in mineral soils normally decreases sharply down the profile,

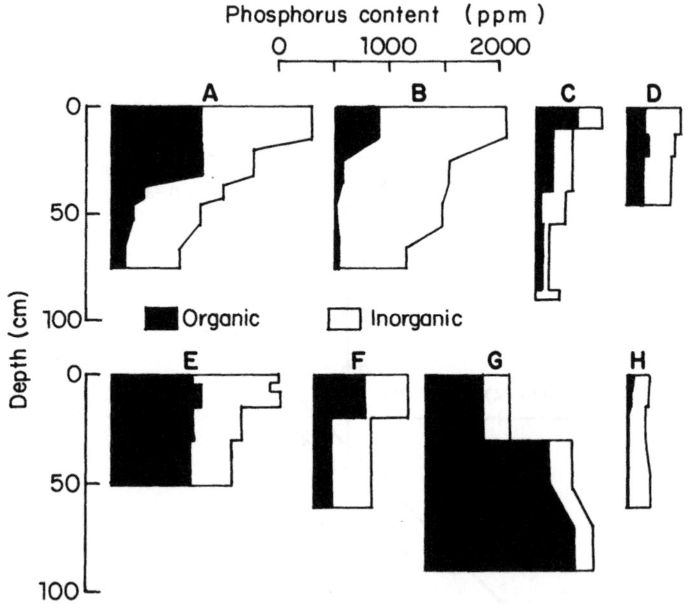

Figure 19. Phosphorus distribution in various soil profiles. A and B, freely and poorly drained cultivated clay loams of the Insch Association, Scotland; C, uncultivated Koputaroa soil developed on windblown sand, New Zealand; D, uncultivated Dawes silt loam, Nebraska, U.S.A., E, uncultivated Pima calcareous clay loam, Arizona, U.S.A., F, cultivated Orthic Deep Black, Melfort, Saskatchewan, Canada; G, uncultivated *Carex globularis* pine bog, Northern Finland; H, leached forest soil, Ibadan, Nigeria, Africa (*Source*: Anderson, 1980).

whereas in the case of peats, organic P often increases with depth. Organic P is generally higher in clay soils than in coarse-textured soils, but lower than in humus soils (Dalal, 1977). Poor drainage characteristics, high soil pH, and cultivation practices also adversely affect the organic P content of soils.

Organic P is located mainly in the fulvic acid fraction (Dalal, 1977; Mukherjee et al., 1979). Brannon and Sommers (1985a,b) also observed that typically more than 40% of the total organic P in soils is associated with the fulvic and humic fractions. However, these authors reported the association of organic P especially with high molecular-weight fractions. Brannon and Sommers (1985a) showed that one mechanism for organic P incorporation into model humic materials involved the linkage of a phosphate ester-amino compound into synthetic humics prepared by oxidative polymerization of polyphenols. They inferred that organic P was covalently bonded to humic compounds with molecular weights exceeding 10,000.

It is commonly believed that organic P has no direct effect on the P nutrition of plants. Organic P has to be mineralized before being absorbed by plants.

2. Analytical Techniques for Soil Organic Phosphorus Characterization

Total soil organic P is determined indirectly by ignition or extraction (Bowman, 1989). In the former, ignited and unignited soil samples are subjected to acid-extraction for soil P. The difference in the amounts of P in the extracts provides a measure of organic P in soil. In the extraction methods, organic P is obtained by taking the difference between total soil P and the inorganic P in soil extracts obtained with appropriate extractants.

Several sequential extraction schemes have been developed (Bowman and Cole, 1978a) that fractionate both inorganic and organic P into fractions based on the solubility of soil P forms in different chemical extractants. Although these schemes do not identify the chemical form of the P compound(s) extracted (Perrott et al., 1989), it is an easy way of quantifying organic P according to its susceptibility or resistance to certain chemical treatments. Data thus generated could form the basis for examining mineralization, microbial turnover, and plant-utilization dynamics of soil organic P (Bowman and Cole, 1978a).

Fractionation schemes have been used to follow depletion of different forms of soil P through various soil and cultural practices (Hedley et al., 1982b,c; Tiessen et al., 1983; Quin et al., 1984; Condron, 1986, cited in Perrott et al., 1989; Perrott and Mansell, 1989). Fractionation procedures have also been used to separate through sequential extraction the organic P in soil extracts into various components (e.g., phospholipids, nucleic acids, phosphoproteins, and acid-soluble esters) that can be further subdivided by chromatography or electrophoresis (Anderson, 1980).

A recently developed approach to the problem of characterizing the soil organic P involves the use of ^{31}P-nuclear magnetic resonance (NMR) spectroscopy to obtain both qualitative and quantitative estimates of the different forms of P in alkaline extracts of soil (Newman and Tate, 1980; Tate and Newman, 1982; Emsley and Niazi, 1983; Hawkes et al., 1984; Condron et al., 1985, 1990; Adams and Byrne, 1989; Gil-Sotres et al., 1990). The major advantage in using ^{31}P-NMR is that it is analytically less complex than the detailed partition chromatography techniques otherwise required for identifying specific organic P compounds. Most ^{31}P-NMR studies of soil organic P have employed a rapid single-extraction technique involving ultrasonic dispersion in 0.5 *M* NaOH, which frequently removes less than half of the total organic P from many of the soils examined (Tate and Newman, 1982; Hawkes et al., 1984). Condron et al. (1985) suggested a combination of sequential acid and alkali extractions and an ultrafiltration technique for nearly quantitatively extracting the soil organic P prior to NMR analysis.

Recently, a chelating, cation-exchange resin, in preference to NaOH-extraction has also been used to prepare the soil extracts for ^{31}P-NMR analysis. It was claimed to be more suitable for use with NMR analysis (Adams and Byrne, 1989).

Figure 20 shows the ^{31}P-NMR spectrum of the alkali extract from an acid soil. Tate (1984) observes that the future prospects for quantitative separation of unaltered organic P from soil are not promising because of the chemical complexity of organic P fractions in soil, the ease of hydrolysis of some of these compounds during extraction, strong organic P sorption in soil by clays, and the interaction of soil organic P with metal cations forming insoluble salts.

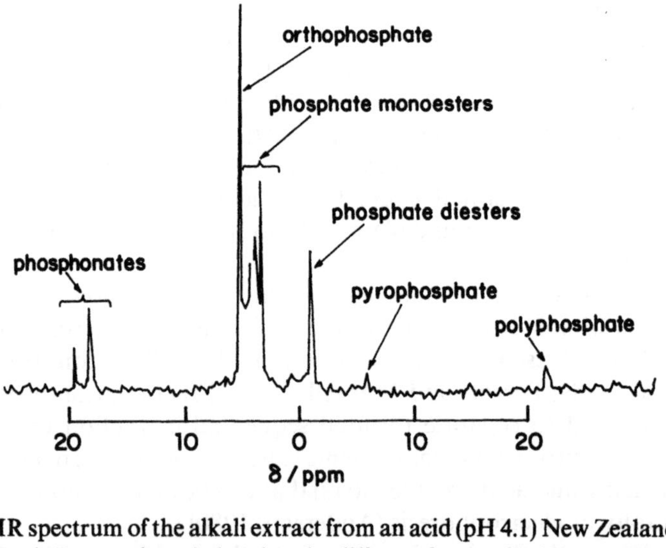

Figure 20. ^{31}P-NMR spectrum of the alkali extract from an acid (pH 4.1) New Zealand high-country soil (Typic Dystrochrept) showing the different forms of P (*Source*: Tate, 1984).

3. Nature of Soil Organic Phosphorus Compounds

According to Anderson (1980), soil organic P compounds can be generally classified into three groups, namely, (1) the inositol phosphates, the major constituent with inositol hexakis- and pentakisphosphates, primarily of plant origin, comprising up to 60% of soil organic P (Tate, 1984), (2) the nucleic acids, and (3) the phospholipids. The presence of phosphoproteins, sugar phosphates, glycerophosphates, and phosphonates has also been reported in soil organic P (Dalal, 1977; Tate, 1984). A large proportion of remaining soil organic P is still chemically unidentified but probably occurs as insoluble complexes with clay minerals and organic matter (Tate, 1984).

Brannon and Sommers (1985a) noted that a portion of unidentified organic P in soil humic substances may result from the incorporation of organic compounds containing both amino and phosphate ester functional groups into humic materials during oxidative polymerization of polyphenols. The organic P thus formed will be resistant to both chemical and enzymatic hydrolysis.

Table 13 gives the relative distribution of organic P compounds in some soils from different countries.

a. Inositol Phosphates

The parent cyclic polyol, inositol, can have several stereoisomers. Phosphate esters of myo-, scyllo-, neo-, and chiro-inositol have been identified in soil (Anderson, 1980). The hexakisphosphate of myo-inositol or the phytic acid, and its Ca–Mg salt, usually called phytin, have been reported in plants although other inositol stereoisomers may also be present in an unphosphorylated form (Dalal, 1977; Mukherjee et al., 1979). Among the inositol stereoisomers, the forms that follow the myo-form in soil organic P in terms of abundance are the scyllo-, chiro-, and neo- in decreasing order (Dalal, 1977).

The first studies of NaOH-soluble organic P compounds by ^{31}P-NMR (Newman and Tate, 1980; Tate and Newman, 1982) suggested that inositol phosphates probably constituted the main group of orthophosphate monoes-

Table 13. Percent distribution of organic phosphorus compounds in some soils

Soils from	Inositol P	Lipid P	Nucleic acid P
Australia	0.4–38	—	—
Bangladesh	9–83	0.5–7.0	0.2–2.3
Britain	24–58	0.6–0.9	0.6–2.4
Canada	11–23	0.9–2.2	—
New Zealand	5–43	0.7–3.1	—
Nigeria	23–30	—	—
U.S.A.	3–52	—	0.2–1.8

Source: Dalal, 1977.

ters. Adams and Byrne (1989), Condron et al. (1985), and Gil-Sotres et al. (1990) also reported that orthophosphate monoesters, which include inositol phosphates, were the dominant organic P components of soil extracts. Condron et al. (1985) found in a long-term field experiment, using superphosphate as the annual P fertilizer, that nearly all the organic P, accumulated in the soil from P fertilizer addition, was detected in the orthophosphate monoester fraction. The fine structure observed in the monoester region of the ^{31}P-NMR spectrum was not resolved sufficiently well to enable identification of individual species, but the peak maxima for 0.1 *N* NaOH and 0.5 *M* NaOH soil extracts did not overlap. This suggested that there are differences in the predominant forms of the monoesters in these two fractions.

Hawkes et al. (1984) also noted that orthophosphate monoesters was the major form of organic P that accumulated in a pasture soil after annual application of superphosphate for 100 years.

The presence of Fe and Al inositol hexakisphosphates has been reported in organic P fraction of soil (Saxena, 1979). A part of inositol phosphates also occurs as complexes of humic and nonhumic substances of high molecular weight in humic and fulvic acid fractions, possibly involving metal bridges. The determination of the inositol phosphates requires alkaline hydrolysis and hypobromite oxidation to release it from the humic substances (Hong and Yamane, 1980a,b). In particular, more drastic procedures are required in determining the inositol phosphate in humic acid (ha) than in fulvic acid (fa). Hong and Yamane (1981) attempted to characterize the inositol phosphate in ha and fa fractions through the ha and fa fractionation by Sephadex gel filtration. Organic P in the two fractions of ha (according to the molecular size) and the high-molecular-size fraction of fa consisted of only inositol phosphate, whereas that in the lower-molecular-size fraction of fa consisted of other organic P compounds as well.

Inositol phosphates have been reported to be much less mobile than, for instance, the phospholipids in soil, thereby tending to accumulate in soil (Cole et al., 1977). Stabilization by interaction with and sorption to clays and the sesquioxides (Stewart and Tiessen, 1987), and the inability of some phosphatase to dephosphorylate penta- and hexaphosphates have caused the accumulation of these phosphates in soils (Tate and Newman, 1982). Rojo et al. (1990) also reported that organic P distribution in 12 soil fractions studied suggested an important association between soil P and humic compounds.

Indeed, the soil fractions containing clay and humic colloids showed the greatest organic P concentrations, although fractions containing plant debris indicate the likely presence of organic P associated with non- or partially humified organic matter. Syers et al. (1969) noted earlier the greatest organic P content in soil fractions with the smallest particle size.

b. Nucleic Acids

The presence and role of P in building up the structure of nucleic acids are well known. Early evidence of the existence of nucleic acids in soil organic matter was obtained by isolation from soil extracts of adenine, guanine, cytosine, xanthine, and hypoxanthine (Saxena, 1979). These nucleic acids may be distinguished from inositol hexakisphosphates by alkaline hypobromite degradation of nucleic acids (Saxena, 1979). Even though the rate of addition of nucleic acids to soils is probably much greater than that of inositol phosphates, nucleic acids are mineralized in most soils much more readily, and incorporated into microbial biomass, therefore occurring only in small quantities in soil (Condron et al., 1985). It has been suggested that the purine and pyrimidine bases obtained from soil humic acid may be present, not in DNA, but in related "limit" polynucleotides left as very stable residues after enzyme attack on DNA (Anderson, 1980).

c. Phospholipids

Phosphoglycerides are the dominant fraction of soil phospholipids, whereas choline phosphoglyceride is the dominant soil phospholipid, followed by ethanolamine phosphoglyceride (Dalal, 1977). An important group of phospholipids, containing myo-inositol, is known as the phosphoinositides. Examples of this can be found in many animal and plant tissues, and in a variety of microorganisms (Anderson, 1980).

There have been some speculations about the origin of soil phospholipids and their contribution to P cycling (Dalal, 1977; Cole et al., 1977). Plants are a major potential source of these compounds, but microbially derived diester-P, including phospholipids, can also make an important contribution. A positive correlation was found between the soil ATP or biomass C content (which are indexes of microbial biomass) and the choline phosphate contents of the soil extracts (Tate and Newman, 1982). It was thus suggested that mineralization of microbially derived diester-P, including phospholipids, could be replenishing the often meager concentration of native inorganic orthophosphate (Molloy and Blakemore, 1974). Tate and Newman (1982) also observed a strong positive correlation between diester-P in the extracts, and annual precipitation. Condron et al. (1985) reported that choline phosphate constituted a significant proportion of the monoester-P found in the acetyl acetone and 0.5 *M* NaOH extracts of soil.

d. Other Phosphate Esters

The presence of sugar phosphates in soil extracts has been reported by some authors (Anderson and Malcolm, 1974; Saxena, 1979). In the ^{31}P-NMR studies of soil organic P compounds, Adams and Byrne (1989) showed that both pentose (ribose-5-phosphate) and hexose (glucose-6-phosphate) sugar phosphates produced P-resonances in the observed monoester region of the NMR spectrum. A trial revealed that when glucose-6-phosphate was added to a soil sample, it could be recovered by a chelating cation-exchange resin and taken through the concentration procedure without appreciable hydrolysis. Muller-Harvey and Wild (1987) also reported the synthesis of a xylose phosphate as a result of microbial activity in forest litter. The sugar phosphate was soluble in water, did not react with acid phosphatase, and accounted for about 5% of the organic P in the litter. Hence, whereas inositol phosphate may be extracted by the chelating cation-exchange resins, sugar phosphates could also contribute to the monoester signal in the ^{31}P-NMR spectra (Adams and Byrne, 1989).

Phosphorylated carboxylic acids other than uronic acids have been isolated from 3*M* NaOH soil extracts, having C/P ratios of approximately 7: or 8:1 (Anderson and Malcolm, 1974). Two esters, each containing glycerol, myo-inositol, chiro-inositol, and an unidentified component, were also detected in the alkaline soil extract (Dalal, 1977).

B. Changes in Soil Organic Phosphorus Due to Cultivation

Depletion of organic P in soil induced by cultivation has been reported by several authors (Chater and Mattingly, 1980; Tiessen et al., 1982, 1983; Tate, 1984). Frequently, cultivation tends to cause a greater degree of aeration that, in turn, encourages the microbial activity, leading to a greater decomposition rate of organic matter (Dalal, 1977). The various forms of soil organic P compounds also differ in their ease of mineralization. The more easily mineralized forms of organic P in the soil environment are the orthophosphate diester-P, which includes phospholipids and nucleic acids rather than the orthophosphate monoester-P, which includes inositol phosphates (Hawkes et al., 1984).

However, the net changes in the soil organic P pool under growing crops are usually very small in relation to total pool sizes, and consequently, observations over several years are required to detect significant changes (Friesen and Blair, 1988). Thus, the annual mineralization rates of organic P determined over periods of 20–50 years at Rothamsted ranged from 0.5–3.2 kg of P/ha in plots not receiving additions of farmyard manure (Chater and Mattingly, 1980). However, this does not imply that organic P is unimportant for although the net changes in this pool are small, considerations from the viewpoint of turnover rates of organic P may become important (White and Ayoub, 1983; Friesen and Blair, 1988).

The presence of a growing plant has been found to enhance P release from the incorporated crop residues (Blair and Boland, 1978). This was, however, contested by Friesen and Blair (1988), who found that cropping only marginally slowed the rates of transfer of inorganic and released residue P into nonlabile pools. Furthermore, McLaughlin et al. (1988a) observed that most of the P taken up by wheat plants grown in a soil, freshly supplied with crop residues, was from soil P, that is, from sources not added that season. On the other hand, the presence of crop residues was found to decrease P uptake by ryegrass derived from a soil poorly supplied with P, whereas in a soil rich in P, higher amounts of P were used and the effects of the residues on absorption of P derived from the soil was less conspicuous (Thiband et al., 1988).

Tiessen et al. (1983) observed that the labile organic P forms in soil, extracted with $NaHCO_3$, rapidly depleted during cultivation. This agrees with the findings of Bowman and Cole (1978b) who found that the rapid hydrolysis of $NaHCO_3$-organic P in soil coincided with the concomitant increase in the plant-available bicarbonate-inorganic P fraction. NaOH-extractable organic P, from the coarse silt and fine clay-associated soil organic matter fractions, was also found to be labile, and therefore suffered significant changes during one growing season (Grindel and Zyrin, 1965), or during several years of cultivation (Batsula and Krivonosova, 1973). In contrast, such P associated with fine silts and coarse clays was quite stable, and therefore corresponds to slower, long-term transformations of soil organic P during soil formation or prolonged cultivation. The latter may also correspond to the highly stable NaOH-extracted organic P, identified by Bowman and Cole (1978a), which may be used to show differences between P forms in different soil types. Tiessen et al. (1983) observed that labile P fractions were greatly reduced during cultivation, indicating a significant reduction in available P, and P fertility of cultivated soils. This reduction was intimately associated with the soil organic matter loss.

Hedley et al. (1982b) studied the changes in inorganic and organic P fractions using a sequential extraction technique in 65 years of continuous cropping in wheat–wheat–fallow rotation. Total P was found to be lower by 29% in the cultivated soil as compared with the adjacent permanent pasture, whereas the major P loss (74% of total P lost) was organic P and residual P. Of the total P lost, 22% was from extractable organic P forms, and 52% originated from stable P. Evidence also showed, from the laboratory incubation studies under simulated fallow with and without residue incorporation, that microbial activity plays a major role in redistributing P into different forms in the soil.

Tiessen et al. (1982) also reported that losses of soil organic matter did not level off even after 60–70 years of cultivation, in contrast with earlier observations (Martel and Paul, 1974). The continued decline was caused by increased erosive losses associated with decreased organic contents of the soil (Voroney et al., 1981; Tiessen et al., 1982), and by an extension of the zone of depletion. Soil P was primarily lost from the organic P fraction until this fraction was depleted sufficiently for dissolution of apatites to occur. In soils with greater water percolation, however, such dissolution of apatite may also occur concurrently with organic P mineralization (Tiessen et al., 1982).

Saunders and Metson (1971), who studied the seasonal variations of P in soils and pastures, suggested that the high available-P status of soils under pasture in spring was due to the release of P from organic residues and soil organic matter by mineralization. A stimulatory effect of liming on the mineralization of organic P has also been observed by several workers (Islam and Mandal, 1977; Haynes, 1984; Perrott and Mansell, 1989), due probably to the creation of a favorable environment for microbial growth, and activity, and probably also increased solubility and availability of some organic phosphate esters (e.g., inositol polyphosphates) (Anderson, 1980).

In a long-term experiment on P fertilizer in an intensively grazed irrigated pasture over a 25-year period (1952–1977), Condron and Goh (1989) reported that the rate of organic P accumulation in soils decreased with time. A marked decline in soil organic P between 1971 and 1974 was attributed to increased mineralization as a result of lime addition in 1972.

However, precipitation on lime particles of P released by mineralization may interfere with its ready availability to plants (Haynes, 1982).

Crop simulation modelling studies on the recycling of soil P have also suggested an increased mineralization of organic P in soils under cultivation, which meets a significant proportion of the total P requirements of the plant (Tate, 1984).

C. Biological Transformations of Soil Phosphorus

1. BuildUp of Microbial Biomass Phosphorus in Soil

Microorganisms are important in the transformations of organic and inorganic forms of P in soils. Renewed interest in the role of microorganisms in P cycling developed through the use of simulation models. The microbial biomass, comprising about 2%–3% of the total organic carbon in soil, is a relatively labile fraction of soil organic matter. It is thus a more important repository of plant nutrients than its small size might indicate (Jenkinson and Ladd, 1981), and particularly a key site for soil organic P mineralization (Anderson and Domsch, 1980; Brookes et al., 1984).

Cole et al. (1977) estimated by simulation technique that soil biomass P uptake was 3–5 times as great as plant P uptake in semiarid grassland. Halm et al. (1972) predicted from microbial biomass estimates and from published values of microbial P contents of laboratory-cultured microorganisms that the soil biomass of a cool, native grassland could contain more P than could the vegetation. However, P concentrations in laboratory-grown microorganisms can vary widely, depending on the age of organisms, P content of the growth medium, and the conditions under which the organisms are grown (Van Veen and Paul, 1979; Tate, 1984). Direct measurement of the soil biomass P is therefore essential for correct assessment of the role of the microbial biomass in P cycling and in crop nutrition.

Subsequently, direct estimates of biomass P were made possible with the development of techniques to extract and measure this pool in soils (Brookes et al., 1982; Hedley and Stewart, 1982). The techniques utilize chloroform ($CHCl_3$) fumigation for the biocidal agent. The P content of the microbial biomass is calculated from the difference in the amounts of P extracted with 0.5 *M* $NaHCO_3$ (pH 8.5) from fumigated and unfumigated portions of soil. A factor (k_P) takes care of the fraction of soil biomass P extracted after fumigation. The values set for k_P are provisional, and Hedley and Stewart (1982) noted that no single k_P factor can be used to calculate accurately the amounts of microbial P in a wide range of soils. However, studies with soils with pH values ranging from 6.2 to 8.2 suggested a value of 0.4 for k_P (Hedley and Stewart, 1982), which agreed with the value assumed by Brookes et al. (1982).

A disadvantage of these techniques for biomass P measurements, when applied to field soils, is that an incubation period is used before fumigation ("preincubation") (McLaughlin et al., 1986). The method due to Hedley and Stewart (1982) used dried, ground, rewetted, and incubated (21 days) soil, whereas that of Brookes et al. (1982) used a 10-day-old incubation of sieved (to remove roots), field-moist soil. Both of these incubation procedures may lead to qualitative and quantitative changes in the soil biota, and hence affect the biomass P estimates. McLaughlin et al. (1986), on the other hand, suggested a modified procedure for the measurement of biomass P in field soils, while testing a wide range of biocides and extractants. They found $CHCl_3$ and hexanol to be the most effective biocides, and suggested that the soils be treated with hexanol or extracted immediately after sampling (with 0.5 *M* $NaHCO_3$, pH 8.5), thereby avoiding preincubation and changes in the biota or biomass. The errors arising from the inclusion of roots in the sample were minimized in this method by removing the bulk of the roots before fumigation.

However, plant roots, if unaccounted for, may cause a serious overestimate of soil biomass P results. Calculations based on published values of rooting densities of field crops have indeed shown that significant errors in the determination of soil biomass P may occur with the $CHCl_3$-fumigation technique unless the extraction time is extended considerably (to 16 h) (McLaughlin and Alston, 1985). Sparling et al. (1985a), on the other hand, proposed that for soil samples with a high viable root density (>6 mg/g), as may be found in dense pastures, greenhouse pot experiments, or rhizosphere soil samples, a preincubation of the soil samples for 7 days prior to fumigation and analysis would considerably minimize the risk that the root materials may be included in the microbial biomass estimates.

Martin and Foster (1985) observed that removal of roots also removes a large population of microorganisms present in the soil adhering to the root surface. In studies of the rhizosphere, where rooting densities are high, this poses a serious problem.

McLaughlin et al. (1987) described a technique, involving the use of porous membranes, to overcome the difficulties of separating the roots from the soil during plant growth while measuring the microbial P in the rhizosphere soil.

It may be noted in this context that some of the P in the 0.5 *M* $NaHCO_3$ (pH 8.5) extract of air-dried soil samples may be derived from the microorganisms killed by drying the soils (Bowman and Cole, 1978b). Consequently, values for plant-available P levels of fresh soils, based on extracts from air-dried soils, such as those used in the Olsen test, may be overestimates, especially for those soils with a high microbial biomass. It is, therefore, necessary to quantify the contribution from the microbial biomass to levels of available-P extracted from soils (Sparling et al., 1985b).

In recent years, microwave irradiation were used as a controlled soil biocidal treatment that could selectively kill microbial biomass (Ferris, 1984;

Speir et al., 1986). The effect on microorganisms of this treatment appears to be entirely thermal.

In a study, Brookes et al. (1983) demonstrated that the soil microbial biomass does not maintain a high adenylate energy charge (AEC), which is a linear measure of the metabolic energy stored in the adenine nucleotide pool (Atkinson, 1977), when air-dried. Once remoistened, the population tends to restore its AEC to the original value. This restoration occurs so fast that it was not attributed to the formation of a new biomass.

Brookes et al. (1984) found a linear relationship between soil biomass C and soil biomass P. The calculated mean annual flux of P through the biomass in grassland soils was more than three times the value through arable soils, suggesting a significant role of biomass P in plant nutrition in grassland. In addition, the decline in biomass P, when an old grassland soil was put into an arable rotation for about 20 years, was sufficient to account for about 50% of the decline in total soil organic P during this period. When an old, arable soil was reverted to woodland soil, organic P doubled in 100 years; biomass P was found to increase 11-fold during the same period.

McLaughlin and Alston (1986) demonstrated that the microbial biomass assimilated a proportion of the applied fertilizer equal to plant uptake, and that a large proportion of the P derived from pasture residues entered and remained in the microbial pool. Incorporation of P derived from fertilizer and plant residues into the soil microbial biomass under field conditions was further studied by using isotopic double labeling (McLaughlin et al., 1988a,b,c). It was found that most of the P taken up by the microbial biomass was derived from native soil P (McLaughlin et al., 1988b). In a study of organic P turnover and P cycling in wheat–pasture rotations (McLaughlin et al., 1988c), an integrated P cycle for the soil under wheat–pasture rotation was developed. It was further shown that fertilization of the pasture phase of the rotation stimulates the buildup of residual inorganic and organic P, whereas fertilization of the wheat phase predominantly stimulates the accumulation of inorganic forms of P in the soil.

2. Phosphorus Transformations in Plant Rhizosphere

The rhizosphere is an especially active site for P transformations because bacteria and fungi, many of which are actively engaged in P transformations in soils, are 20–50 times more abundant in the rhizosphere than in the bulk soil (Newman, 1978; Rovira, 1979), and because roots withdraw P from the soil solution and may also exude or leak P back into the soil. Indeed, the growing plants largely influence the reactions taking place in soil, which affect P transformations (Helal and Sauerbeck, 1984; Jungk and Classen, 1986). The competition for the small amount of P in the soil solution between plant roots and microorganisms is most intense in the rhizosphere. Although rhizosphere microorganisms benefit plant growth, the mechanisms are far from being clearly understood (Tate, 1984). Several mechanisms of plant or microbial origin have been proposed by which the P concentration in the soil solution of the rhizosphere may be increased. These have been summarized by Grinsted et al. (1982) as follows:

1. Low-molecular-weight organic acids are secreted by, or exude from plant roots or microorganisms, and lower the rhizosphere pH, or accelerate the dissolution of sparingly soluble phosphate minerals by complexing the metal cation of the mineral.
2. Organic acid anions accumulate in sufficient concentrations in the rhizosphere to compete effectively with orthophosphate for adsorption sites on Fe or Al oxides.
3. The plant alters the rhizosphere pH, and hence modifies soil P solubility by net excretion of H^+ or HCO_3^- to maintain a balance of electric charge associated with cations and anions crossing the root membrane.

Phosphorus concentration in rhizosphere soil solution may also be increased by the hydrolytic cleavage of soil organic P forms, notably the phosphate monoesters, through the action of extracellular phosphohydrolases, which are generated by a demand for P (Chhonkar and Tarafdar, 1981; Sharma and Saxena, 1982; Tate, 1984). Speir and Ross (1978) proposed soil phosphatase activity as an index of soil organic P mineralization. Tarafdar and Jungk (1987) measured the distribution of phosphatase activity in the rhizosphere soil around plant roots, and reported up to a sevenfold higher concentration of acid phosphatase within the same zone. The alkaline phosphatases, which are solely of microbial origin, also showed similar increases. Furthermore, both the phosphatase fractions were found to increase with the age of seedlings of gram, Egyptian clover, wheat, and mustard (Chhonkar and Tarafdar, 1981). Tarafdar and Roy (1981) also reported a higher acid phosphatase concentration in the rhizosphere of nine mesta (*Hibiscus* spp.) genotypes in a soil in India. Cole and Sanford (1989) commented that several findings of this nature confirmed the conjecture of Sharpley (1985) that the enrichment of phosphatases at the soil–root interface renders soil organic P available to plants.

Mycorrhizal fungi are an important component of the rhizosphere of some plant species (Cole and Sanford, 1989). Thus, when the P supply in soil limits growth, plants infected with vesicular–arbuscular mycorrhizal (VAM) fungi are able to take up more P from soil and grow better than uninfected plants (Mosse, 1973; Tinker, 1980; Abbott and Robson, 1982; Ramanie et al., 1986; Raju et al., 1987; Eivazi and Weir, 1989). Radioactive p (^{32}P) has been used to test the possibility whether or not mycorrhizal fungi can help plants take up P that would otherwise be chemically unavailable to non-mycorrhizal plants (Sanders and Tinker, 1971; Bolan et al., 1984). However, similar values of specific activity of P in the mycorrhizal and non-mycorrhizal plants have been obtained signifying that both types meet their P requirements from the same P sources in soil (Sanders and Tinker, 1971; Gianinazzi-

Pearson et al., 1981; Brechet and Le Tacon, 1984). However, Bolan et al. (1983, 1984) suggested that this conclusion does not necessarily follow in view of the fact that forms of P in soils differ in their availability to mycorrhizal and nonmycorrhizal plants and are uniformly labeled by the addition of ^{32}P.

In addition, increased plant growth due to mycorrhizal infection has been found (Bolan et al., 1987b) to be greater with a poorly soluble P source than with soluble P. But Pairunan et al. (1980) and Barrow et al. (1977) found, by using the complete response curve, that although mycorrhizal plants were better than nonmycorrhizal plants, as users of P from poorly soluble PR reacted with soils, the improvement was comparable to that obtained with soluble P or freshly added P. Bolan et al . (1983), on the other hand, demonstrated that the addition of Fe hydroxide to soil decreased the growth and P uptake by nonmycorrhizal plants while having no effect on mycorrhizal plants. These authors (1987b) further showed, by using Fe phosphates of varied solubilities, that VAM plants can obtain P from different sources than can nonmycorrhizal plants. Bolan et al. (1987b) hypothesized that VAM fungi can form siderophores that would chelate Fe and release P from Fe–phosphate complexes in soils of low pH. This was confirmed by Reid et al. (1985) who found that the synthetic chelate, ethylene-diamine [di(*o*-hydroxy-phenylacetic)] acid (EDDHA) mobilized Fe in a high-pH soil and Fe and P in a low-pH soil. Indeed, Jayachandran et al. (1989) used EDDHA to simulate the effect of siderophores in soil to demonstrate that the former amendment to soil increases P-availability in soils of low-P fertility. Results indicate that siderophore production by mycorrhizae or other soil microbes can significantly increase P-availability in low-pH soils. This may be a feasible mechanism by which mycorrhizal plants could assess P sources unavailable to nonmycorrhizal plants.

While studying the response of chickpea to VAM infection in relation to the level of P application to an Ando-soil, Hirata et al. (1988) suggested that VAM infection may promote the photosynthetic activity at the pod-filling stage of chickpea, presumably due to the continuous uptake of P from soil.

Jurinak et al. (1986) presented a thermodynamic model that would explain the enhanced P uptake by mycorrhizal plants in semiarid soils. They presented evidence for the existence of Ca–oxalate crystals at the soil–hyphae interface of mycorrhizal *Pascopyrum smithii*. This leads to an increased solubility of Ca–apatite in both calcite–apatite (calcareous) and exchangeable Ca–apatite (noncalcareous) system. The result is a marked increase of soluble P in soil solution.

In mixed plant stands, such as legumes and grasses growing in mixed communities, the main mechanism of VAM activity has been attributed to the extraradical hyphae of the fungal symbiont, increasing the absorbing surface of the root, particularly for the less-mobile nutrients like P (Barea et al., 1989).

Dodd et al. (1987) reported that the phosphatase activity of roots and rhizosphere of plants infected with VAM fungi was higher for onions and wheat plants than that for control plants. This may explain for the higher P uptake by the infected plants.

The VAM association often varies according to the abundance of available soil P. A recent study by Rajapakse et al. (1989) demonstrated that plant-growth parameters were affected differently by inoculation with a VAM fungus under increasing soil P levels. Thus, in addition to the observed cultivar variability for beneficial soil P level (Powell, 1982), several parameters were found to exhibit different optimal P requirements for derivation of the greatest benefit from VAM fungi. Indeed, Sainz and Arines (1988) reported that the plant growth of red clover increased, whereas VAM infection (by native VAM endophytes) decreased significantly with increasing P levels.

Reduced light supply (photon irradiance or photoperiod) to plant shoots often reduces the growth enhancement of mycorrhizal plants over nonmycorrhizal controls (Bishop, 1979; Bethlenfalvay and Pacovsky, 1983). This is usually attributed to an increased carbohydrate transfer to the mycorrhizal fungi in plants grown with a low light supply (Bethlenfalvay and Pacovsky, 1983). Tester et al. (1985), on the other hand, suggested that this could be due to a "concentration effect" (Jarrell and Beverly, 1981), with plants grown under low light supply growing slowly, and still being influenced by high P levels stored in the seed. It was suggested that the decreased growth response to mycorrhizal infection with decreased photon irradiance is due to an increased significance of carbohydrate drain by the fungus.

Grinsted et al. (1982) and Hedley et al. (1982c) studied the mechanism by which nonmycorrhizal species, such as fodder rape (*Brassica napus*) also proves to be particularly efficient in absorbing soil and fertilizer P. These authors found that rape plants could solubilize soil P when grown at high root densities (>90 cm/cm^3) in a P-deficient soil. Hedley et al. (1982c) further showed that H^+ released from the roots, during periods when cation uptake by the plant exceeded that of anions, was the most likely cause of soil P solubilization. No evidence was presented for significant hydrolysis of soil organic P in these experiments.

Dodd et al. (1987) also observed that higher phosphatase activities were associated with roots and rhizospheres of nonmycorrhizal rape than that with mycorrhizal onions or mycorrhizal wheat. This was attributed to the possibility of an inverse relationship between phosphatase activity and the mycorrhizal dependence of a crop species.

3. Phosphorus Cycle in Soil

Several studies have examined the aspects of the P cycle in soil–plant systems (Cole et al., 1977; Chauhan et al., 1979, 1981; Tiessen et al., 1983; Tate, 1984; Smeck, 1985; Stevenson, 1986), including the mathematical simulation models that incorporate the various conceptual frameworks developed (Cole et al., 1977; Jones et al., 1984a; Parton et al., 1987, 1988). Recent reviews, which discussed these concepts, have been published (Smeck, 1985; Stewart and Tiessen, 1987; Cole and Sanford, 1989; Sanford et al., 1989). Thus, Chauhan et al. (1979, 1981) measured the rate of P movement between soil inorganic P, organic P, and biomass P compartments following regular additions of grass and cellulose. Results suggest that continued addition of cellulose without P for a long period of time (>9 months) would eventually have exhausted the reserve of labile (inorganic) P, leaving the microbial population dependent on the mineralization rate of organic P forms.

Tate (1984) emphasized the need to understand better P rates and pathways through organic matter in soil (including microbial biomass) and of the interaction between the biological and physicochemical processes controlling the P cycle to refine the predictive models. Smeck (1985), on the other hand, distinguished between the pedologic soil P transformations and the P dynamics in rapid biological pathways. Stevenson (1986) stressed the interlinking of P cycling to cycles for other elements.

Harrison (1982a,b) studied the mineralization rates of labile organic P in woodland soils, and found that nearly half the variation in mineralization rates were accounted for by plant-available P contents of soils. Harrison (1987) further presented the statistical interrelationships of soil organic P with soil properties, effects of climate, vegetation, parent material, distribution in soil profiles, and the effects of land management practices.

Stewart and Tiessen (1987) proposed that only a small portion of the total soil organic matter and, hence, soil organic P may be biologically active. They also discussed the effects of various factors controlling organic P transformations in soil–plant systems.

Elliott et al. (1984) observed a flush of mineralization in fallow wheat plots in wet and dry seasons, and attributed this to an increase in the protozoan biomass.

The agronomic implications of P cycling can be understood by following P transformations in laboratory incubation studies (Hedley et al., 1982b), P depletion patterns in plant rhizosphere (Hedley et al., 1982c), and changes of P fertility status during cultivation (Tiessen et al., 1982, 1983). This type of approach has been used along with soil biomass measurements to study the altered pattern of nutrient cycling in soils where different management of agricultural residues significantly changed the pattern of nutrient cycling (Stewart and McKercher, 1983).

Sanford et al. (1989) presented a P submodel describing the soil P cycle, similar to that in Figure 21. The primary P source in soil is the weathering of apatitic minerals, which contributes to the labile P pool. Phosphorus uptake by plants and soil microorganisms from soil solution causes a close linking with C and N cycles. Thus, as more P gets fixed via plant and soil microbial uptake, larger amounts of P become immobilized in organic matter, leading to an accumulation of soil organic matter (Cole and Heil, 1981). The plant residue P is subdivided into structural and metabolic materials, depending upon C/P ratios. Mineralization of soil organic P by microorganisms contributes to the labile P pool.

In the model (Fig. 21), the structural and the metabolic products are subdivided into three fractions, based on the turnover time of the various fractions. It has been shown that the C/P ratio of active, slow, and passive soil organic matter pools (Fig. 21) varies as a function of the labile P level (McGill

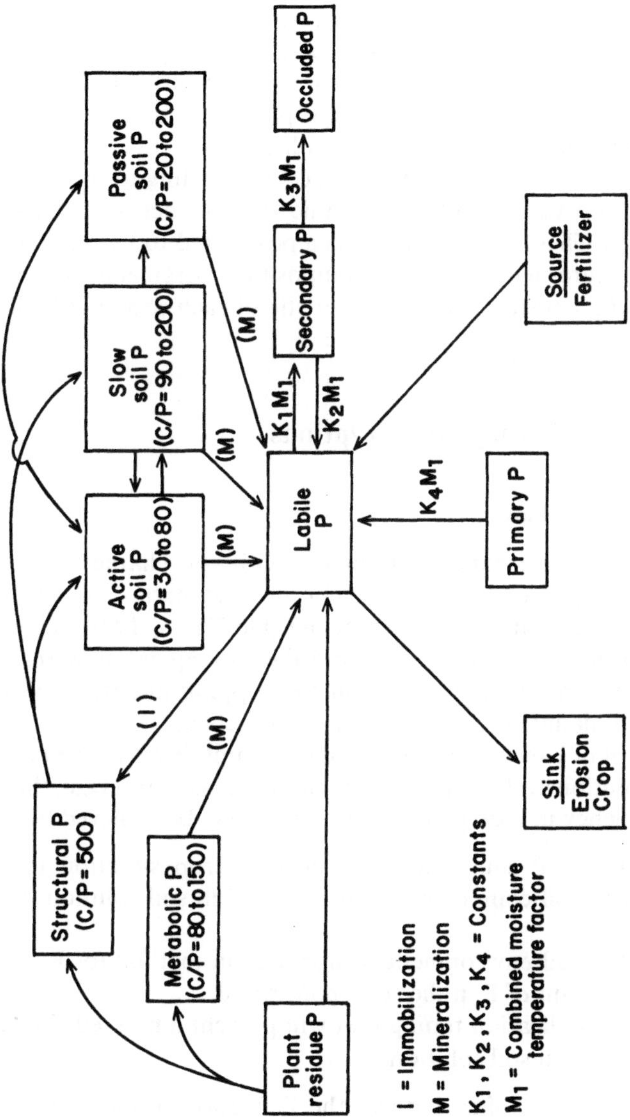

Figure 21. Flow diagram for the P submodel (After Parton et al., 1988) (*Source:* Sanford et al., 1989).

and Cole, 1981; Sanford et al., 1989) . The decomposition of metabolic plant materials, as well as active, slow, and passive soil organic matter (with low C/P ratios) leads to a mineralization contribution to labile P. On the other hand, P immobilization from the labile P pool is needed for decomposition of the structural plant materials having a high C/P ratio. Several additional P fluxes have also been defined (Fig. 21), corresponding to the redistribution of primary P into various other inorganic P forms between soil solids and solution phases. These changes have been postulated to be controlled by the combined moisture–temperature decomposition parameter (M_1) multiplied by the value of a constant (K), characteristic of each process (Sanford et al., 1989).

VI. Phosphorus Management Options

A. General Observations

Management of P is strongly interlinked with soil characteristics such as soil reaction, degree of weathering, amount and nature of clay minerals, organic matter content, and water regime (De Datta, 1981). The extent of initial P adsorption reaction in soil and the subsequent slow reaction rate should also be considered, in addition to cropping intensity and cropping patterns, when planning for P management practices.

Management of P fertilizers should aim to maintain a sufficient available-P level in soil solution at the appropriate time at a reasonable cost, thus increasing P use efficiency in crop production. These may be achieved by

1. using a suitable P source for a given soil–crop situation to minimize reactions with soil components that render P in soil solution unavailable to crop,
2. modifying the soil environment or application method (of P fertilizer) to reduce the amount of P in the solid phase; and
3. selecting a P application timing that will prevent a marked rise and fall of P concentration in soil solution.

Liming an acid soil often reduces the P fixation problem. However, the effect of lime on P-availability seemed to be influenced by two opposing factors. An increase in pH decreases P adsorption by amphoteric adsorbing surfaces, whereas a high exchangeable Al content of the soil may generate fresh P-adsorbing surfaces in soil through the precipitation of hydroxy–Al polymers (Haynes, 1984).

A rise in pH may also reduce the extent of P desorption. Haynes (1984) suggested that compared with that in unlimed soil, P adsorption can be greater immediately following liming, with such P adsorption capacity falling, however, with time. The presence of inorganic and organic anions, that inhibit the transformation of hydroxy–Al polymers to crystalline products of reduced P sorption capacity would also tend to maintain a high adsorption capacity of limed soil. Liming appears to benefit greatly an acid soil that has a very low-P status, and that is about to be fertilized with P (Barrow, 1989).

Use of sparingly soluble P sources in place of soluble P fertilizers has been considered where improved efficiency of soluble fertilizers is difficult to attain. Ground PR per unit of P costs about one-quarter the price of simple superphosphate (Tian-ren et al., 1989). It has a moderately high-P content and is known to be effective on the P-deficient acid soils of the humid tropics (Hammond et al., 1986). PAPR also costs less than the fully acidulated PR products, and is intermediate in solubility between superphosphate and ground PR. Use of PAPR has often been found to lead to comparable yields as soluble P sources (De Datta et al., 1989).

As explained earlier, for dissolution of PR to continue, the soil should provide a source of H^+ ions, and also a sink, particularly for Ca^{2+} and $H_2PO_4^-$ ions (reaction e). Sanchez and Uehara (1980) suggested that PR should be applied several months ahead of liming so that an increase in exchangeable Ca^{2+} and pH does not interfere with PR dissolution.

Effectiveness of PR fertilizers increases with a decrease in particle size. However, it is difficult to handle and spread finely ground PR, and even then, it is doubtful whether PR, including finely ground, highly reactive materials, could be economic substitutes for superphosphates for annual crops and pastures (Hammond et al., 1986; Bolland and Barrow, 1988; Bolland and Gilkes, 1989).

To achieve high efficiency for a longer period, researchers have suggested using a very high level of PR application, especially in acid soils having high P fixing capacity. The main advantage of such application is that, apart from satisfying the P fixing capacity of the soil, the fixed P is gradually released over a period of several years at rates sufficient to support adequate crop growth. A further advantage of massive P applications is an increase in cation-exchange capacity of the soils (Sanchez and Uehara, 1980). However, Barrow (1989) pointed out that the diffusion zones around each particle are more likely to overlap at a higher level of application, causing the rate of solution to fall. As a result, even very high doses of PR application may fail to attain the same maximum yields as those obtainable from soluble P sources (Bolland and Barrow, 1988).

B. Phosphorus Management Practices for Lowland Rice

1. P Source

Single and triple superphosphate, diammonium phosphate, and ammonium phosphate are among the P fertilizer materials most commonly used for lowland rice. Fused Ca–Mg phosphates, urea ammonium phosphate sulfate, hyperphosphate, nitric phosphate, and PR have also been used (De Datta et al., 1989). De Datta (1978, 1981) considered that there was no significant difference in P availability to rice from various kinds of P fertilizers, except on very strongly acid or alkaline soils.

PR application to lowland rice has been evaluated in many Asian countries (Hammond et al., 1986). Such application meets with two difficulties: (1) pH of an acidic soil will rise following submergence, and this may adversely affect the solubility and rate of PR dissolution in soil; and (2) the ability of rice to derive P from PR is relatively low (Ru-kun et al., 1982). Thus PRs available to other crops are often less available to rice.

However, Tian-ren et al. (1989) argues that a considerable amount of PR could dissolve before the soil pH rises appreciably on flooding, particularly in highly acidic soils. They further emphasized the residual value of PR relative to water-soluble sources, which calls for a greater understanding of the continuing slow reactions between soil components and P in flooded soils. The latter need not be correlated with the extent of the initial P adsorption reactions in the soil.

Results with PR in flooded soils generally suggest a high potential for use of PR (relative to the soluble sources), even though the rate of application must be high, especially so for less-reactive PR sources indigenous to many Asian countries (De Datta et al., 1989). Thus even though the indigenous or imported commercial PR may offer low-cost P fertilizers, the relative agronomic value of the materials must be determined to identify their potential role as P sources for rice. A procedure has been developed to allow for a choice of P source on the basis of both agronomic effectiveness, and prices of P in TSP and PRs (De Datta, 1983). This is shown in Figure 22.

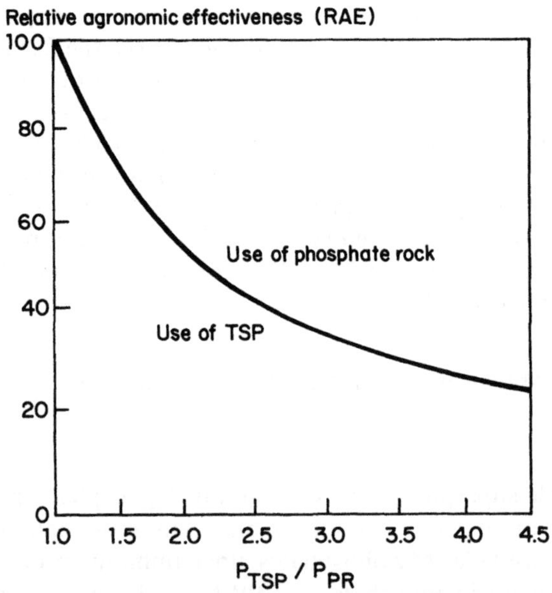

Figure 22. Relationship between the relative agronomic effectiveness (RAE) values and price ratios of TSP (P_{TSP}) and PR (P_{PR}) (*Source*: Engelstad et al., 1974.)

2. Application Methods

Food and Agriculture Organization of the United Nations (FAO) and the International Atomic Energy Agency (IAEA) conducted field experiments with ^{32}P-labeled P fertilizers on different soils in the Philippines, Thailand, Burma, Pakistan, Egypt, and Hungary (Ru-kun et al., 1982). Results indicated that surface broadcasting or incorporation of fertilizer before transplanting was more effective than other methods such as deep placement of P at 10- or 20-cm depth either in planting hills or between rows (Ru-kun et al., 1982; De Datta et al., 1989). As a possible cause for this observation, Ru-kun et al. (1982) suggested that the rice root system may remain mostly confined to the surface soil layer. The latter, coupled with a probable higher P diffusion rate under submergence, may contribute to the superiority of the surface broadcasting method.

Dipping rice seedlings into P fertilizer slurry before transplanting has also been reported to be useful. In China, this practice has resulted in a 40%–60% saving on P fertilizer in irrigated rice soils (Ru-kun et al., 1982).

3. Application Time

De Datta (1978, 1981) suggested that P fertilizer for rice should generally be applied at transplanting, but may also be applied later before the vigorous tillering stage. He did not consider split P application necessary in view of the high mobility of P from old to younger leaves, increase in available soil P with time of submergence, and low leaching losses. Experiments on P fertilizer application methods for rice in seven countries demonstrated that P applied 2 weeks before panicle initiation is as effective as that applied at transplanting (De Datta et al., 1989).

For rice, the best time and method of P fertilization thus appear to be the application of the total amount as a basal dose at transplanting (Ru-kun et al., 1982). Patrick et al. (1974) recommended an early application of P because of the following (Ru-kun et al., 1982; De Datta et al., 1989):

1. more P is required by the rice plant during the early growth stages;
2. increased available P after flooding the soil falls short of the P requirement at this early stage;
3. sufficient P supply encourages the development of a root system and tillering;
4. in areas of low temperature, more P is required in the early stages of rice; and
5. application at transplanting is more convenient than topdressing later.

A judicious fertilization scheme is for the entire rotation within a cropping system. The variation of soil P under the rotation of rice and upland crops influences the direct as well as residual effects of P fertilizers (Willett, 1979; Willett and Higgins, 1980; Bradley et al., 1984). Generally, fertilizer P recovery by rice ranges from 8% to 20% (De Datta et al., 1966; Goswami and Banerjee, 1978; Ru-kun et al., 1982), with 80%–90% of the applied P remaining in the

Table 14. Effect of phosphorus application methods on total crop yield in a wheat–rice rotation (Average of 8 Soils).

	Grain yield (g/pot)			P uptake (mg/pot)		
	Wheat	Rice	Total	Wheat	Rice	Total
All P applied to wheat	8.8 ± 1.3	17.5 ± 2.4	26.3 ± 2.6	17.6 ± 8.9	35.0 ± 11.8	55.0 ± 15.5
All P applied to rice	2.1 ± 1.0	12.5 ± 2.1	14.6 ± 2.7	2.2 ± 1.4	30.4 ± 12.8	31.7 ± 13.5

Source: Ru-ken et al., 1982.

soil for the succeeding crop. As P availability changes with alternate drying and submergence (Patrick and Mahapatra, 1968; Willett, 1982, 1986; Sah and Mikkelsen, 1986a,c), the P applied to the upland crop may have a greater residual effect on the succeeding rice, whereas P applied to rice may have a less residual value for the succeeding upland crop (Ru-kun et al., 1982). Bradley et al. (1984) suggested that fertilizer P should be applied when water-logging is least likely to occur to minimize fixation before being used by the plant.

In a study using eight lowland rice soils of contrasting properties, Ru-kun et al. (1982) reported that the total P uptake by crops and the total yield of wheat and rice grown in rotation doubled when all of P was applied to the upland wheat crop as compared with that when all of P was applied to rice. Table 14 illustrates the point.

In another study, Willett and Higgins (1980) reported that the depressing effects of a lowland rice crop on subsequent upland crop growth may have declined by the third year, after drainage of a rice crop. Thus, a decline in soil P sorptivity, and a corresponding increase in P-availability would be expected during the 2 years after drainage of the rice crop. The authors suggested that this could be important in the management of soils undergoing rice–upland-crop rotations, and for the efficient use of P fertilizer.

Recycling of fertilizer P through a preceding green manure crop may also be beneficial for the succeeding lowland rice. Singhabutra et al. (1987) observed that green manuring with *Sesbania* spp. in light-textured acid soils of Thailand significantly improved the soil-P status, and also significantly increased rice yield. Ru-kun et al. (1982) also reported that PR applied to radish (*Raphanus sativus*) preceding rice led to a substantial yield increase of the succeeding rice. Radish is cultivated in some parts of China as a green manure crop. Beri and Meelu (1980) found that applying P to the green manure crop was beneficial to N accumulation, and increased rice yield (on a soil low in available P) more than did P applied directly to rice.

From the results of experiments at nine sites in five countries in South and Southeast Asia, De Datta and Gomez (1982) suggested that rice responses to P and K should be evaluated with N, and in continuous rice cropping in several seasons. Large interaction effects between P and N and between K and N were observed.

VII. Unresolved Challenges

Despite sustained efforts to understand the intricate chemistry of P transformations, several issues are still open to questions. Further intensive research is imperative to bridge these knowledge gaps and so obtain a more comprehensive understanding of the different processes involved. This would help foment better management practices, thus deriving maximum benefits from P inputs to soils, while minimizing the adverse effects on the environment. In this context, the following areas of research needs sustained attention.

1. A better predictive understanding of P sorption and release behavior of soils, including the mechanisms of these processes is needed, in particular, more information on the rate and amount of desorption of inorganic P from different soils. This information would be useful in adopting the suitable P management practices under different soil–crop–environment situations.
2. The analytical methods used for soil testing for P should be carefully assessed. Possible improvements of P soil tests in terms of combination of intensity, quantity, and buffering capacity data need to be considered. The possibility of including in the soil test values of the contribution from organic P to the pool of available-P in soil may also be explored.

   Interpretation of fertilizer response data based on soil testing values is also inadequate, especially so for flooded soils. The influence of environmental factors on such response data should be given due consideration.

   A complementary approach to soil testing may be the modeling approach, wherein previous P applications are considered while assessing the current P status of soils. For this to be operative, more intense research is necessary to characterize further the rates of continuing reactions of P with soil, and the effect of soil properties on such rates.
3. The turnover rate of organic P is much less well understood, especially so for lowland rice soils. Information should be generated on the quantification of pool size and turnover of biomass P in the soil.
4. There should be more definitive studies to understand better the root development, structure, and function in various soil–plant systems that would provide a basis for manipulation. Indeed, the organic acids production by the roots (e.g., citric, malic, polygalacturonic acids) may have an important bearing on P desorption in rhizosphere soil.
5. For soil mycorrhizal research, there is a need to explore the possibility of matching VAM and plant genotypes in a variable P environment.
6. The beneficial effects of green manuring with regard to the P and N supply to soils and crops, as well as the recycling value of crop residues,

deserves more attention in view of their significance in sustaining crop yields at minimal input levels.

7. The efficiency of sparingly soluble P sources, such as PR and PAPR, should be examined further. There is also a need for an increase in the geological exploration for finding fresh sedimentary PR deposits, as reserves of high-quality deposits are being rapidly depleted.
8. There is a need to formulate P recommendations based on the cropping system as a whole rather than for each separate crop in the rotation sequence. This is especially true for a rotation that includes lowland rice and an upland crop wherein soil P dynamics vary considerably between flooded and drained phases of rotation.
9. More research should be directed to pinpoint the reasons for yield declines of lowland rice observed in some adverse rice-growing soils, for example, acid sulfate soils, peat soils, saline soils, or Fe-toxic soils. Indeed, the kinetics of reactions of P in these soils often vary considerably from those in nonadverse soils, and should be studied in greater depth to explore the possibility whether P is a contributing factor to such yield reverses.
10. The residual effects of continued fertilizer P application over many years should be studied in greater depth for planning the long-term P management strategies. It is interesting to note in this context that Barrow (1989) proposed an integration of concepts of available and unavailable P in soil in a scheme of continuum of P supply rate. The latter ranges from instantaneous for freshly added P to a rate that is far too slow for adequate growth of plants. The availability of previously applied P in this continuum depends on how long ago it was applied, and also on the characteristics of the particular soil.
11. Losses of P through erosion, runoff, or leaching from different upland situations need to be appropriately quantified.
12. Information should be generated to fill the numerous gaps to understand the processes that govern the P cycle. Such knowledge would enable one to move to the mathematical models from the conceptual ones.

References

Abbott, L.K., and A.D. Robson. 1982. The role of vesicular–arbuscular mycorrhizal fungi in agriculture and the selection of fungi for inoculation. *Aust. J. Agric. Res.* 33:389–408.

Adams, J.F., and J.W. Odom. 1985. Effects of pH and phosphorus rates on soil-solution phosphorus and phosphorus availability. *Soil Sci.* 140:202–205.

Adams, M.A., and L.T. Byrne. 1989. ^{31}P-NMR analysis of phosphorus compounds in extracts of surface soils from selected Karri (*Eucalyptus diversicolor* F. muell.) forests. *Soil Biol. Biochem.* 21:523–528.

Adams, W.A., S.N. Gafoor, and M.I. Karim. 1987. Composition and properties of poorly ordered minerals in Welsh soils. II. Phosphate adsorption and reactivity towards NaF solution. *J. Soil Sci.* 38:95–103.

Agboola, A.A., and O.J. Ayodele. 1983. An attempt to evaluate plant available phosphorus in Western Nigeria savannah soils under traditional fallow systems. *In Proc. 3rd Int. Congr. on Phosphorus Compounds, (Brussels, Belgium).* Institut Mondial du Phosphate, Casablanca, Morocco, pp. 261–267.

Al-Khateeb, I.K., M.J. Raihan, and S.R. Asker. 1986. Phase equilibria and kinetics of orthophosphate in some Iraqi soils. *Soil Sci.* 141:31–37.

Alva, A.K., S. Larsen, and S.W. Bille. 1980. The influence of rhizosphere in rice crop on resin extractable phosphate in flooded soils at various levels of phosphate applications. *Plant Soil* 56:17–25.

Amarasiri, S.L., and S.R. Olsen. 1973. Liming as related to solubility of phosphorus and plant growth in an acid tropical soil. *Soil Sci. Soc. Am. Proc.* 37:716–721.

Anderegg, J.C., and D.V. Naylor. 1988. Phosphorus and pH relationships in an Andic soil with surface and incorporated organic amendment. *Plant Soil* 107:273–278.

Anderson, G. 1980. Assessing organic phosphorus in soils. *In* F.E. Khasawneh, E.C. Sample, and E.J. Kamprath, (Eds.). *The Role of Phosphorus in Agriculture.* Soil Sci. Soc. Am., Madison, Wisconsin, pp. 411–431.

Anderson, G., and R.E. Malcolm. 1974. The nature of alkali-soluble soil organic phosphates. *J. Soil Sci.* 25:282–297.

Anderson, J.P.E., and K.H. Domsch. 1980. Quantities of bacteria and fungi in chloroform-fumigated soils. *Soil Sci.* 130:211–216.

Aoba, T., and E.C. Moreno. 1985. Adsorption of phosphoserine onto hydroxyapatite and its inhibitory activity on crystal growth. *J. Colloid Interface Sci.* 106:110–121.

Apthorp, J.N., M.J. Hedley, and R.W. Tillman. 1987. The effects of nitrogen fertilizer form on the plant availability of phosphate from soil, phosphate rock and monocalcium phosphate. *Fert. Res.* 12:269–284.

Araki, S., H. Hirai, and K. Kyuma. 1986. Phosphate absorption of red and/or yellow colored soil materials in relation to the characteristics of free oxides. *Soil Sci. Plant Nutr.* 32:609–616.

Atkinson, D.E. 1977. *Cellular Energy Metabolism and its Regulation.* Acad. Press, New York.

Atkinson, R.J., A.M. Posner, and J.P. Quirk. 1972. Kinetics of isotopic exchange of phosphate at the L-FeOOH-aqueous solution interface. *J. Inorg. Nucl. Chem.* 34:2201–2211.

Ayodele, O., and A.A. Agboola. 1981. Evaluation of phosphorus fixation capacity of tropical savannah soils of Western Nigeria. *Soil Sci. Soc. Am. J.* 45:462–464.

Bache, B.W. 1963. Aluminum and iron phosphate studies relating to soils. I. Solution and hydrolysis of variscite and strengite. *J. Soil Sci.* 14:113–123.

Bache, B.W. 1964. Aluminum and iron phosphate studies relating to soils. II. Reactions between phosphate and hydrous oxides. *J. Soil Sci.* 15:110–116.

Bache, B.W., and C. Ireland. 1980. Desorption of phosphate from soils using anion exchange resins. *J. Soil Sci.* 31:297–306.

Bache, B.W., and E.G. Williams. 1971. A phosphate sorption index for soils. *J. Soil Sci.* 22:289–301.

Bangor, K.C., K.S. Yadav, and M.M. Mishra. 1985. Transformation of rock phosphate during composting and the effect of humic acid. *Plant Soil* 85:259–266.

Barea, J.M., F. El-Atrach, and R. Azcon. 1989. Mycorrhiza and phosphate interactions as affecting plant development, nitrogen fixation, nitrogen transfer and nitrogen uptake from soil in legume–grass mixtures by using a ^{15}N dilution technique. *Soil Biol. Biochem.* 21:581–589.

Barekzai, A., and K. Mengel. 1985. Atterung von wasserlöslichem Dungerphosphat bei verschiedenen Bodentypen. *Z. Pflanzenernaehr. Bodenkd.* 148:365–378.

Barrow, N.J. 1974. Effect of previous additions of phosphate on phosphate adsorption by soils. *Soil Sci.* 118:82–89.

Barrow, N.J. 1978. The description of phosphate adsorption curves. *J. Soil Sci.* 29:447–462.

Barrow, N.J. 1979a. The description of desorption of phosphate from soil. *J. Soil Sci.* 30:259–270.

Barrow, N.J. 1979b. Three effects of temperature on the reactions between inorganic phosphate and soil. *J. Soil Sci.* 30:271–279.

Barrow, N.J. 1980. Evaluation and utilization of residual phosphorus in soils. *In* F.E. Khasawneh, E.C. Sample, and E.J. Kamprath (Eds.). *The Role of Phosphorus in Agriculture.* Soil Sci. Soc. Am., Madison, Wisconsin, pp. 333–359.

Barrow, N.J. 1983a. A mechanistic model for describing the sorption and desorption of phosphate by soil. *J. Soil Sci.* 34:733–750.

Barrow, N.J. 1983b. On the reversibility of phosphate sorption by soils. *J. Soil Sci.* 34:751–758.

Barrow, N.J. 1983c. A discussion of the methods for measuring the rate of reaction between soil and phosphate. *Fert. Res.* 4:51–61.

Barrow, N.J. 1984. Modelling the effects of pH on phosphate sorption by soils. *J. Soil Sci.* 35:283–297.

Barrow, N.J. 1985. Reaction of anions and cations with variable-charge soils. *Adv. Agron.* 38:183–230.

Barrow, N.J. 1987. *Reactions with Variable-Charge soils.* Martinus Nijhoff Pub., Dordrecht.

Barrow, N.J. 1989. Relating chemical processes to management systems. Paper presented at the *Symp. on Phosphorus Requirements for Sustainable Agriculture in Asia and Oceania,* 6–10 March 1989. International Rice Research Institute, Los Banos, Laguna, Philippines.

Barrow, N.J., J.W. Bowden, A.M. Posner, and J.P. Quirk. 1980. Describing the effects of electrolyte on adsorption of phosphate by a variable-charge surface. *Aust. J. Soil Res.* 18:395–404.

Barrow, N.J., and A.S. Ellis. 1986. Testing a mechanistic model. V. The points of zero salt effect for phosphate retention, for zinc retention and for acid/alkali titration of a soil. *J. Soil Sci.* 37:303–310.

Barrow, N.J., N. Malajczuk, and T.C. Shaw. 1977. A direct test of the ability of vesicular-arbuscular mycorrhiza to help plants take up fixed soil phosphate. *New Phytol.* 78:269–276.

Barrow, N.J., and T.C. Shaw. 1975a. The slow reactions between soil and anions. 2. Effect of time and temperature on the decrease in phosphate concentration in the soil solution. *Soil Sci.* 119:167–177.

Barrow, N.J., and T.C. Shaw. 1975b. The slow reactions between soil and anions. 3. The effects of time and temperature on the decrease in isotopically exchangeable phosphate. *Soil Sci.* 119:190–197.

Barrow, N.J., and T.C. Shaw. 1979. Effects of ionic strength and nature of the cation on desorption of phosphate from soil. *J. Soil Sci.* 30:53–65.

Batsula, A.A., and S.M. Krivonosova. 1973. Phosphorus in the humic and fulvic acids of some Ukrainian soils. *Soviet Soil Sci.* 5:347–350.

Bennoah, E.O., and D.K. Acquaye. 1989. Phosphate sorption characteristics of selected major Ghanian soils. *Soil Sci.* 148:114–123.

Beri, V., and O.P. Meelu. 1980. Add nitrogen through green manure. *Prog. Farming* 16:8–9.

Bethlenfalvay, G.J., and R.S. Pacovsky. 1983. Light effects in mycorrhizal soybeans. *Plant Physiol.* 73:969–972.

Bhujbal, B.M., and K.B. Mistry. 1985. Studies on the dissolution of major Indian phosphate rocks in an acid soil. *J. Indian Soc. Soil Sci.* 33:568–573.

Bishop, C.L. 1979. Effects of light intensity on vesicular-arbuscular mycorrhizal infection in *Zea mays. J. Colorado-Wyoming Acad. Sci.* 11:28.

Blair, G.J., and O.W. Boland. 1978. The release of phosphorus from plant material added to soil. *Aust. J. Soil Res.* 16:101–111.

Blanchar, R.W., and G.K. Stearman. 1984. Ion products and solid-phase activity to describe phosphate sorption by soils. *Soil Sci. Soc. Am. J.* 48:1253–1258.

Blanchar, R.W., and G.K. Stearman. 1985. Prediction of phosphate sorption in soils from regular solid solution theory. *Soil Sci. Soc. Am. J.* 49:578–583.

Bloom, P.R. 1981. Phosphorus adsorption by an aluminum–peat complex. *Soil Sci. Soc. Am. J.* 45:267–272.

Bolan, N.S., and N.J. Barrow. 1984. Modelling the effect of adsorption of phosphate and other anions on the surface charge of variable-charge oxides. *J. Soil Sci.* 35:273–281.

Bolan, N.S., N.J. Barrow, and A.M. Posner. 1985. Describing the effect of time on sorption of phosphate by iron and aluminum hydroxides. *J. Soil Sci.* 36:187–197.

Bolan, N.S., and M.J. Hedley. 1989. Dissolution of phosphate rocks in soils. 1. Evaluation of extraction methods for the measurement of phosphate rock dissolution. *Fert. Res.* 19:65–75.

Bolan, N.S., M.J. Hedley, J.K. Syers, and R.W. Tillman. 1987a. Single superphosphate-reactive phosphate rock mixtures. 1. Factors affecting chemical composition. *Fert. Res.* 13:223–229.

Bolan, N.S., A.D. Robson, and N.J. Barrow. 1983. Plant and soil factors including mycorrhizal infection causing sigmoidal response of plants to applied phosphorus. *Plant Soil* 73:187–201.

Bolan, N.S., A.D. Robson, and N.J. Barrow. 1987b. Effects of vesicular–arbuscular mycorrhiza on the availability of iron phosphates to plants. *Plant Soil* 99:401–410.

Bolan, N.S., A.D. Robson, N.J. Barrow, and L.A.G. Aylmore. 1984. Specific activity of phosphorus in mycorrhizal and non-mycorrhizal plants in relation to the availability of phosphorus to plants. *Soil Biol. Biochem.* 16:299–304.

Bolan, N.S., J.K. Syers, and R.W. Tillman. 1986. Ionic strength effects on surface charge and adsorption of phosphate and sulfate by soils. *J. Soil Sci.* 37:379–388.

Bolan, N.S., J.K. Syers, and R.W. Tillman. 1988. Effect of pH on adsorption of phosphate and potassium in batch and in column experiments. *Aust. J. Soil Res.* 26:165–170.

Bolland, M.D.A., and N.J. Barrow. 1988. Effect of level of application on the relative effectiveness of rock phosphate. *Fert. Res.* 15:181–192.

Bolland, M.D.A., and J.W. Bowden. 1982. Long-term availability of phosphorus from calcined rock phosphate compared with superphosphate. *Aust. J. Agric. Res.* 33:1061–1071.

Bolland, M.D.A., and J.W. Bowden. 1984. The initial and residual value for subterranean clover of phosphorus from Candallite rock phosphates, apatite rock phosphates and superphosphate. *Fert. Res.* 5:295–307.

Bolland, M.D.A., J.W. Bowden, M.F. D'Antuono, and R.J. Gilkes. 1984. The current and residual value of superphosphate, Christmas Island C-grade ore, and Calci-phos as fertilizers for a subterranean clover pasture. *Fert. Res.* 5:335–354.

Bolland, M.D.A., and R.J. Gilkes. 1989. Reactive rock phosphate fertilizers and soil testing for phosphorus: The effect of particle size of the rock phosphate. *Fert. Res.* 21:75–93.

Bolland, M.D.A., A.J. Wheatherley, R.J. Gilkes, and J.W. Bowden. 1986. Granular reactive apatite rock phosphate is not an effective phosphorus fertilizer in the short term on lateritic soils in south-western Australia. *Aust. J. Exp. Agric.* 26:217–225.

Borggaard, O.K. 1983. The influence of iron oxides on phosphate adsorption by soil. *J. Soil Sci.* 34:333–341.

Bowden, J.W., S. Nagarajah, N.J. Barrow, A.M. Posner, and J.P. Quirk. 1980. Describing the adsorption of phosphate, citrate, and selenite on a variable-charge mineral surface. *Aust. J. Soil Res.* 18:49–60.

Bowden, J.W., A.M. Posner, and J.P., Quirk. 1977. Ionic adsorption on variable-charge mineral surfaces. Theoretical charge development and titration curves. *Aust. J. Soil Res.* 15:121–136.

Bowman, R.A. 1989. A sequential extraction procedure with concentrated sulfuric acid and dilute base for soil organic phosphorus. *Soil Sci. Soc. Am. J.* 53:362–366.

Bowman, R.A., and C.V., Cole. 1978a. An exploratory method for fractionation of organic phosphorus from grassland soils. *Soil Sci.* 125:95–101.

Bowman, R.A., and C.V. Cole. 1978b. Transformations of organic phosphorus substrates in soils as evaluated by $NaHCO_3$ extraction. *Soil Sci.* 125:49–54.

Bowman, R.A., and S.R. Olsen. 1985. Assessment of phosphate-buffering capacity. 2. Greenhouse methods. *Soil Sci.* 140:387–392.

Bradley, J., I. Vimpany, and P.J. Nicholls. 1984. Effects of waterlogging and subsequent drainage of a pasture soil on phosphate sorption, extractable phosphate, and oxalate-extractable iron. *Aust. J. Soil Res.* 22:455–461.

Braithwaite, A.C. 1987. Processing aspects of production of partially acidulated phosphate rock fertilizers using phosphoric acid. *Fert. Res.* 13:87–96.

Braithwaite, A.C., A.C. Eaton, and P.S. Groom. 1989. Some factors associated with the use of the extractants 2% citric acid and 2% formic acid as estimators of available phosphorus in fertilizer products. *Fert. Res.* 19:175–181.

Brandon, A.M., and D.S. Mikkelsen. 1979. Phosphorus transformations in alternately flooded California soils. I. Cause of plant phosphorus deficiency in rice rotation crops and correctional methods. *Soil Sci. Soc. Am. J.* 43:989–994.

Brannon, C.A., and L.E. Sommers. 1985a. Preparation and characterization of model humic polymers containing organic phosphorus. *Soil Biol. Biochem.* 17:213–219.

Brannon, C.A., and L.E. Sommers. 1985b. Stability and mineralization of organic phosphorus incorporated into model humic polymers. *Soil Biol. Biochem.* 17:221–227.

Brechet, C., and F. Le Tacon. 1984. Response of endomycorrhizal and nonmycorrhizal plants of *Acer pseudoplatanus* to different levels of soluble and nonsoluble phosphorus. *Eur. J. For. Pathol.* 14:68–77.

Broeshart, H., E. Haunold, and M. Fried. 1965. The effect of water conditions and oxidation–reduction status of rice soils on the availability of soil and fertilizer phosphorus. *Plant Soil* 23:305–313.

Brookes, P.C., D.S. Powlson, and D.S. Jenkinson. 1982. Measurement of microbial biomass phosphorus in soil. *Soil Biol. Biochem* 14:319–329.

Brookes, P.C., D.S. Powlson, and D.S. Jenkinson. 1984. Phosphorus in the soil microbial biomass. *Soil Biol. Biochem.* 16:169–175.

Brookes, P.C., , K.R. Tate, and D.S. Jenkinson. 1983. The adenylate-energy charge of the soil microbial biomass. *Soil Biol. Biochem.* 15:9–16.

Buchter, B., B. Davidoff, M.C. Amacher, C., Hinz, I.K. Iskandar, and H.M. Selim. 1989. Correlation of Freundlich K_d and n retention parameters with soils and elements. *Soil Sci.* 148:370–379.

Burnham, C.P., and D. Lopez-Hernandez. 1982. Phosphate retention in different soil taxonomy. *Soil Sci.* 134:376–380.

Chang, S.C. 1978. Evaluation of the fertility of rice soils. *In Soils and Rice.* International Rice Research Institute, Los Baños, Laguna, Philippines, pp. 521–541.

Chang, S.C., and M.L. Jackson. 1957. Fractionation of soil phosphorus. *Soil Sci.* 84:133–144.

Chang, S.C., and K. Maleewan. 1972. Correlation between availability indices of soil phosphorus determined on aerobic and anaerobic soil samples. *In Compilation of Reports on Rice Fertilization Research.* Presented at the 4th meeting of the Advisory Committee on rice fertilization, 15–16 May 1972, Tennessee Valley Authority and Rice Department, Ministry of Agriculture, Bangkok, Thailand.

Chater, M., and G.E.G. Mattingly. 1980. Changes in organic phosphorus contents of soil from long-continued experiments at Rothamsted and Saxmundham. *Rothamsted Experiment Station Report for 1979*, Part 2, pp. 41–61.

Chatterjee, S.C., A.N. Datta, and S.K. Ghosh. 1983. Evaluation of low-grade phosphate rock deposits in Indian context. *Fert. News.* 28:27–32.

Chauhan, B.S., J.W.B. Stewart, and E.A. Paul. 1979. Effect of carbon additions on soil labile inorganic, organic and microbially held phosphate. *Can. J. Soil Sci.* 59:387–396.

Chauhan, B.S., J.W.B. Stewart, and E.A. Paul. 1981. Effect of labile inorganic phosphate status and organic carbon additions on the microbial uptake of phosphorus in soils. *Can. J. Soil Sci.* 61:373–385.

Chaverri, J.G., and C.A. Black. 1966. Theory of the solubility of phosphate rock. *Iowa State J. Sci.* 41:77–95.

Chhonkar, P.K., and J.C. Tarafdar. 1981. Characteristics and location of phosphatases in soil-plant system. *J. Indian Soc. Soil Sci.* 29:215–219.

Chien, S.H. 1977a. Dissolution rates of phosphate rocks. *Soil Sci. Soc. Am. J.* 41:656–657.

Chien, S.H. 1977b. Thermodynamic considerations on the solubility of phosphate rock. *Soil Sci.* 123:117–121.

Chien, S.H. 1979a. *Dissolution of phosphate rocks in soils and solutions.* Special Publication IFDC-Sl. International Fertilizer Development Center, Muscle Shoals, Alabama.

Chien, S.H. 1979b. Dissolution of phosphate rocks in acid soils as influenced by nitrogen and potassium fertilizers. *Soil Sci.* 127:371–376.

Chien, S.H., F. Adams, F.E. Khasawneh, and J. Henao. 1987b. Effects of combinations of triple superphosphate and a reactive phosphate rock on yield and phosphate uptake by corn. *Soil Sci. Soc. Am. J.* 51:1656–1658.

Chien, S.H., and C.A. Black. 1976. Free energy of formation of carbonate apatites in some phosphate rock. *Soil Sci. Soc. Am. J.* 40:234–239.

Chien, S.H., and W.R. Clayton. 1980. Application of Elovich equation to the kinetics of phosphate release and sorption in soils. *Soil Sci. Soc. Am. J.* 44:265–268.

Chien, S.H., W.R. Clayton, and G.H. McClellan. 1980a. Kinetics of dissolution of phosphate rocks in soil. *Soil Sci. Soc. Am. J.* 44:260–264.

Chien, S.H., and L.L. Hammond. 1978a. Comparison of various laboratory methods for predicting the agronomic potential of phosphate rocks for direct application. *Soil Sci. Soc. Am. J.* 42:935–939.

Chien, S.H., and L.L. Hammond. 1978b. A simple chemical method for evaluating the agronomic potential of granulated phosphate rock. *Soil Sci. Soc. Am. J.* 42:615–617.

Chien, S.H., and L.L. Hammond. 1989. Agronomic effectiveness of partially acidulated phosphate rock as influenced by soil phosphorus-fixing capacity. *Plant Soil* 120:159–164.

Chien, S.H., L.L. Hammond, and L.A. Leon. 1987a. Long-term reactions of phosphate rocks with an Oxisol in Colombia. *Soil Sci.* 144:257–265.

Chien, S.H., L.A. Leon, and H.R. Tejeda. 1980b. Dissolution of North Carolina phosphate rock in acid Colombian soils as related to soil properties. *Soil Sci. Soc. Am. J.* 44:1267–1271.

Chien, S.H., N.K. Savant, and U. Mokwunye. 1982. Effect of temperature on phosphate sorption and desorption in two acid soils. *Soil Sci.* 133:160–166.

Cholitkul, W., and E.H. Tyner. 1971. Inorganic phosphorus fractions and their relation to some chemical indices of phosphate availability for some lowland rice soils of Thailand. *In Proc. Int. Symp. on Soil Fertility Evaluation.* Indian Soc. Soil Sci., New Delhi, pp. 7–20.

Cole, C.V., and R.D. Heil. 1981. Phosphorus effects on terrestrial nitrogen cycling. *Ecol. Bull.* 33:363–374.

Cole, C.V., G.S. Innis, and J.W.B. Stewart. 1977. Simulation of phosphorus cycling in semiarid grasslands. *Ecology* 58:1–15.

Cole, C.V., and R.L. Sanford, Jr. 1989. Biological aspects of the phosphorus cycle. Paper presented at the *Symp. on Phosphorus Requirements for Sustainable Agriculture in Asia and Oceania,* 6–10 March 1989. International Rice Research Institute, Los Baños, Laguna, Philippines.

Condron, L.M., E. Frossard, H. Tiessen, R.H. Newman, and J.W.B. Stewart. 1990. Chemical nature of organic phosphorus in cultivated and uncultivated soils under different environmental conditions. *J. Soil Sci.* 41:41–50.

Condron, L.M., and K.M. Goh. 1989. Effects of long-term phosphatic fertilizer applications on amounts and forms of phosphorus in soils under irrigated pasture in New Zealand. *J. Soil Sci.* 40:383–395.

Condron, L.M., K.M. Goh, and R. H. Newman. 1985. Nature and distribution of soil phosphorus as revealed by a sequential extraction method followed by ^{31}P nuclear magnetic resonance analysis. *J. Soil Sci.* 36:199–207.

Cook, R.L. 1935. Divergent influence of degree of base saturation of soils on the availability of native, soluble, and rock phosphates. *Am. Soc. Agron.* 27:297–311.

Cooke, G.W. 1956. The value of rock phosphate for direct application. *Emp. J. Exp. Agric.* 24:295–306.

Dalal, R.C. 1977. Soil organic phosphorus. *Adv. Agron.* 29:83–117.

Dalal, R.C. 1985. Comparative prediction of yield response and phosphorus uptake from soil using anion- and cation-anion exchange resins. *Soil Sci.* 139:227–231.

Dash, R.N., S.K. Mohanty, and S. Patnaik. 1981a. Chemical composition and reactivity of some Indian rock phosphates. *J. Indian Sci. Soil Sci.* 29:559–562.

Dash, R.N., S.K. Mohanty, and S. Patnaik. 1981b. Efficiency of HCl- and H_2SO_4-acidulated rock phosphates for a rice based cropping system. *Fert. Res.* 2:109–118.

Dash, R.N., S.K. Mohanty, and S. Patnaik. 1982a. Comparative study of reactivity and agronomic effectiveness of Indian, U.S., African, and Middle East rock phosphates for growing rice. *Fert. Res.* 3:177–189.

Dash, R.N., S.K. Mohanty, and S. Patnaik. 1982b. Note on the effects of HCl- and H_2SO_4-acidulated rock phosphates in relation to the time of application to rice. *Indian J. agric. Sci.* 52:252–254.

Datta, N.R., and N.P. Datta. 1963. Response to phosphate in rice and wheat in different soils. *J. Indian Soc. Soil Sci.* 11:117–128.

Debnath, N.C., S.K. Banerjee, and S.K. Mandal. 1974. Influence of humic and fulvic acids on the availability and inorganic transformation of phosphorus in soil. *J. Indian Chem. Soc.* 51:811–813.

Debnath, N.C., and R.K. Basak. 1986. Effect of rock phosphate and basic slag on available phosphorus in acid soils in relation to soil characteristics, seasons, and moisture regimes. *J. Indian Soc. Soil Sci.* 34:464–470.

De Datta, S.K. 1978. Fertilizer management for efficient use in wetland rice soils. *In Soils and Rice.* International Rice Research Institute, Los Baños, Laguna, Philippines, pp. 671–701.

De Datta, S.K. 1981. *Principles and Practices of Rice Production.* John Wiley & Sons, New York.

De Datta, S.K. 1983. Phosphorus requirements and phosphorus fertilization of lowland rice. *In Proc. 3rd Int. Congr. on Phosphorus Compounds (Brussels, Belgium).* Institut Mondial du Phosphate, Casablanca, Morocco, pp. 401–421.

De Datta, S.K., T.K. Biswas, and C. Charoenchamratcheep. 1989. Phosphorus requirements and management for lowland rice. Paper presented at the *Symp. on Phosphorus Requirements for Sustainable Agriculture in Asia and Oceania,* 6–10 March 1989, International Rice Research Institute, Los Baños, Laguna, Philippines.

De Datta, S.K., and K.A. Gomez. 1982. Changes in phosphorus and potassium responses in wetland rice soils in South and Southeast Asia. *In* E. Pushparajah and H.A.H. Sharifuddin (Eds.). *Phosphorus and Potassium in the Tropics.* The Malaysian Soc. Soil Sci., Kuala Lumpur, pp. 127–146.

De Datta, S.K., J.C. Moomaw, V.V. Racho, and G.V. Simsiman. 1966. Phosphorus supplying power of lowland rice soils. *Soil Sci. Soc. Am. Proc.* 30:613–617.

Diamond, R.B. 1985. Availability and management of phosphorus in wetland soils in relation to soil characteristics. *In Wetland Soils: Characterization, Classification, and Utilization.* International Rice Research Institute, Los Baños, Laguna, Philippines, pp. 269–283.

Dodd, J.C., C.C. Burton, R.G. Burns, and P. Jeffries. 1987. Phosphatase activity associated with the roots and the rhizosphere of plants infected with vesicular-arbuscular mycorrhizal fungi. *New Phytol.* 107:163–172.

Dolui, A.K., and S.K. Gangyopadhyay. 1984. Fixation of phosphorus in relation to properties of some red and lateritic soils of West Bengal. *Indian J. Agric. Chem.* 17:177–182.

Earl, K.D., J.K. Syers, and J.R. McLaughlin. 1979. Origin of the effects of citrate, tartarate, and acetate on phosphate sorption by soils and synthetic gels. *Soil Sci. Soc. Am. J.* 43:674–678.

Eivazi, F., and C.C. Weir. 1989. Phosphorus and mycorrhizal interaction on uptake of phosphorus and trace elements by maize. *Fert. Res.* 21:19–22.

Elliot, E.T., K. Horton, J.C. Moore, D.C. Coleman, and C.V. Cole. 1984. Mineralization dynamics in fallow dryland wheat plots, Colorado. *Plant Soil* 76:149–155.

Emsley, J., and S. Niazi. 1983. The analysis of soil phosphorus by ICP and ^{31}P-NMR spectroscopy. *Phosphorus and Sulfur* 16:303–312.

Enfield, C.G., C.C. Harlin, Jr., and B.E. Bledsoe. 1976. Comparison of five kinetic methods for orthophosphate reactions in mineral soils. *Soil Sci. Soc. Am. J.* 40:243–249.

Engelstad, O.P., A. Jugsujinda, and S. K. De Datta. 1974. Response by flooded rice to phosphate rocks varying in citrate solubility. *Soil Sci. Soc. Am. Proc.* 38:524–529.

Engelstad, O.P., and G.L. Terman. 1980. Agronomic effectiveness of phosphate fertilizers. *In* F.E. Khasawneh, E.C. Sample, and E.J. Kamprath (Eds.). *The Role of Phosphorus in Agriculture.* Soil Sci Soc. Am., Madison, Wisconsin, pp. 311–332.

Evans, A. Jr. 1985. The adsorption of inorganic phosphate by a sandy soil as influenced by dissolved organic compounds. *Soil Sci.* 140:251–255.

Ferris, R.S. 1984. Effects of microwave oven treatment on microorganisms in soil. *Phytopathol.* 74:121–126.

Fischer, W.R. 1983. Theoretische Betrachtungen zur reduktiven Ausflösung von Eisen (III)-oxiden. *Z. Pflanzenernaehr. Bodenkd.* 146:611–622.

Fixen, P.E., and A.E. Ludwick. 1982a. Residual available phosphorus in near neutral and alkaline soils. I. Solubility and capacity relationships. *Soil Sci. Soc. Am. J.* 46:332–334.

Fixen, P.E., and A.E. Ludwick. 1982b. Residual available phosphorus in near-neutral and alkaline soils. II. Persistence and quantitative estimation. *Soil Sci. Soc. Am. J.* 46:335–338.

Fixen, P.E., A.E. Ludwick, and S.R. Olsen. 1983. Phosphorus and potassium fertilization of irrigated alfalfa on calcareous soils. Soil phosphorus solubility relationships. *Soil Sci. Soc. Am. J.* 47:112–117.

Fox, R.L. 1974. Examples of anion and cation adsorption by soils of tropical America. *Tropic. Agric.* 51:200–210.

Fox, R. L., and E.J. Kamprath. 1970. Phosphate sorption isotherms for evaluating the phosphate requirements of soils. *Soil Sci. Soc. Am. Proc.* 34:902–907.

Fox, R.L., and B.T. Kang. 1978. Influence of phosphorus fertilizer placement and fertilization rate on maize nutrition. *Soil Sci.* 125:34–40.

Freeman, J.S., and D.L. Rowell. 1981. The adsorption and precipitation of phosphate onto calcite. *J. Soil Sci.* 32:75–84.

Friesen, D.K., and G.J. Blair. 1988. A dual radiotracer study of transformations of organic, inorganic and plant residue phosphorus in soil in presence and absence of plants. *Aust. J. Soil Res.* 26:355–366.

Friesen, D.K., A.S.R. Juo, and M.H. Miller. 1980. Liming and lime–phosphorus–zinc interactions in two Nigerian Ultisols. 1. Interactions in the soil. *Soil Sci. Soc. Am. J.* 44:1221–1226.

Garbouchev, I.P. 1981. The manufacture and agronomic efficiency of a partially acidulated phosphate rock fertilizer. *Soil Sci. Soc. Am. J.* 45:970–974.

Gebhardt, H., and N.T. Coleman. 1974. Anion adsorption by allophanic tropical soils. III. Phosphate adsorption. *Soil Sci. Soc. Am. Proc.* 38:263–266.

Gianinazzi-Pearson, V., J. Fardeau, S. Asimi, and S. Gianinazzi. 1981. Source of additional phosphorus absorbed from soil by vesicular–arbuscular mycorrhizal soybeans. *Physiol. Veg.* 19:33–43.

Gil-Sotres, F., W. Zech, and H.G. Alt. 1990. Characterization of phosphorus fractions in surface horizons of soil from Galicia (N.W. Spain) by ^{31}P-NMR spectroscopy. *Soil Biol. Biochem.* 22:75–79.

Goldberg, S., and G. Sposito. 1984a. A chemical model of phosphate adsorption by soils. I. Reference oxide minerals. *Soil Sci. Soc. Am. J.* 48:772–778.

Goldberg, S., and G. Sposito. 1984b. A chemical model of phosphate adsorption by soils. II. Noncalcareous soils. *Soil Sci. Soc. Am. J.* 48:779–783.

Goswami, N.N., and N.K. Banerjee. 1978. Phosphorus, potassium, and other macroelements. *In Soils and Rice.* International Rice Research Institute, Los Baños, Laguna, Philippines, pp. 561–580.

Graham, E.R. 1955. Availability of natural phosphates according to energy changes. *Soil Sci. Soc. Am. Proc.* 19:26–29.

Greenland, D.J., and S.K. De Datta. 1985. Constraints to rice production and wetland soil characteristics. *In Wetland Soils: Characterization, Classification, and Utilization.* International Rice Research Institute, Los Baños, Laguna, Philippines, pp. 23–36.

Greenland, D.J., J.M. Oades, and T.W. Sherwin. 1968. Electron-microscope observations of iron oxides in some red soils. *J. Soil Sci.* 19:123–126.

Gregg, P.E.H., J.K. Syers, and A.D. MacKay. 1981. Agronomic effectiveness of reactive phosphate rocks in hill country pastures. *In Proc. Tech. Workshop on Potential of Phosphate Rock as a Direct Application Fertilizer in New Zealand. Occasional Report No.3.* Massey University, Palmerston North, New Zealand, pp. 4–10.

Gregory, P.J. 1988. Growth and functioning of plant roots. *In* A. Wild (Ed.). *Russell's Soil Conditions and Plant Growth.* 11th Ed. Longman, Harlow, England, pp. 155–163.

Grindel, N.M., and N.G. Zyrin. 1965. Method of determination and dynamics of organic phosphate compounds in the plow horizon of slightly cultivated sod-podzolic soils. *Soviet Soil Sci.* 12:1391–1401.

Grinsted, M.J., M.J. Hedley, R.E. White, and P.H. Nye. 1982. Plant-induced changes in the rhizosphere of rape (*Brassica napus* var. Emerald) seedlings. I. pH change and the increase in phosphorus concentration in the soil solution. *New Phytol.* 91:19–29.

Gunary, D. 1970. A new adsorption isotherm for phosphate in soil. *J. Soil Sci.* 21:72–77.

Gunjigake, N., and K. Wada. 1981. Effects of phosphorus concentration and pH on phosphate retention by active aluminum and iron of Ando soils. *Soil Sci.* 132:347–352.

Hagin, J. 1985. *Partially Acidulated Phosphate Rock. A Review.* Water and Soils Research Centre, Technion-Israel Institute of Technology.

Halm, B.J., J.W.B. Stewart, and R.L. Halstead. 1972. The phosphorus cycling in a natural grassland eco-system. In: *Isotopes and Radiation in Soil Plant Relationships Including Forestry.* IAEA, Vienna, pp. 571–586.

Hammond, L.L., S.H. Chien, and A.U. Mokwunye. 1986. Agronomic value of unacidulated and partially acidulated phosphate rocks indigenous to the tropics. *Adv. Agron.* 40:89–140.

Hammond, L.L., S.H. Chien, and J.R. Polo. 1980. Phosphorus availability from partial acidulation of two phosphate rocks. *Fert. Res.* 1:37–49.

Hammond, L.L., S.H. Chien, A.H. Roy, and A.U. Mokwunye. 1989. Solubility and agronomic effectiveness of partially acidulated phosphate rocks as influenced by their iron and aluminum oxide content. *Fert. Res.* 19:93–98.

Hammond, L.L., and L.A. Leon. 1983. Agronomic effectiveness of natural and altered phosphate rocks from Latin America. *In Proc. 3rd Int. Congr. on Phosphorus Compounds, (Brussels, Belgium).* Institut Mondial du Phosphate, Casablanca, Morocco, pp. 503–518.

Harrison, A.F. 1982a. ^{32}P-method to compare rates of mineralization of labile organic phosphorus in woodland soils. *Soil Biol. Biochem.* 14:337–341.

Harrison, A.F. 1982b. Labile organic phosphorus mineralization in relationship to soil properties. *Soil Biol. Biochem.* 14:343–351.

Harrison, A.F. 1987. *Soil Organic Phosphorus. A review of World Literature.* C.A.B. International Wallingford, United Kingdom.

Harrison, R.B. and F. Adams. 1987. Solubility characteristics of residual phosphate in a fertilized and limed Ultisol. *Soil Sci. Soc. Am. J.* 51:963–969.

Harter, R.D. 1969. Phosphorus adsorption sites in soils. *Soil Sci. Soc. Am. Proc.* 33:630–632.

Harter, R.D. 1984. Curve-fit errors in Langmuir adsorption maxima. *Soil Sci. Soc. Am. J.* 48:749–752.

Havlin, J.L., and D.G. Westfall. 1984. Soil test phosphorus and solubility relationships in calcareous soils. *Soil Sci. Soc. Am. J.* 48:327–330.

Hawkes, G.E., D.S. Powlson, E.W. Randall, and K.R. Tate. 1984. A ^{31}P-nuclear magnetic resonance study of the phosphorus species in soils from long continued field experiments. *J. Soil Sci.* 35:35–45.

Haynes, R.J. 1982. Effects of liming on phosphate availability in acid soils. A critical review. *Plant Soil* 68:289–308.

Haynes, R.J. 1983. Effect of lime and phosphate applications on the adsorption of phosphate, sulfate, and molybdate by a Spodosol. *Soil Sci.* 135:221–227.

Haynes, R.J. 1984. Lime and phosphate in the soil-plant system. *Adv. Agron.* 37:249–315.

Haynes, R.J., and T.E. Ludecke. 1981. Effect of lime and phosphorus applications on concentrations of available nutrients and on P, Al and Mn uptake by two pasture legumes in an acid soil. *Plant Soil* 62:117–128.

Hedley, M.J., P.H. Nye, and R.E. White. 1982a. Plant-induced changes in the rhizosphere of rape (*Brassica napus* var. Emerald) seedlings. II. Origin of the pH change. *New Phytol.* 91:31–44.

Hedley, M.J., and J.W.B. Stewart. 1982. Method to measure microbial phosphate in soils. *Soil Biol. Biochem.* 14:377–385.

Hedley, M.J., J.W.B. Stewart, and B.S. Chauhan. 1982b. Changes in inorganic and organic soil phosphorus fractions induced by cultivation practices and by laboratory incubations. *Soil Sci. Soc. Am. J.* 46:970–976.

Hedley, M.J., R.E. White, and P.H. Nye. 1982c. Plant-induced changes in the rhizosphere of rape (*Brassica napus* var. Emerald) seedlings. III. Changes in *L* value, soil phosphate fractions and phosphatase activity. *New Phytol.* 91:45–56.

Helal, H.M., and D. Sauerbeck. 1984. Influence of plant roots on C and P metabolism in soil. *Plant Soil* 76:175–182.

Hellums, D.T., S.H. Chien, and J.T. Touchton. 1989. Potential agronomic value of calcium in some phosphate rocks from south America and west Africa. *Soil Sci. Soc. Am. J.* 53:459–462.

Hingston, F.J. 1982. A review of anion adsorption. *In* M.A. Anderson and A.J. Rubin (Eds.). *Adsorption of Inorganics at Solid–Liquid Surfaces.* Ann Arbor Sci. Pub., Inc., Ann Arbor, Michigan, pp. 51–90.

Hirata, H., M. Toshihisa, and H. Koiwa. 1988. Response of chickpea grown on Ando-soil to vesicular-arbuscular mycorrhizal infection in relation to the level of phosphorus application. *Soil Sci. Plant Nutr.* 34:441–449.

Holford, I.C.R. 1979. Evaluation of soil phosphate-buffering indices. *Aust. J. Soil Res.* 17:495–504.

Holford, I.C.R. 1988. Buffering of phosphate in the soil solution during growth of ryegrass compared with buffering by sorption. *Plant Soil* 111:3–9.

Holford, I.C.R., and W.H. Patrick, Jr. 1979. Effects of reduction and pH changes on phosphate sorption and mobility in an acid soil. *Soil Sci. Soc. Am. J.* 43:292–297.

Hong, J.K., and I. Yamane. 1980a. Inositol phosphate and inositol in humic and fulvic acid fractions extracted by three methods. *Soil Sci. Plant Nutr.* 26:491–496.

Hong, J.K., and I. Yamane. 1980b. Examination of the conventional method and a proposal for a new method of determining inositol phosphate in the humic acid fraction. *Soil Sci. Plant Nutr.* 26:497–505.

Hong, J.K., and I. Yamane. 1981. Distribution of inositol phosphate in the molecular size fractions of humic and fulvic acid fractions. *Soil Sci. Plant Nutr.* 27:295–303.

Howe, D.O., and E.R. Graham. 1957. Salt concentration: A factor in the availability of phosphorus from rock phosphate as revealed by the growth and composition of alfalfa. *Soil Sci. Soc. Am. Proc.* 21:25–28.

Hsu, P.H. 1965. Fixation of phosphate by aluminum and iron in acidic soils. *Soil Sci.* 99:398–402.

Hughes, J.C., and R.J. Gilkes. 1984. The effect of chemical extractant on the estimation of rock phosphate fertilizer dissolution. *Aust. J. Soil Res.* 22:475–481.

Hughes, J.C., and R.J. Gilkes. 1986a. The effect of soil properties and level of fertilizer application on the dissolution of Sechura rock phosphate in some soils from Brazil, Colombia, Australia and Nigeria. *Aust. J. Soil Res.* 24:219–227.

Hughes, J.C., and R.J. Gilkes. 1986b. The effect of rock phosphate properties on the extent of fertilizer dissolution in soils. *Aust. J. Soil Res.* 24:209–217.

Ibrahim, H.S., and P.F. Pratt. 1982. Effects of rate of application and time on phosphorus sorption by soils. *Soil Sci. Soc. Am. J.* 46:926–928.

Imai, H., W.T. Goulding, and O. Talibudeen. 1981. Phosphate adsorption in allophanic soils. *J. Soil Sci.* 32:555–570.

Inskeep, W., and J.C. Silvertooth. 1988. Inhibition of hydroxyapatite precipitation in the presence of fulvic, humic and tannic acids. *Soil Sci. Soc. Am. J.* 52:941–946.

Islam, A., and R. Mandal. 1977. Amounts and mineralization of organic phosphorus compounds and derivatives in some surface soils of Bangladesh. *Geoderma* 17:57–68.

Jarrell, W.M., and R.B. Beverly. 1981. The dilution effect in plant nutrition studies. *Adv. Agron.* 34:197–224.

Jayachandran, K., A.P. Schwab, and B.A.D. Hetrick. 1989. Mycorrhizal mediation of phosphorus availability: Synthetic iron chelate effects on phosphorus solubilization. *Soil Sci. Soc. Am. J.* 53:1701–1706.

Jenkinson, D.S., and J.N. Ladd. 1981. Microbial biomass in soil: Measurement and turnover. *In* E.A. Paul and J.N. Ladd (Eds.). *Soil Biochemistry.* Marcel Dekker, New York, pp. 415–471.

Jensen, H.E. 1971. Phosphate solubility in Danish soils equilibrated with solutions of differing phosphate concentrations. *J. Soil Sci.* 22:261–266.

Jones, C.A., C.V. Cole, A.N. Sharpley, and J.R. Williams. 1984a. A simplified soil and plant phosphorus model. I. Documentation. *Soil Sci. Soc. Am. J.* 48:800–805.

Jones, U.S., J.C. Katyal, C.P. Mamaril, and C.S. Park. 1982. Wetland rice nutrient deficiencies other than nitrogen. *In Rice Research Strategies for the Future.* International Rice Research Institute, Los Baños, Laguna, Philippines, pp. 327–378.

Jones, C.A., A.N. Sharpley, and J.R. Williams. 1984b. A simplified soil and plant phosphorus model. III. Testing. *Soil Sci. Soc. Am. J.* 48:810–813.

Joos, L.L., and C.A. Black. 1950. Availability of phosphate rock as affected by particle size and contact with bentonite and soil of different pH values. *Soil Sci. Soc. Am. Proc.* 15:69–75.

Junge, A., and W. Werner. 1989. Investigations on interactions of phosphorus compounds in partially acidulated phosphate rock and fertilizer effectiveness. *Fert. Res.* 20:129–134.

Jungk, A., and N. Classen. 1986. Availability of phosphate and potassium as the result of interactions between root and soil in rhizosphere. *Z. Pflanzenernaehr Bodenkd.* 149:411–427.

Juo, A.S.R. 1981. Chemical characteristics. *In* D.J. Greenland (ed.) *Characterization of Soils.* Clarendon Press, Oxford, pp. 51–79.

Jurinak, J.J., L.M. Dudley, M.F. Allen, and W.G. Knight. 1986. The role of calcium oxalate in the availability of phosphorus in soils of semiarid regions. A thermodynamic study. *Soil Sci.* 142:255–261.

Kamprath, E.J., and M.E. Watson. 1980. Conventional soil and tissue tests for assessing the phosphorus status of soils. *In* F.E. Khasawneh, E.C. Sample, and E.J. Kamprath (Eds.). *The Role of Phosphorus in Agriculture.* Soil Sci. Soc. Am., Madison, Wisconsin, pp. 433–469.

Kanabo, I.A.K., and R.J. Gilkes. 1987a. A comparison between plant response and chemical measurements of the dissolution of reactive phosphate rock in soils of different pH and phosphorus retention. *Aust. J. Soil Res.* 25:451–460.

Kanabo, I.A.K., and R.J. Gilkes. 1987b. The role of soil pH in the dissolution of phosphate rock fertilizers. *Fert. Res.* 12:165–179.

Kanabo, I.A.K., and R.J. Gilkes. 1987c. The influence of the addition of goethite to soil on the dissolution of North Carolina phosphate rock. *Aust. J. Soil Res.* 25:313–322.

Kanabo, I.A.K., and R.J. Gilkes. 1988. The effects of moisture regime and incubation period on the dissolution of North Carolina phosphate rock in soil. *Aust. J. Soil Res.* 26:153–163.

Kato, N., and N. Owa. 1989. Kinetics of phosphate adsorption by sandy and clayey soils. *Soil Sci. Plant Nutr.* 35:119–129.

Khasawneh, F.E., and E.C. Doll. 1978. The use of phosphate rock for direct application to soils. *Adv. Agron.* 30:159–206.

Khasawneh, F.E., E.C. Sample, and I. Hashimoto. 1974. Reactions of ammonium ortho- and polyphosphate fertilizers in soil. I. Mobility of phosphorus. *Soil Sci. Soc. Am. Proc.* 38:446–451.

Kirk, G.J.D., and P.H. Nye. 1986. The dissolution and dispersion of dicalcium phosphate dihydrate in soils. IV. Experimental evaluation of the model for particles. *J. Soil Sci.* 37:525–528.

Kittrick, J.A., and M.L. Jackson. 1955a. Rate of phosphate reaction with soil minerals and electron microscrope observations on the reaction mechanism. *Soil Sci. Soc. Am. Proc.* 19:292–295.

Kittrick, J.A., and M.L. Jackson. 1955b. The common ion effect on phosphate solubility. *Soil Sci.* 79:415–421.

Kittrick, J.A., and M.L. Jackson. 1956. Electron microscope observation of the reaction of phosphate with minerals leading to a unified theory of phosphate fixation. *J. Soil Sci.* 7:81–89.

Klages, M.G., R.A. Olsen, and V.A. Haby. 1988. Relationship of phosphorus isotherms to $NaHCO_3$-extractable phosphorus as affected by soil properties. *Soil Sci.* 146:85–91.

Kodama, H., and M.D. Webber. 1975. Clay-inorganic studies. III. Hydroxy aluminum phosphate–montmorillonite complex. *Can. J. Soil Sci.* 55:225–233.

Kumar, P., and B. Mishra. 1986. Dissolution pattern of Indian rock phosphates in acid soils. *J. Indian Soc. Soil Sci.* 34:611–613.

Kuo, S. 1988. Application of a modified Langmuir isotherm to phosphate sorption by some acid soils. *Soil Sci. Soc. Am. J.* 52:97–102.

Kuo, S., and E.J. Jellum. 1987. Influence of soil characteristics and environmental conditions on seasonal variations of water-soluble phosphate in soils. *Soil Sci.* 143:257–263.

Kuo, S., E.J. Jellum, and W.L. Pan. 1988. Influence of phosphate sorption parameters of soils on the desorption of phosphate by various extractants. *Soil Sci. Soc. Am. J.* 52:974–979.

Kuo, S., and E.G. Lotse. 1974a. Kinetics of phosphate adsorption and desorption by hematite and gibbsite. *Soil Sci.* 116:400–406.

Kuo, S., and E.G. Lotse. 1974b. Kinetics of phosphate adsorption and desorption by lake sediments. *Soil Sci. Soc. Am. Proc.* 38:50–54.

Kuo, S., and B.L. McNeal. 1984. Effects of pH and phosphate on cadmium sorption by a hydrous ferric oxide. *Soil Sci. Soc. Am. J.* 48:1040–1044.

Kuo, S., and D.S. Mikkelsen. 1979. Distribution of iron and phosphorus in flooded and unflooded soil profiles and their relation to phosphorus adsorption. Soil Sci. 127:18–25.

Kwong, K.F. Kee Ng, and P.M. Huang. 1979. Surface reactivity of aluminum hydroxides precipitated in the presence of low molecular weight organic acids. *Soil Sci. Soc. Am. J.* 43:1107–1113.

Lajtha, K., and S.H. Bloomer. 1988. Factors affecting phosphate sorption and phosphate retention in a desert ecosystem. *Soil Sci.* 146:160–167.

Lal, M., and I.C. Mahapatra. 1979. Phosphate transformation under continuous submergence in rice–barley rotation. *J. Indian Soc. Soil Sci.* 27:375–382.

Lehr, J.R. 1972. Chemical reactions of micronutrients in fertilizers. *In* J.J. Mortvedt, P.M. Giardano, and W.L. Lindsay, (Eds.). *Micronutrients in Agriculture.* Soil Sci. Soc. Am., Madison, Wisconsin, pp. 459–503.

Lehr, J. R. 1980. Phosphate raw materials and fertilizers. Part I—A look ahead. *In* F.E. Khasawneh, E.C. Sample, and E.J. Kamprath (Eds.). *The Role of Phosphorus in Agriculture.* Soil Sci. Soc. Am., Madison, Wisconsin, pp. 81–128.

Lehr, J.R., and G.H. McClellan. 1972. A revised laboratory reactivity scale for evaluating phosphate rocks for direct application. *In TVA Bull. Y-43.* Tennessee Valley Authority, Muscle Shoals, Alabama, p. 36.

Le Mare, P. H. 1982. Sorption of isotopically exchangeable and nonexchangeable phosphate by some soils of Colombia and Brazil, and comparisons with soils of southern Nigeria. *J. Soil Sci.* 33:691–707.

Le Mare, P.H., and L.A. Leon. 1989. The effects of lime on adsorption and desorption of phosphate in five Colombian soils. *J. Soil Sci.* 40:59–69.

Leon, L. A., W.E. Fenster, and L.L. Hammond. 1986. Agronomic potential of eleven phosphate rocks from Brazil, Colombia, Peru and Venezuela. *Soil Sci. Soc. Am. J.* 50:798–802.

Lin, C., W.J. Busscher, and L.A. Douglas. 1983a. Multifactor kinetics of phosphate reactions with minerals in acidic soils. I. Modeling and simulation. *Soil Sci. Soc. Am. J.* 47:1097–1103.

Lin, C., L.A. Douglas, H.L. Motto, and W.J. Busscher. 1986. Phosphate reaction models for the Lakehurst soil in the New Jersey Pine barrens. *Soil Sci.* 142:125–131.

Lin, C., H.L. Motto, L.A. Douglas, and W.J. Busscher. 1983b. Multifactor kinetics of phosphate reactions with minerals in acid soils. II. Experimental curve fitting. *Soil Sci. Soc. Am. J.* 47:1103–1109.

Lindsay, W.L., J.R. Lehr, and H.F. Stephenson. 1959a. Nature of the reactions of monocalcium phosphate monohydrate in soils. III. Studies with metastable triple-point solution. *Soil Sci. Soc. Am. Proc.* 23:342–345.

Lindsay, W.L., M. Peech, and J.S. Clark. 1959b. Solubility criteria for the existence of variscite in soils. *Soil Sci. Soc. Am. Proc.* 23:357–360.

Loganathan, P., N.O. Isirimah, and D.A. Nwachuku. 1987. Phosphorus sorption by Ultisols and Inceptisols of the Niger delta in southern Nigeria. *Soil Sci.* 144:330–338.

Logan, T.J., and E.O. McLean. 1977. Diffusion of ^{32}P from partially acidulated rock phosphate. *Soil Sci.* 123:203–206.

Lopez-Hernandez, D. 1987. Phosphate adsorption vaiability within soil series and in diverse soil populations. *Soil Sci.* 144:408–411.

MacKay, A.D., and J.K. Syers. 1986. Effect of phosphate, calcium, and pH on the dissolution of a phosphate rock in soil. *Fert. Res.* 10:175–184.

MacKay, A.D., J.K. Syers, P.E.H. Gregg, and R.W. Tillman. 1984. A comparison of three soil-testing procedures for estimating the plant availability of phosphorus in soils receiving either superphosphate or phosphate rock. *New Zeal. J. Agr. Res.* 27:231–245.

MacKay, A.D., J.K. Syers, R.W. Tillman, and P.E.H. Gregg. 1986. A simple model to describe the dissolution of phosphate rock in soils. *Soil Sci. Soc. Am. J.* 50:291–296.

Madrid, L., and A.M. Posner. 1979. Desorption of phosphate from goethite. *J. Soil Sci.* 30:697–707.

Mahapatra, I.C., and W.H. Patrick, Jr. 1969. Inorganic phosphate transformation in waterlogged soils. *Soil Sci.* 107:281–288.

Mahapatra, I.C., and W.H. Patrick, Jr. 1971. Evaluation of phosphate fertility in waterlogged soils. *In Proc. Int. Symp. on Soil Fertility Evaluation.* Indian Soc. Soil Sci., New Delhi, pp. 53–62.

Mandal, L.N. 1964. Effect of time, starch, and lime on the transformation of inorganic phosphorus in waterlogged rice soil. *Soil Sci.* 97:127–132.

Mandal, L.N. 1979. Transformation of phosphorus in waterlogged soil. *Bull. Indian Soc. Soil Sci.* 12:73–80.

Mandal, L.N. 1984. Soil research in relation to rice. *J. Indian Soc. Soil Sci.* 32:575–582.

Mandal, L.N., and K.C. Mandal. 1973. Influence of organic matter and lime on the transformation of applied phosphate in acidic lowland rice soils. *J. Indian Soc. Soil Sci.* 21:57–62.

Mandal, L.N., and S.K. Khan. 1975. Influence of soil moisture regime on transformation of inorganic phosphorus in rice soils. *J. Indian Soc. Soil Sci.* 23:31–37.

Mandal, L.N., and S.K. Khan. 1977a. Transformation of phosphorus in soils under waterlogged condition. *J. Indian Soc. Soil Sci.* 25:122–128.

Mandal, L.N., and S.K. Khan. 1977b. Influence of moisture regimes on the transformation of the recently applied phosphate in soil and its availability to rice crop. *J. Indian Soc. Soil Sci.* 25:379–383.

Manikandan, P., and T.G. Sastry. 1988. Role of different soil components in phosphate adsorption in some soils of Mysore plateau. *J. Indian Soc. Soil Sci.* 36:228–234.

Mansell, R.S., and H.M. Selim. 1981. Mathematical models for predicting reactions and transport of phosphorus applied to soils. *In* I.K. Iskandar (Ed.). *Modeling Wastewater Renovation. Wiley Interscience.*, New York, pp. 600–646.

Martel, Y.A., and E.A. Paul. 1974. Effects of cultivation on the organic matter of grassland soils as determined by fractionation and radiocarbon dating. *Can. J. Soil Sci.* 54:419–426.

Martin, J.K., and R.C. Foster. 1985. A model system for studying the biochemistry and biology of the root soil interface. *Soil Biol. Biochem.* 17:261–269.

Martin, R.R., R.St.C. Smart, and Q. Tazaki. 1988. Direct observation of phosphate precipitation in the goethite/phosphate system. *Soil Sci. Soc. Am. J.* 52:1492–1500.

Marwaha, B.C. 1983. Partially acidulated rock phosphate as a source of fertilizer phosphorus with special reference to high P-fixing acid soils—A review. *Proc. Indian Natn. Sci. Acad.* B49:436–446.

Marwaha, B.C., B.S. Kapoor, and B.R. Tripathi. 1983. Responsiveness of some Indian rock phosphates to acidulation with phosphoric and nitric acids. *J. Indian Soc. Soil Sci.* 31:328–330.

Mattson, S., E. Alvsaker, E. Koutler-Anderson, E. Barkoff, and K. Vahtras. 1950. Phosphate relationships of soil and plant. VI. The salt effect on phosphate solubility in pedalfer soils. *Ann. Roy. Agric. Coll. Swed.* 17:141–160.

McClellan, G.H., and L.R. Gremillion. 1980. Evaluation of phosphatic raw materials. *In* F.E. Khasawneh, E.C. Sample, and E.J. Kamprath (Eds.). *The Role of Phosphorus in Agriculture.* Soil Sci. Soc. Am., Madison, Wisconsin, pp. 43–80.

McGill, W.B., and C.V. Cole. 1981. Comparative aspects of cycling of organic C, N, S, and P through soil organic matter. *Geoderma* 26:267–286.

McLaughlin, M.J., and A.M. Alston. 1985. Measurement of phosphorus in the soil microbial biomass: Influence of plant material. *Soil Biol. Biochem.* 17:271–274.

McLaughlin, M.J., and A.M. Alston. 1986. The relative contribution of plant residues and fertilizer to the phosphorus nutrition of wheat in a pasture/cereal system. *Aust. J. Soil Res.* 24:517–526.

McLaughlin, M.J., A.M. Alston, and J.K. Martin. 1986. Measurement of phosphorus in the soil microbial biomass: A modified procedure for field soils. *Soil Biol. Biochem.* 18:437–443.

McLaughlin, M.J., A.M. Alston, and J.K. Martin. 1987. Transformations and movement of P in the rhizosphere. *Plant Soil* 97:391–399.

McLaughlin, M.J., A.M. Alston, and J.K. Martin. 1988a. Phosphorus cycling in wheat–pasture rotations. I. The source of phosphorus taken up by wheat. *Aust. J. Soil Res.* 26:323–331.

McLaughlin, M.J., A.M. Alston, and J.K. Martin. 1988b. Phosphorus cycling in wheat–pasture rotations. II. The role of the microbial biomass in phosphorus cycling. *Aust. J. Soil Res.* 26:333–342.

McLaughlin, M.J., A.M. Alston, and J.K. Martin. 1988c. Phosphorus cycling in wheat-pasture rotations. III. Organic phosphorus turnover and phosphorus cycling. *Aust. J. Soil Res.* 26:343–353.

McLaughlin, J.R., J.C. Ryden, and J.K. Syers. 1981. Sorption of inorganic phosphate by iron- and aluminum-containing components. *J. Soil Sci.* 32:365–377.

McLaughlin, J.R., and J.K. Syers. 1978. Stability of ferric phosphates. *J. Soil Sci.* 29:499–504.

McLean, E.O., and B.S. Balam. 1967. Partially acidulated rock phosphate as a source of phosphorus to plants. III. Uptake by corn from soils of different calcium status. *Soil Sci. Soc. Am. Proc.* 31:811–814.

McLean, E.O., and T.J. Logan. 1970. Sources of phosphorus for plants grown in soils with different phosphorus fixation tendencies. *Soil Sci. Soc. Am. Proc.* 34:906–911.

McLean, E.O., and R.W. Wheeler. 1964. Partially acidulated rock phosphate as a source of phosphorus to plants. I. Growth chamber studies. *Soil Sci. Soc. Am. Proc.* 28:545–550.

McLean, E.O., R.W. Wheeler, and J.D. Watson. 1965. Partially acidulated rock phosphate as a source of phosphorus to plants. Growth chamber and field corn studies. *Soil Sci. Soc. Am. Proc.* 29:625–628.

McSweeney, G., and A.G. Charleston. 1985. Partially acidulated phosphate rocks—reactions with water. *Fert. Res.* 8:75–83.

Mead, J.A. 1981. A comparison of the L, F, and T equations to describe phosphate adsorption properties of soils. *Aust. J. Soil Res.* 19:333–342.

Mehadi, A.A., and R.W. Taylor. 1988. Phosphate adsorption by two highly weathered soils. *Soil Sci. Soc. Am. J.* 52:627–632.

Mendoza, R.E., and N.J. Barrow. 1987a. Characterizing the rate of reaction of some Argentinian soils with phosphate. *Soil Sci.* 143:105–112.

Mendoza, R.E., and NJ. Barrow. 1987b. Ability of three soil extractants to reflect the factors that determine the availability of soil phosphate. *Soil Sci.* 144:319–329.

Mishra, B., P. Kumar, and G.K. Dwivedi. 1985. Effectiveness of Indian rock phosphates in acid soils. *J. Indian Soc. Soil Sci.* 33:574–580.

Mishra, M.M., K.K. Kapoor, and K.S. Yadav. 1982. Preparation of P-enriched compost with rock phosphate and its effect on crop yield. *Indian J. agric. Sci.* 52:674–678.

Mizota, C., M.A. Carrasco, and K. Wada. 1982. Clay mineralogy and some chemical properties of Ap horizons of Ando soils used for paddy rice in Japan. *Geoderma* 27:225–237.

Mokwunye, U. 1975. The influence of pH on the adsorption of phosphate by soils from the Guinea and Sudan savannah zones of Nigeria. *Soil Sci. Soc. Am. Proc.* 39:1100–1102.

Mokwunye, A.U., and S.H. Chien. 1980. Reactions of partially acidulated phosphate rock with soils from the tropics. *Soil Sci. Soc. Am. J.* 44:477–482.

Molloy, L.F., and L.C. Blackemore. 1974. Studies on a climosequence of soils in tussock grasslands. I. Introduction, sites, and soils. *New Zeal. J. Sci.* 17:233–255.

Moody, P.W., and D.J. Radcliffe. 1986. Phosphorus sorption by Andepts from the southern highlands of Papua New Guinea. *Geoderma* 37:137–147.

Morel, F.M.M., J.G. Yeasted, and J.C. Westall. 1981. Adsorption models: A mathematical analysis in the framework of general equilibrium calculations. *In* M.A. Anderson and A.J. Rubin (Eds.). *Adsorption of Inorganics at Solid–Liquid Interfaces.* Ann Arbor Sci. Pub., Inc., Ann Arbor, Michigan, pp. 263–294.

Mosse, B. 1973. Advances in the study of vesicular-arbuscular mycorrdhiza. *Annu. Rev. Phytopathol.* 11:171–196.

Mouat, M.C.H. 1983. Release of phosphate from soil measured by sequential desorption. *New Zeal. J. Agr. Res.* 26:321–326.

Muchovej, R.M.C., J.J. Muchovej, and V.H. Alvarez V. 1989. Temporal relations of phosphorus fractions in an Oxisol amended with rock phosphate and *Thiobacillus thiooxidans*. *Soil Sci. Soc. Am. J.* 53:1096–1100.

Mukherjee, S.K., S.K. Ghosh, and K. Ghosh. 1979. Mineralogy and chemistry of phosphorus in the soil. *Bull. Indian Soc. Soil Sci.* 12:9–22.

Muller-Harvey, I., and A. Wild. 1987. Isolation of a new sugar phosphate from forest leaf litter in Nigeria. *Soil Biol. Biochem.* 19:323–327.

Munch, J.C., T. Hillebrand, and J.C.G. Ottow. 1978. Transformations in the Fe_0/Fe_d ratio of pedogenic iron oxides affected by iron-reducing bacteria. *Can. J. Soil Sci.* 58:475–486.

Munch, J.C., and J.C.G. Ottow. 1980. Preferential reduction of amorphous to crystalline iron oxides by bacterial activity. *Soil Sci.* 129:15–21.

Munns, D.N., and R.L. Fox. 1976. The slow reaction which continues after phosphate adsorption: Kinetics and equilibrium in some tropical soils. *Soil Sci. Soc. Am. J.* 40:46–51.

Murrmann, R.P., and M. Peech. 1968. Reaction products of applied phosphate in limed soils. *Soil Sci. Soc. Am. Proc.* 32:493–496.

Murrmann, R.P., and M. Peech. 1969. Relative significance of labile and crystalline phosphates in soil. *Soil Sci.* 107:249–255.

Naidu, R., J.K. Syers, R.W. Tillman, and J.H. Kirkman. 1990. Effect of liming on phosphate sorption by acid soils. *J. Soil Sci.* 41:165–175.

Nanzyo, M. 1984. Diffuse reflectance infrared spectra of phosphate sorbed on alumina gel. *J. Soil Sci.* 35:63–69.

Nanzyo, M. 1986. Infrared spectra of phosphate sorbed on iron hydroxide gel and the sorption products. *Soil Sci. Plant Nutr.* 32:51–58.

Newman, E.J. 1978. Root microorganisms: Their significance in the ecosystem. *Biol. Rev.* 53:511–554.

Newman, R.H., and K.R. Tate. 1980. Soil phosphorus characterization by ^{31}P nuclear magnetic resonance. *Commun. Soil Sci. Plant Anal.* 11:835–842.

Nordengren, S. 1957. New theories of phosphate reactions in the soil. *Fert. and Feeding Stuffs J.* 47:348–352.

Notholt, A.J.G. 1975. Phosphate rock: World production, trade, and resources. *Proc. 1st Ind. Miner, Int. Congr., London.* Metals Bull. Ltd., Surrey, England, pp. 104–120.

Novak, L.T., and J.W. Petschauer. 1979. Kinetics of reaction between orthophosphate ions and Muskegon dune sand. *J. Environ. Qual.* 8:312–318.

O'Connor, G.A., K.L. Knudtsen, and G.A. Connel. 1986. Phosphorus solubility in sludge-amended calcareous soils. *J. Environ. Qual.* 15:308–312.

Okajima, H., H. Kubota, and T. Sakuma. 1983. Hysteresis in the phosphorus sorption and desorption processes of soils. *Soil Sci. Plant Nutr.* 29:271–283.

Olsen, R.A. 1975. Rate of dissolution of phosphate from minerals and soils. *Soil Sci. Soc. Am. Proc.* 29:634–639.

Olsen, R.G., and M.N. Court. 1982. Effect of wetting and drying of soils on phosphate adsorption and resin extraction of soil phosphate. *J. Soil Sci.* 33:709–717.

Olsen, S.R., and F.E. Khasawneh. 1980. Use and limitations of physical–chemical criteria for assessing the status of phosphorus in soils. *In* F.E. Khasawneh, E.C. Sample, and E.J. Kamprath (Eds.). *The Role of Phosphorus in Agriculture. Soil Sci. Soc. Am.*, Madison, Wisconsin, pp. 361–410.

Onken, A.B., and R.L. Matheson. 1982. Dissolution rate of EDTA-extractable phosphate from soils. *Soil Sci. Soc. Am. J.* 46:276–279.

Pairunan, A.K., A.D. Robson, and L.K. Abbott. 1980. The effectiveness of vesicular–arbuscular mycorrhizas in increasing growth and phosphorus uptake of subterranean clover from phosphorus sources of different solubilities. *New Phytol.* 84:327–338.

Palmer, B., M.D.A. Bolland, and R.J. Gilkes. 1979. A re-evaluation of the effectiveness of calcined Christmas Island C-grade rock phosphate. *Aust. J. Expt. Agric. Anim. Husb.* 19:605–610.

Parfitt, R.L. 1977. Phosphate adsorption on an Oxisol. *Soil Sci. Soc. Am. J.* 41:1064–1067.

Parfitt, R.L. 1978. Anion adsorption by soils and soil materials. Adv. Agron. 30:1–50.

Parfitt, R.L., L.J. Hume, and G.P. Sparling. 1989. Loss of availability of phosphate in

New Zealand soils. *J. Soil Sci.* 40:371–382.

Parton, W.J., D.S. Schimel, C.V. Cole, and D.S. Ojima. 1987. Analysis of factors controlling soil organic matter levels in great plains grasslands. *Soil Sci. Soc. Am. J.* 51:1173–1179.

Parton, W.J., J.W.B. Stewart, and C.V. Cole. 1988. Dynamics of C, N, P and S in grassland soils: A model. *Biogeochemistry* 5:109–131.

Pathiratna, L.S.S., U.P. De, S. Waidyantha, and O.S. Peries. 1989. The effect of apatite and elemental sulfur mixtures on growth and P content of *Centrocema pubescens. Fert. Res.* 21:37–43.

Patrick, W.H., Jr., and I.C. Mahapatra. 1968. Transformation and availability to rice of nitrogen and phosphorus in waterlogged soils. *Adv. Agron.* 20:323–359.

Patrick, W.H., Jr., D.S. Mikkelsen, and B.R. Wells. 1985. Plant nutrient behavior in flooded soil. *In Fertilizer Use and Technology*, 3d ed. Soil Sci. Soc. Am., Madison, Wisconsin, pp. 197–228.

Patrick, W.H., Jr., and F.J. Peterson, and F.E. Wilson. 1974. Response of lowland rice to time and method of application of phosphate. *Agron. J.* 66:459–460.

Pavlatou, A., and N.A. Polyzopoulos. 1988. The role of diffusion in the kinetics of phosphate desorption: The relevance of the Elovich equation. *J. Soil Sci.* 39:425–436.

Pena, F., and J. Torrent. 1984. Relationships between phosphate sorption and iron oxides in Alfisols from a river terrase sequence of Mediterranean Spain. *Geoderma* 33:283–296.

Perrott, K.W., F.M. Maher, and B.S. Thorrold. 1989. Accumulation of phosphorus fractions in yellow-brown pumice soils with development. *New Zeal. J. Agr. Res.* 32:53–62.

Perrott, K.W., and G.P. Mansell. 1989. Effect of fertilizer phosphorus and liming on inorganic and organic soil phosphorus fractions. *New Zeal. J. Agr. Res.* 32:63–70.

Polyzopoulos, N.A., V.Z., Keramidas, and H. Kiosse. 1985. Phosphate sorption by some Alfisols of Greece as described by commonly used isotherms. *Soil Sci. Soc. Am. J.* 49:81–84.

Polyzopoulos, N.A., V.Z. Keramidas, and A. Pavlatou. 1986. On the limitations of the simplified Elovich equation in describing the kinetics of phosphate sorption and release from soils. *J. Soil Sci.* 37:81–87.

Ponnamperuma, F.N. 1965. Dynamic aspects of flooded soils and the nutrition of rice plant. *In The Mineral Nutrition of the Rice Plant.* Johns Hopkins, Baltimore, Maryland, pp. 295–328.

Ponnamperuma, F.N. 1972. The chemistry of submerged soils. *Adv. Agron.* 24:29–97.

Ponnamperuma, F.N. 1985. Chemical kinetics of wetland rice soils relative to soil fertility. *In Wetland Soils: Characterization, Classification, and Utilization.* International Rice Research Institute, Los Baños, Laguna, Philippines, pp. 71–89.

Ponnamperuma, F.N., E.M. Tianco, and T.A. Loy. 1967. Redox equilibria in flooded soils: I. The iron hydroxide systems. *Soil Sci.* 103:374–382.

Posner, A.M., and N.J. Barrow, 1982. Simplification of a model for ion adsorption on oxide surfaces. *J. Soil Sci.* 33:211–217.

Powell, C.L.T. 1982. Phosphate response curves of mycorrhizal and non-mycorrhizal plants. III. Cultivar effects in *Lotus pedunculatus* Cav. and *Trifolium repens* L. *New Zeal. J. Agr. Res.* 25:217–222.

Prabhudesai, S.S., and S.B. Kadrekar. 1984. Reaction products from fertilizer phosphorus in lateritic soils of Konkan region. *J. Indian Soc. Soil Sci.* 32:52–56.

Probert, M.E. 1980. Growth responses to various calcium sources in a yellow earth soil with low calcium status. *Aust. J. Exp. Agric. Anim. Husb.* 20:240–246.

Quin, B.F. 1981. Performance of reactive phosphate rocks on irrigated and non-irrigated pasture. *In Proc. Tech. Workshop on the Potential of Phosphate Rock as a Direct Application Fertilizer in New Zealand. Occasional Report No. 3.* Massey Univ., Palmerston North, New Zealand, pp. 13–20.

Quin, B.F., L.M. Condron, D.S. Richard, and K.M. Goh. 1984. Changes in soil organic phosphorus. *In Proc. Workshop on Lime in Agriculture. MAF Workshop,* Ruakura, pp. 43–46.

Racz, G.J. 1979. Release of P in organic soils under aerobic and anaerobic conditions. *Can. J. Soil Sci.* 59:337–339.

Rajan, S.S.S. 1982a. Influence of phosphate rock reactivity and granule size on the effectiveness of "biosuper." *Fert. Res.* 3:3–12.

Rajan, S.S.S. 1982b. Availability to plants of phosphate from "biosupers" and partially acidulated phosphate rock. *New Zeal. J. Agric. Res.* 25:355–361.

Rajan, S.S.S. 1983a. Effect of sulfur content of phosphate rock/sulfur granules on the availability of phosphate to plants. *Fert. Res.* 4:287–296.

Rajan, S.S.S. 1983b. Rotokawa sulfur in phosphate rock/sulfur granules—a greenhouse study. *New Zeal. J. Agric. Res.* 26:233–236.

Rajan, S.S.S. 1985. Partial acidulation of an "underground" phosphate rock: 1. Preparation and characteristics. *Fert. Res.* 8:147–155.

Rajapakse, S., D.A. Zuberer, and J.C. Miller, Jr. 1989. Influence of phosphorus level on VA mycorrhizal colonization and growth of cowpea cultivars. *Plant Soil* 114:45–52.

Raju, P.S., R.B. Clark, J.R. Ellis, and J.W. Maranville. 1987. Vesicular-arbuscular mycorrhizal infection effects on sorghum growth, phosphorus efficiency, as mineral element uptake. *J. Plant Nutr.* 10:1331–1339.

Ram, M., R.N. Prasad, and P. Ram. 1987. Studies on phosphate adsorption and phosphate fixation in Alfisols and Entisols occurring in different altitudes of Meghalaya. *J. Indian Soc. Soil Sci.* 35:207–216.

Ram, P., and R.N. Rai. 1987. Effect of liming on phosphate potential and solubility in acid soils of Sikkim. *J. Indian Soc. Soil Sci.* 35:369–374.

Ramanie, S.K., J.H. Baker, and A.V. Barker. 1986. Phosphorus uptake by mycorrhizal and non-mycorrhizal roots of soybean. *J. Plant Nutr.* 9:1303–1313.

Rao, A.S., R. Hasan, and A.B. Ghosh. 1983. Phosphate adsorption behavior under continuous cropping and fertilizer use in an alluvial soil. *J. Indian Soc. Soil Sci.* 31:606–607.

Ratkowsky, D.A. 1986. A statistical study of seven curves for describing the sorption of phosphate by soil. *J. Soil Sci.* 37:183–189.

Reddy, K.R., and P.S.C. Rao. 1983. Nitrogen and phosphorus fluxes from a flooded organic soil. *Soil Sci.* 136:300–307.

Reid, R.K., C.P.P. Reid, and P.J. Szanislo. 1985. Effects of synthetic and microbially produced chelates on the diffusion of iron and phosphorus to a simulated root in soil. *Biol. Fertil. Soils* 1:45–52.

Resseler, H., and W. Werner. 1989. Properties of unreacted rock residues in partially acidulated phosphate rocks affecting their reactivity. *Fert. Res.* 20:135–142.

Rojo, M.J., S.G. Carcedo, and M.P. Mateos. 1990. Distribution and characterization of phosphatase and organic phosphorus in soil fractions. *Soil Biol. Biochem.* 22:169–174.

Roy, A.C., and S.K. De Datta. 1985. Phosphate sorption isotherms for evaluating phosphorus requirement of wetland rice soils. *Plant Soil* 86:185–196.

Rovira, A.D. 1979. Biology of soil-root interface. *In* J.L. Harley and R.S. Russell (Eds.). *The Soil Root Interface*. Acad. Press, New York, pp. 145–160.

Ru-kun, L., J. Bai-fan, and L. Ching-kwei. 1982. Phosphorus management for submerged soil. *Proc. 12th Int. Cong. Soil Sci., Symposia Paper II*. Indian Soc. Soil Sci., New Delhi, pp. 182–191.

Russell, J.S., E.J. Kamprath, and C.S. Andrew. 1988. Phosphorus sorption of subtropical acid soils as influenced by the nature of the cation suite. *Soil Sci. Soc. Am. J.* 52:1407–1410.

Ryden, J.C., J.R. McLaughlin, and J.K. Syers. 1977a. Mechanisms of phosphate sorption by soils and hydrous ferric oxide gel. *J. Soil Sci.* 28:72–92.

Ryden, J.C., J.R. McLaughlin, and J.K. Syers. 1977b. Time-dependent sorption of phosphate by soils and hydrous ferric oxides. *J. Soil Sci.* 28:585–595.

Ryden, J.C., and P.F. Pratt. 1980. Phosphorus removal from wastewater applied to land. *Hilgardia* 48:1–36.

Ryden, J.C., and J.K. Syers. 1977. Desorption and isotopic exchange relationships of phosphate sorbed by soils and hydrous ferric oxide gel. *J. Soil Sci.* 28:596–609.

Ryden, J.C., J.K. Syers, and R.W. Tillman. 1987. Inorganic anion sorption and interactions with phosphate sorption by hydrous ferric oxide gel. *J. Soil Sci.* 38:211–217.

Sah, R.N., and D.S. Mikkelsen. 1986a. Transformations of inorganic phosphorus during the flooding and draining cycles of soil. *Soil Sci. Soc. Am. J.* 50:62–67.

Sah, R.N., and D.S. Mikkelsen. 1986b. Effects of anaerobic decomposition of organic matter on sorption and transformations of phosphate in drained soils: 2. Effects on amorphous iron content and phosphate transformation. *Soil Sci.* 142:346–351.

Sah, R.N., and D.S. Mikkelsen. 1986c. Effects of temperature and prior flooding on intensity and sorption of phosphate in soil. I. Effects on the kinetics of soluble P in soil. *Plant Soil* 95:163–171.

Sah, R.N., and D.S. Mikkelsen. 1986d. Effects of temperature and prior flooding on intensity and sorption of phosphate in soil. II. Effects on P sorption. *Plant Soil* 95:173–181.

Sah, R.N., and D.S. Mikkelsen. 1986e. Effects of anaerobic decomposition of organic matter on sorption and transformations of phosphate in drained soils: 1. Effects on phosphate sorption. *Soil Sci.* 142:267–274.

Sah, R.N., and D.S. Mikkelsen. 1986f. Sorption and bioavailability of phosphorus during the drainage period of flooded-drained soils. *Plant Soil* 92:265–278.

Sah, R.N., and D.S. Mikkelsen. 1989. Phosphorus behavior in flooded-drained soils. I. Effects on phosphorus sorption. *Soil Sci. Soc. Am. J.* 53:1718–1722.

Sah, R.N., D.S. Mikkelsen, and A.A. Hafez. 1989a. Phosphorus behavior in flooded-drained soils. II. Iron transformation and phosphorus sorption. *Soil Sci. Soc. Am. J.* 53:1723–1729.

Sah, R.N., D.S. Mikkelsen, and A.A. Hafez. 1989b. Phosphorus behavior in flooded-drained soils. III. Phosphorus desorption and availability. *Soil Sci. Soc. Am. J.* 53:1729–1732.

Sahrawat, K.L., and G.P. Warren. 1989. Sorption of labelled phosphate by a Vertisol and an Alfisol of the semi-arid zone of India. *Fert. Res.* 20:17–25.

Sainz, M.J., and J. Arines. 1988. Phosphorus absorbed from soil by mycorrhizal red clover plants as affected by soluble P fertilization. *Soil Biol. Biochem.* 20:61–67.

Salih, H.M., A.I. Yahya, A.M. Abdul-Rahem, and B.H. Munam. 1989. Availability of phosphorus in a calcareous soil treated with rock phosphate or superphosphate as affected by phosphate-dissolving fungi. *Plant Soil* 120:181–185.

Sample, E.C., R.J. Soper, and G.J. Racz. 1980. Reactions of phosphate fertilizers in soils. *In* F.E. Khasawneh, E.C. Sample, and E.J. Kamprath (Eds.). *The Role of Phosphorus in Agriculture.* Soil Sci. Soc. Am., Madison, Wisconsin, pp. 263–310.

Sanchez, P.A., and J. Salinas. 1981. Low-input technology for managing Oxisols and Ultisols in tropical America. *Adv. Agron.* 34:279–406.

Sanchez, P.A., and G. Uehara. 1980. Management considerations for acid soils with high phosphorus fixation capacity. *In* F.E. Khasawneh, E.C. Sample, and E.J. Kamprath (Eds.). *The Role of Phosphorus in Agriculture.* Soil Sci. Soc. Am., Madison, Wisconsin. pp. 471–514.

Sanders, F.E., and P.B. Tinker. 1971. Mechanism of absorption of phosphate from soil by *Endogone mycorrhizas. Nature* 233:278–279.

Sanford, R.L., Jr., W.J. Parton, and C.V. Cole, 1989. Soil phosphorus interrelationships with carbon, nitrogen and sulfur: A modeling approach. *In* H. Tiessen (ed.). *Phosphorus Cycles in Terrestrial and Aquatic Ecosystem, Regional Workshop 1: Europe.* Saskatchewan Institute of Pedology, Univ. of Saskatchewan, Saskatoon, Canada, pp. 30–41.

Sanyal, S.K., P.Y. Chan, and S.K. De Datta. 1990. Phosphate sorption–desorption behavior of some acidic soils in South and Southeast Asia. Paper presented at the *6th Philippine Chemistry Congress*, Cebu City, Philippines, 24–26 May, 1990.

Sarkar, S.K., S.N. Basu, and D. Guha. 1986. Phosphate equilibria in acid soil. *J. Indian Soc. Soil Sci.* 34:471–475.

Satyanarayana, T., and B.P. Ghildyal. 1970. Influence of soil water regimes on the growth and nutrient uptake by rice (*Oryza sativa*). *J. Indian Soc. Soil Sci.* 18:41–44.

Saunders, W.M.H. 1965. Phosphate retention by New Zealand soils and its relationship to free sesquioxides, organic matter, and other soil properties. *New Zeal. J. Agric. Res.* 8:30–57.

Saunders, W.M.H., and A.J. Metson. 1971. Seasonal variation of phosphorus in soil and pasture. *New Zeal. J. Agric. Res.* 14:307–328.

Savant, N.K., D.R. Kene, and N.N. Kibe. 1970. Influence of alternate submergence and drying of rice soils prior to resubmergence on available phosphorus. *Plant Soil* 32:521–525.

Saxena, S.N. 1979. Biochemistry of soil phosphorus. *Bull. Indian Soc. Soil Sci.* 12:42–57.

Schwab, A.P., and W.L. Lindsay. 1983. Effect of redox on the solubility and availability of iron. *Soil Sci. Soc. Am. J.* 47:201–205.

Sharma, J.P., and S.N. Saxena. 1982. Dynamics of soil phosphorus in the rhizosphere of *Pisum sativum. J. Indian Soc. Soil Sci.* 30:230–231.

Sharpley, A.N. 1985. Phosphorus cycling in unfertilized and fertilized agricultural soils. *Soil Sci. Soc. Am. J.* 49:905–911.

Sharpley, A.N., and L.R. Ahuja. 1983. A diffusion interpretation of soil phosphorus desorption. *Soil Sci.* 135:322–326.

Sharpley, A.N., L.R. Ahuja, M. Yamamoto, and R.G. Menzel. 1981. The kinetics of phosphorus desorption from soil. *Soil Sci. Soc. Am. J.* 45:493–496.

Sharpley, A.N., C.A. Jones, C. Gray, and C.V. Cole. 1984. A simplified soil and plant phosphorus model. II. Prediction of labile, organic and sorbed phosphorus. *Soil Sci. Soc. Am. J.* 48:805–809.

Sharpley, A.N., U. Singh, G. Uehara, and J. Kimble. 1989. Modeling soil and plant phosphorus dynamics in calcareous and highly weathered soils. *Soil Sci. Soc. Am. J.* 53:153–158.

Shaviv, A., and N. Sachar. 1989. A kinetic-mechanistic model of phosphorus sorption in calcareous soils. *Soil Sci.* 148:172–178.

Shinde, B.N., P.A. Sarangamath, and S. Patnaik. 1978. Efficiency of HCl-and H_2SO_4-acidulated rock phosphates for rice (*Oryza sativa* L.) on acid soils. *Plant Soil* 50:575–584.

Sibanda, H.M., and S.D. Young. 1986. Competitive adsorption of humus acids and phosphate on goethite, gibbsite and two tropical soils. *J. Soil Sci.* 37:197–204.

Sibbesen, E. 1981. Some new equations to describe phosphate sorption by soils. *J. Soil Sci.* 32:67–74.

Sims, T.J., and B.G. Ellis. 1983a. Adsorption and availability of phosphorus following the application of limestone to an acid, aluminous soil. *Soil Sci. Soc. Am. J.* 47:888–893.

Sims, T.J., and B.G. Ellis. 1983b. Changes in phosphorus adsorption associated with aging of aluminum hydroxide suspensions. *Soil Sci. Soc. Am. J.* 47:912–916.

Singh, B.B., and M.A. Tabatabai. 1977. Effects of soil properties on phosphate sorption. *Commun. Soil Sci. Plant Anal.* 8:97–107.

Singhabutra, N., S. Arunin, and Y. Anuluxtipan. 1987. Experiment on two Rhizobium strains for inoculation into *Sesbania* spp. used as a green manure on the reclamation of saline soil. Paper presented at the 25th Ann. Tech. Mtg., Kasetsart Univ., Bangkok, pp. 25.

Singhania, R.A., and N.N. Goswami. 1978. Transformation of applied phosphorus under simulated conditions of growing rice and wheat in a sequence. *J. Indian Soc. Soil Sci.* 26:193–197.

Smeck, N.E. 1985. Phosphorus dynamics in soils and landscapes. *Geoderma* 36:185–199.

Smillie, G.W., D. Curtin, and J.K. Syers. 1987. Influence of exchangeable calcium on phosphate retention by weakly acid soils. *Soil Sci. Soc. Am. J.* 51:1169–1172.

Smyth, T.J., and P.A. Sanchez. 1982. Phosphate rock dissolution and availability in Cerrado soils as affected by phosphorus sorption capacity. *Soil Sci. Soc. Am. J.* 46:339–345.

Solis, P., and J. Torrent. 1989. Phosphate sorption by calcareous Vertisols and Inceptisols of Spain. *Soil Sci. Soc. Am. J.* 53:456–459.

Sparks, D.L. 1986. Kinetics of reactions in pure and mixed systems. *In* D.L. Sparks (Ed.). *Soil Physical Chemistry*. C.R.C. Press, Florida, pp. 83–145.

Sparling, G.P., A.W. West, and K.N. Whale. 1985a. Interference from plant roots in the estimation of soil microbial ATP, C, N, and P. *Soil Biol. Biochem.* 17:275–278.

Sparling, G.P., K.N. Whale, and A.J. Ramsay. 1985b. Quantifying the contribution from the soil microbial biomass to the extractable P levels of fresh and air-dried soils. *Aust. J. Soil Res.* 23:613–621.

Speir, T.W., J.C. Cowling, G.P. Sparling, A.W. West, and D.M. Corderoy. 1986. Effects of microwave radiation on the microbial biomass, phosphatase activity and levels of extractable N and P in a low-fertility soil under pasture. *Soil Biol. Biochem.* 18:377–382.

Speir, T.W., and D.J. Ross. 1978. Soil phosphatase and sulfatase. *In* R.G. Burns (Ed.). *Soil Enzymes*. Acad. Press, London, pp. 197–250.

Sposito, G. 1980. Derivation of the Freundlich equation for ion exchange reactions in soils. *Soil Sci. Soc. Am. J.* 44:652–654.

Sposito, G. 1981. The operational definition of the zero-point of charge in soils. *Soil Sci. Soc. Am. J.* 45:292–297.

Sposito, G. 1982. On the use of the Langmuir equation in the interpretation of "adsorption" phenomena. II. The "two-surface" Langmuir equation. *Soil Sci. Soc. Am. J.* 46:1147–1152.

Sposito, G. 1984. *The Surface Chemistry of Soils.* Clarendon Press, Oxford.

Stephen, R.C. 1985. The agronomic value of products resulting from the partial acidulation of North Carolina phosphate rock with phosphoric acid. *Fert. Res.* 8:67–73.

Stephen, R.C., and L.M. Condron. 1986. An assessment of the agronomic efficiency of partially acidulated phosphate rock fertilizers. *Fert. Res.* 10:269–282.

Stevenson, F.J. 1986. The phosphorus cycle. *In Cycles of Soil C, N, P, and S Micronutrients.* John Wiley and Sons, New York, pp. 231–284.

Stewart, J.W.B., and F.B. McKercher. 1983. Phosphorus cycling in soils. Agronomic considerations. *In Proc. 3rd Int. Congr. on Phosphorus Compounds (Brussels, Belgium).* Institut Mondial du Phosphate, Casablanca, Morocco, pp. 551–565.

Stewart, J.W.B., and H. Tiessen. 1987. Dynamics of soil organic phosphorus. *Biogeochemistry* 4:41–60.

Stumm, W., R. Kummert, and L. Sigg. 1980. A ligand-exchange model for the adsorption of inorganic and organic ligands at hydrous oxide interfaces. *Croat. Chem. Acta* 53:291–312.

Syers, J.K., M.G. Browman, G.W. Smillie, and R.B. Corey. 1973. Phosphate sorption by soils evaluated by the Langmuir adsorption equation. *Soil Sci. Soc. Am. Proc.* 37:358–363.

Syers, J.K., and A.D. MacKay. 1986. Reactions of Sechura phosphate rock and single superphosphate in soil. *Soil Sci. Soc. Am. J.* 50:480–485.

Syers, J.K., A.D. MacKay, M.W. Brown, and L.D. Currie. 1986. Chemical and physical characteristics of phosphate rock minerals of varying reactivity. *J. Sci. Food Agric.* 37:1057–1064.

Syers, J.K., and L. Ru-kun. 1989. Inorganic reactions influencing phosphorus cycling. Paper presented at the *Symp. on Phosphorus Requirements for Sustainable Agriculture in Asia and Oceania,* 6–10 March 1989. International Rice Research Institute, Los Baños, Laguna, Philippines.

Syers, J.K., R. Shah, and T.W. Walker. 1969. Fractionation of phosphorus in two alluvial soils and particle-size separates. *Soil Sci.* 103:283–289.

Talibudeen, O. 1974. The nutrient potential of soils. *Soils Ferts.* 37:41–45.

Tambe, K.N., and N.K. Savant. 1978. Kinetics of sorption of orthophosphate and pyrophosphate by ammoniated tropical soils. *Commun. Soil Sci. Plant Anal.* 9:745–754.

Tarafdar, J.C., and A. Jungk. 1987. Phosphatase activity in the rhizosphere and its relation to the depletion of soil organic phosphorus. *Biol. Fert. Soils* 3:199–204.

Tarafdar, J.C., and A.B. Roy. 1981. Urease and phosphatases in the rhizosphere of different mesta (*Hibiscus spp*) genotypes. *J. Indian Soc. Soil Sci.* 29:469–472.

Tate, K.R. 1984. The biological transformation of phosphorus in soil. *Plant Soil* 76:245–256.

Tate, K.R., and R.H. Newman. 1982. Phosphorus fractions of a climosequence of soils in New Zealand turrock grassland. *Soil Biol. Biochem.* 14:191–196.

Terman, G.L., and S.E. Allen. 1967. Response of corn to phosphorus in underacidulated phosphate rock and rock-superphosphate fertilizer. *J. Agric. Food Chem.* 15:354–358.

Terman, G.L., S.E. Allen, and O.P. Engelstad. 1970. Response of paddy rice to rates and sources of applied phosphorus. *Agron. J.* 62:390–394.

Terman, G.L., E.C. Moreno, and G. Osborn. 1964. Acidulation of phosphate rock in soil. *Soil Sci. Soc. Am. Proc.* 28:104–107.

Tester, M., F.A. Smith, and S.E. Smith. 1985. Phosphate inflow into *Trifolium subterraneum* L: Effects of photon irradiance and mycorrhizal infection. *Soil Biol. Biochem.* 17:807–810.

Thiband, M.C., C. Morel, and J.C. Fardeau. 1988. Contribution of phosphorus issued from crop residues to plant nutrition. *Soil Sci. Plant Nutr.* 34:481–491.

Tian-ren, Y. 1985. *Physical Chemistry of Paddy Soils.* Science Press, Beijing, and Springer-Verlag, Berlin, Heidelberg, New York, Tokyo.

Tian-ren, Y., G.J.D. Kirk, and F.A. Chaudhury. 1989. Phosphorus chemistry in relation to water regime. Paper presented at the *Symp. on Phosphorus Requirements for Sustainable Agriculture in Asia and Oceania,* 6–10 March 1989. International Rice Research Institute, Los Baños, Laguna, Philippines.

Tiessen, H., J.W.B. Stewart, and J.R. Bettany. 1982. Cultivation effects on the amounts and concentration of carbon, nitrogen, and phosphorus in grassland soils. *Agron. J.* 74:831–835.

Tiessen, H., J.W.B. Stewart, and J.O. Moir. 1983. Changes in organic and inorganic phosphorus composition of two grassland soils and their particle size fractions during 60–90 years of cultivation. *J. Soil Sci.* 34:815–823.

Timmermann, F. 1972. Chemisch–physikalische Untersuchungen zur Bewertung von teilaufgeschlossenen Phosphaten. *Landwirtsch. Forsch.* 25:71–83.

Tinker, P.B. 1980. Role of rhizosphere microorganisms in phosphorus uptake. *In* F.E. Khasawneh, E.C. Sample, and E.J. Kamprath (Eds.). *The Role of Phosphorus in Agriculture.* Soil Sci. Soc. Am., Madison, Wisconsin, pp. 617–654.

Tiwari, V.N., A.N. Pathak, and L.K. Lehri. 1988. Manurial value of compost enriched with rock phosphate and microbial inoculants to greengram. *J. Indian Soc. Soil Sci.* 36:280–283.

Torrent, J. 1987. Rapid and slow phosphate sorption by Mediterranean soils: Effect of iron oxides. *Soil Sci. Soc. Am. J.* 51:78–82.

Tsutsuki, K., and F.N. Ponnamperuma. 1987. Behavior of anaerobic decomposition products in submerged soils. *Soil Sci. Plant Nutr.* 33:13–33.

Uwasawa, M., P. Sangtong, and W. Cholitkul. 1988a. Behavior of phosphorus in paddy soils of Thailand. I. Contents of inorganic, organic, and available soil phosphorus in relation to rice plant P-nutrition. *Soil Sci. Plant Nutr.* 34:41–53.

Uwasawa, M., P. Sangtong, and W. Cholitkul. 1988b. Behavior of phosphorus in paddy soils of Thailand. II. Fate of phosphorus during rice cultivation in some representative soils. *Soil Sci. Plant Nutr.* 34:183–194.

Van Riemsdijk, W.H., L.J.M. Boumans, and F.A.M. de Haan. 1984. Phosphate sorption by soils: I. A model for phosphate reaction with metal-oxides in soil. *Soil Sci. Soc. Am. J.* 48:537–541.

Van Veen, J.A., and E.A. Paul. 1979. Conversion of biovolume measurements of soil organisms, grown under various moisture tensions, to biomass and their nutrient content. *Applied Environ. Microbiol.* 37:686–692.

Veith, J.A. 1978. Formation of X-ray amorphous aluminum o-phosphate from precipitation and secondary precipitation. *Z. Pflanzenernaehr. Bodenkd.* 141:29–42.

Veith, J.A., and G. Sposito. 1977. Reactions of aluminosilicates, aluminum hydrous oxides, and aluminum oxide with o-phosphate: The formation of X-ray amorphous analogs of variscite and montebrasite. *Soil Sci. Soc. Am. J.* 41:870–876.

Verma, T.S., and B.R. Tripathi. 1982. Evaluation of chemical methods for the determination of available phosphorus in waterlogged Alfisols: 2. Suitability of soil test methods in relation to plant growth parameters. *Soil Sci.* 134:294–299.
Vig, A.C., and G. Dev. 1984. Phosphorus adsorption characteristics of some acid and alkaline soils. *J. Indian Soc. Soil Sci.* 32:235–239.
Vijayachandran, P.K., and R.D. Harter. 1975. Evaluation of phosphorus adsorption by a cross section of soil types. *Soil Sci.* 119:119–126.
Voroney, R.P., J.A. Van Vean, and E.A. Paul. 1981. Organic C dynamics in grassland soils. 2. Model validation and simulation of the long-term effects of cultivation and rainfall erosion. *Can. J. Soil Sci.* 61:211–224.
Wada, K. 1985. The distinctive properties of Andosols. *Adv. Soil Sci.* 2:173–229.
Wada, K., and N. Gunjigake. 1979. Active aluminum and iron and phosphate sorption in Ando soils. *Soil Sci.* 128:331–336.
Wada, K., L. Xue-yuan, and P.W. Moody. 1989. Chemistry of adverse upland soils. Paper presented at the *Symp. on Phosphorus Requirements for Sustainable Agriculture in Asia and Oceania*, 6–10 March 1989, International Rice Research Institute, Los Baños, Laguna, Philippines.
Webber, M.D. 1978. Effects of temperature and time on hydroxy aluminum phosphate. *Soil Sci.* 125:107–114.
Welp, G., U. Herms, and G. Brümmer. 1983. Einfluß von Bodenreaktion, Redoxbedingungen und organischer Substanz auf die Phosphatgehalte der Bodenlösung. *Z. Pflanzenernaehr. Bodenkd.* 146:38–52.
Werner, W., and A. Solle. 1983. Untersuchungen zur Phosphatlöslichkeit und -wirkung von kalciniertem Aluminum–Calciumphosphat. *Landwirtsch. Forsch.* 36:308–319.
White, R.E., and A.T. Ayoub. 1983. Decomposition of plant residues of variable C/P ratio and the effect on phosphate availability. *Plant Soil* 74:163–173.
White, R.E., and A.W. Taylor. 1977. Effect of pH on phosphate adsorption and isotopic exchange in acid soils at low and high additions of soluble phosphate. *J. Soil Sci.* 28:48–61.
Willett, I.R. 1979. The effects of flooding for rice culture on soil chemical properties and subsequent maize growth. *Plant Soil* 52:373–383.
Willett, I.R. 1982. Phosphorus availability in soils subjected to short periods of flooding and drying. *Aust. J. Soil Res.* 20:131–138.
Willett, I.R. 1985. The reductive dissolution of phosphated ferrihydrite and strengite. *Aust. J. Soil Res.* 23:237–244.
Willett, I.R. 1986. Phosphorus dynamics in relation to redox processes in flooded soils. *13th Int. Congr. Soil Sci. Trans. (Hamburg)* 6:748–755.
Willett, I.R. 1989. Causes and prediction of changes in extractable phosphorus during flooding. *Aust. J. Soil Res.* 27:45–54.
Willett, I.R., C.J. Chartres, and T.T. Nguyen. 1988. Migration of phosphate into aggregated particles of ferrihydrite. *J. Soil Sci.* 39:275–282.
Willett, I.R., and R.B. Cunningham. 1983. Influence of sorbed phosphate on the stability of ferric hydrous oxide under controlled pH and E_h conditions. *Aust. J. Soil Res.* 21:301–308.
Willett, I.R., and M.L. Higgins. 1978. Phosphate sorption by reduced and reoxidized rice soils. *Aust. J. Soil Res.* 16:319–326.
Willett, I.R., and M.L. Higgins. 1980. Phosphate sorption and extractable iron in soils during rice-upland crop rotations. *Aust. J. Exp. Agric. Anim. Husb.* 20:346–353.
Willett, I.R., W.A. Muirhead, and M.L. Higgins. 1978. The effects of rice growing on soil phosphorus immobilization. *Aust. J. Exp. Agric. Anim. Husb.* 18:270–275.

Williams, J.D.H., J.K. Syers, and T.W. Walker. 1967. Fractionation of soil inorganic phosphate by a modification of Chang and Jackson's procedure. *Soil Sci. Soc. Am. Proc.* 31:736–739.

Wilson, M.A., and B.G. Ellis. 1984. Influence of calcium solution activity and surface area on the solubility of selected rock phosphates. *Soil Sci.* 138:354–359.

Yuan, T. 1980. Adsorption of phosphate and water-extractable soil organic material by synthetic aluminum silicates and soils. *Soil Sci. Soc. Am. J.* 44:951–955.

Equilibrium-Based Modeling of Chemical Sorption on Soils and Soil Constituents

C.P. Schulthess and D.L. Sparks

Advances in Soil Science, Volume 16

I. Introduction

A. What's a Model?

Models are tools used to describe any phenomena in nature. One should not be misled to believe that if a given set of data can be accurately described or predicted, then the model used must therefore be correct. This conclusion, though it is often made, should be questioned as several different models often predict a given set of data equally well. A model that is always correct, however, eventually becomes recognized as a law of nature. The research emphasis, therefore, should be in testing models (or hypotheses) with various data sets from independent experiments. The more accurate the model, the broader the range of data it will predict.

In the field of adsorption-related phenomena, several models have been advanced that can be classified as (1) empirical models or (2) mechanistic models. All these models are expressed mathematically; this is inevitable because quantitative, rather than merely qualitative, predictions are generally desired. Mechanistic models are based on thermodynamic concepts that include the law of mass action, the law of chemical equilibrium, the principle of mass balance, and the principle of electroneutrality. An empirical model, however, may be developed using mathematical parameters that lack any apparent physical or chemical meaning. Parts of a given mechanistic model may also be considered empirical if it contains a parameter whose magnitude cannot be definitively determined or whose physicochemical correlation is based on an unproven assumption.

In this chapter, we will discuss the similarities and differences between several models commonly encountered in the literature. Particular emphasis will be placed on how the law of chemical equilibrium and the principle of mass balance are applied with each model; these are fundamental components of most mechanistic models. Whenever applicable, the term *equilibrium-based model* will be used because it gives more information as to the kind of model being discussed. Another term commonly encountered in the literature is *surface complexation model*, which is very broad in meaning (i.e., it is analogous to the term *surface adsorption model*). The use of this term emphasizes that the aqueous components interact with the solid phase, but it does not describe how the model is to be developed. In this chapter, modern surface complexation models (which are also equilibrium-based models) that are commonly encountered in the literature will be referred to as *potential determining ion (PDI) models* because this term is more descriptive of their theoretical development.

Of course, none of the equilibrium-based models that are presented provide definitive proof as to the physical characteristics of the adsorbed species. This limitation is inevitable when using a macroscopic approach with these models. At best, the physical characteristics (e.g., degree of hydration of the adsorbed ions) are merely implied. They also do not provide information on the adsorption rates, which is essential in order to understand fully the adsorption behavior in a dynamic environment such as that of natural soils. One questions if equilibrium conditions are ever achieved in most soil environments. These equilibrium-based models, however, are easily modified for use in kinetic studies because of the close interdependence of the law of chemical equilibrium with the law of mass action. To obtain information on the physical and rate characteristics, one must also collect spectroscopic and kinetic data. Accordingly, the use of equilibrium-based models in combination with spectroscopic and kinetic approaches is also discussed in this chapter.

B. The Fundamental Components of Equilibrium-Based Models

The first observation that the quantity of a chemically reacting substance can affect the course of a reaction is attributed to Berthollet in 1801 (Silbermann, 1961). The law of mass action and the law of chemical equilibrium connected therewith are attributed to the Norwegians, Guldberg and Waage, who first presented their results in 1864 (Kauffman, 1972, 1976).

The law of mass action has played a central historical role in our understanding of the parameters that affect chemical reactions. Given the general schematic reaction

$$aA + bB + \cdots \rightarrow \text{products}, \tag{1}$$

the law of mass action states that at any time (t) the rate of chemical change (v) is proportional to the product of the effective concentrations (activities) of the reactants:

$$v = k(A)^a(B)^b\cdots, \tag{2}$$

where the parentheses denote the activity of reactants, k = rate constant, and v_i = d(activity of product i)/dt = $-$d(activity of reactant i)/dt (Sparks, 1989b).

The law of chemical equilibrium is easily derived for a reversible reaction that has a dynamic balance between the rate of formation and the rate of disappearance of each chemical species. Given the reversible reaction

$$aA + bB + \cdots \rightleftharpoons xX + yY + \cdots, \tag{3}$$

the instantaneous rates are

$$v_f = k_f(A)^a(B)^b\cdots \tag{4}$$

for the forward (subscript f) reactions, and

$$v_b = k_b(X)^x(Y)^y\cdots \tag{5}$$

for the backward (subscript b) reactions. At equilibrium $v_f = v_b$, and

$$\frac{k_f}{k_b} = K = \frac{(X)^x(Y)^y\cdots}{(A)^a(B)^b\cdots}, \tag{6}$$

where K = equilibrium constant. Equation (6) is a special form of the law of chemical equilibrium. The derivation of Eq. (6) depends on the validity of the rate equations used to describe the reaction kinetics (Silbermann, 1961). The validity of the rate equations, in turn, are based on extensive experimental observations, particularly in homogeneous aqueous reactions.

Although it is common to describe a chemical reaction using either equilibrium or kinetic experimental techniques, the close dependence of equilibrium-based models to kinetic theories should not be overlooked. Modern applications of kinetic theories to soil science research will be discussed in the final section of this chapter. For now, the brief introduction given above should provide an adequate background on the law of chemical equilibrium for the equilibrium-based models discussed in this chapter.

In addition to the law of chemical equilibrium, equilibrium-based models also rely on the conservation principles of mass balance and electroneutrality, which are somewhat obvious statements on the mass and charge balance of a system. These principles are similar in form to the first law of thermodynamics on the conservation of energy.

The principle of mass balance states that the total mass of each element in the system must always be constant, but may be distributed through various phases present within the boundaries of the system. The concept of a finite number of reactive sites on solid phases is based on this principle of mass balance.

The principle of electroneutrality states that the total positive charge in a system must equal the total negative charge. The electroneutrality principle is basic to the data interpretation of the adsorption of aqueous charged ions by solid phases (e.g., proton adsorption isotherms) and affects the subsequent development of the proposed mechanistic models (Schulthess and Sparks, 1986, 1988).

II. Selection of Solid and Aqueous Phase Components for Validating a Proposed Model

A. Comments on Modeling Adsorption of Organics

Natural organic matter (OM) are mixtures of hydrophilic and hydrophobic macromolecules (Leenheer and Huffman, 1979). Hydrophilic compounds exhibit ion exchange (more commonly called *ligand-exchange*) mechanisms, which are described based on common principles applied to inorganic ion exchange models (e.g., Tipping, 1981a). Organic cations are also subject to competitive ion exchange; Zachara et al. (1987) suggest that aromatic nitrogen heterocyclic organic cations adsorb competitively at pH values near the pK_a of the compounds (circa pH 5). Adsorption of OM [molecular weight (MW) = 5–13,000 g/mol] can still exhibit a Langmuir type adsorption (Tipping, 1981b). This suggests that the OM adsorption is reversible and competitive with the other anions in the system (perhaps OH^-). With larger molecules, Nabzar et al. (1986) showed a pH dependence and Langmuir adsorption of an acrylamide-acrolein copolymer ($MW = 1.2 \times 10^6$ g/mol) on kaolinite and synthetic aluminosilicates.

Chiou and Shoup (1985), working with vapor phase adsorption of nonionic (hydrophobic) organics, clearly showed decreased organic adsorption with increased relative humidity. The greater displacement of organics by polar compounds (e.g., water or acetone versus hexane) emphasizes the significance of competitive solute-solid interactions.

Due to their size, a given organic molecule may display several adsorption phenomena and are, therefore, much more difficult to predict than inorganic compounds. These include the "octopus" effect and metal bridging mechanisms discussed below.

1. The "Octopus" Effect

High molecular weight organics with multiple ionizable functional groups often have high adsorption affinities that are at best only weakly affected by competitive anions (e.g., Cl^- or HCO_3^-). Podoll et al. (1987) attributed the strong adsorption of polymers to an "octopus" effect. The larger a molecule, the greater the number of segments (functional groups) that will attach to the surface, and the more difficult it becomes to desorb the molecule. The competitive nature of other aqueous anions may affect the total number of attached segments; however, unless all the segments are desorbed, the oligomer will remain adsorbed with each of its desorbed functional groups always relatively close to the surface for competitive readsorption.

2. Metal Bridging Mechanisms

Adsorption of OM to clay particles through a polyvalent cation link (or bridge) is a common occurrence. Mortland (1970) explains that based on infrared absorption studies (the lowering of the OH stretching frequencies of water due to the H–bond formation) the adsorption mechanism should include a water bridge, namely clay–metal–H_2O–OM, particularly for cations of high solvation energy.

The presence of aqueous polyvalent cations is particularly significant in sandy and montmorillonitic soils because OM is usually inert toward Si at near neutral pH values (Greenland, 1971; Schulthess and Huang, 1991). The cation–OM link is effectively an energy barrier bridge; that is, because OM does not interact with silicon (Si) surface sites, it will prefer to complex with the polyvalent cation, which, in turn, is easily adsorbed by the Si surface sites.

Conversely, the OM may complex with the aqueous cations to the extent of causing a caging phenomenon around the central cation. Organic-metal complex formation could result in a surface-blocking effect (Ho and Miller, 1985) and, thus, hinder the adsorption of the metal. The degree of OM-metal complexation probably affects the Lewis acid strength of the metal and its ability to further interact with the surface sites (Schulthess and Huang, 1991). That is, a surface-blocking effect may correspond either to a physical (or steric) caging phenomenon or to an electron sharing effect on the reactivity of the metal toward the surface sites.

The authors at present do not know of any quantitative adsorption modeling of organics that includes a metal bridging term. In its simplest form, the adsorbed metal may be viewed as a new (layered) adsorbent surface; that is, the bridging metal may "modify" the solid surface as viewed by the complexing OM.

B. Effects of Adsorption on the Surface

Any compound adsorbed will change the subsequent chemical interactions of the "new" solid phase with the remaining compounds in the aqueous phase. Boyd et al. (1988a,b) observed that adsorption of organic cations on the surface of smectites will enhance the sorptive capacity of the clay toward aqueous hydrophobic organics. Boyd et al. (1988c) also modified the organic adsorptive capacity of soils by pretreatment with organic cations. The adsorption of hydrophilic compounds, therefore, can encourage organic–organic adsorption mechanisms (e.g., partitioning-type mechanisms) for the hydrophobic molecules.

Weiss (1966) showed that a complete coverage of a surface with alkyl chains is not required to cause the surface to behave hydrophobically as long as a stable water structure forms on the surface. Lagaly et al. (1983) speculate that if sufficiently close-packed chains exceeding a critical length exist, then the resulting strong aggregation of the ordered water molecules around the chains exhibit a discontinuity with the random orientation of the bulk water; that is, if an ordered structure is formed, then hydrophobic behavior will dominate even if an organic monolayer coverage is not complete. Additionally, hydrophobic interaction chromatography (HIC) research shows that the chain length of the hydrophobic solid phase affects the partitioning of organic compounds in the mobile phase (Schmuck et al., 1986).

Adsorption of inorganic ions also affects the behavior of the solid phase. This is easily demonstrated through electrokinetic studies. As is illustrated in Figure 1, the electrophoretic mobility (EM) (under an applied electric field) of a given particle resembles the mobility of the surface metals rather than the metals beneath the outermost (adsorbed) surface (James and Healy, 1972; Elliott and Sparks, 1981). Consequently, studies on surface behavior and adsorption modeling should weigh heavily on the chemical composition of the outermost layer of the solid phase.

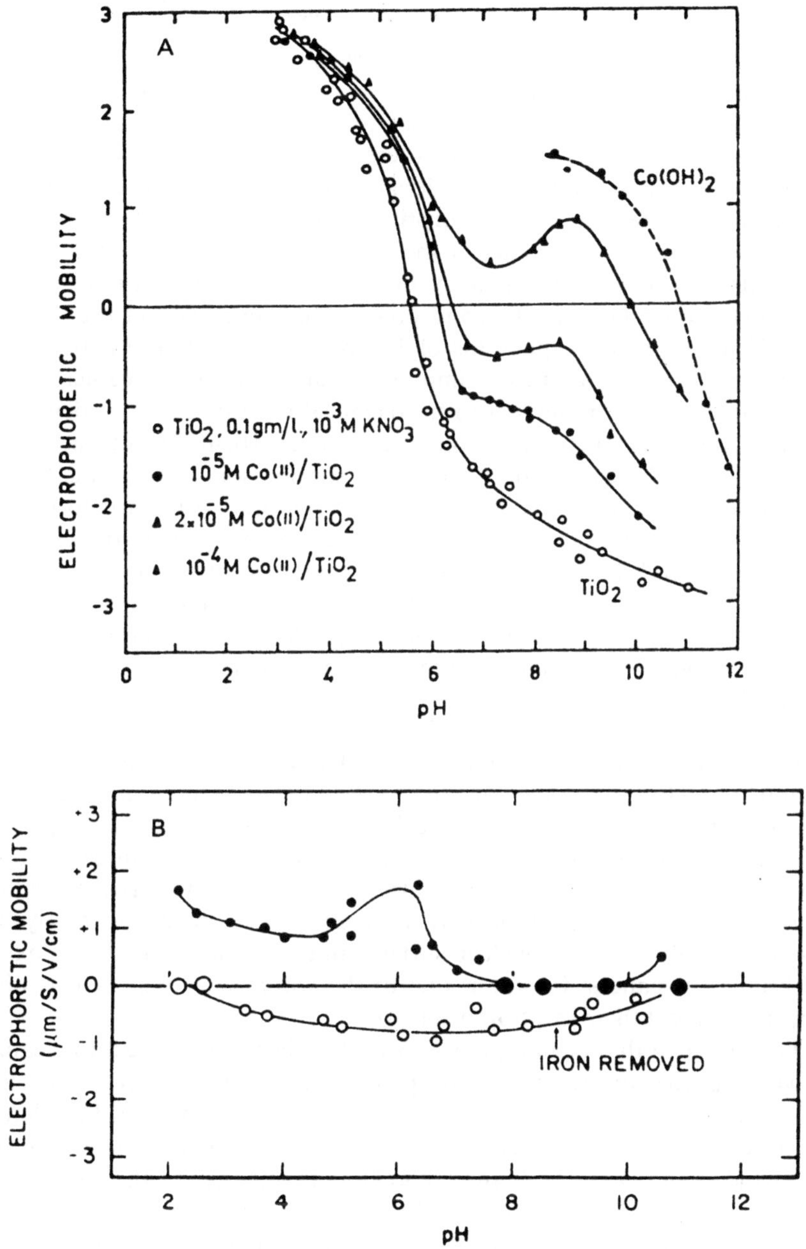

Figure 1. A, The electrophoretic mobility (EM) of TiO_2 as a function of pH in 10^{-3} *M* HNO_3–KOH electrolyte with and without the presence of various concentrations of $Co(NO_3)_2$. (*Source*: James and Healy, 1972). B, The effect on the EM of a paleudult soil sample as a function of pH in the presence and absence of a natural Fe coating. (*Source:* Elliot, H.A. and D.L. Sparks, *Soil Science* 132, 402–409. © by Williams & Wilkens, 1981).

C. Discussion of Criteria Used in Selecting Sorbates and Sorbents

Selecting the proper solid and aqueous phase components is particularly important to soil chemists who study adsorption phenomena in natural environments. From the previous discussion on the diverse sorption mechanisms of organics, the inclusion of these compounds in the experimental procedure should be carefully considered if the primary research objective is to test (or validate) a proposed sorption model. Likewise, the presence of multiple inorganic components in the aqueous phase should also be considered as this may form an array of monolayer "modifications" on the outermost surface layer of the solid phase. The complexity of the aqueous phase should reflect the capabilities of the proposed model. Alternatively, one may assume that the mixture of components can be grouped together, or that they are inert and/or indifferent to the chemical reaction being modeled.

Typical questions encountered on selecting the sorbents include the following: (1) Should the whole soil rather than specific soil components (oxides or minerals) be studied? and (2) Should the samples be washed to remove organics, inorganics, and/or amorphous oxides that may be partially coating the solid surfaces? For the purposes of testing adsorption models, the first question can be restated as follows: How complex can the system be? If the system is too complex (e.g., whole soil samples), then the mathematical parameters of a given model may become meaningless in terms of their intended physicochemical significance.

Most investigators have used soil constituents, such as hydrous oxides, humic substances, and clay minerals, rather than whole soils to model sorption phenomena. This is in part due to the less heterogeneous nature of soil constituents. However, it should be realized that they may contain impurities and frequently do not mimic constituents in field soils. For example, attempts to model sorption on clay minerals with the objective of simulating sorption on soils that contain these same clay minerals is often unrealistic. Many clay minerals in soils do not have clean surfaces but rather are coated with oxides and humic substances. However, due to the complexity of soils, it is often difficult to model their sorption behavior.

The answer to the second question asked earlier, with respect to cleaning (or washing) the surface of the solid phase, is somewhat subjective. A clean surface is often desired for modeling purposes because of the need to identify the physicochemical characteristics of the adsorption site(s). The presence of "impurities" (i.e., undesirable or unknown components) on the solid phase, however, is very difficult to remove. These may include the inorganic carbon species, which are very prominent in nature, and other strongly held anions.

Additionally, washing a clay surface free of amorphous material may not guarantee that the resulting surface will remain crystalline throughout the adsorption experiment. The processes of dissolution and reprecipitation occur continuously at all pH values in aqueous solutions. These unavoidable processes tend to reshape and restructure the outermost surface layer(s) on a solid particle. These phenomena, therefore, should caution one from prematurely concluding that a particular surface characteristic was present at the time that a surface reaction actually occurred (e.g., adsorption of aqueous compounds or ions).

In summary, selection of solid and aqueous phase components for validating a proposed model is often initially restricted to very simple systems. The adsorption of protons by a synthetic oxide, for example, is a common initial proof-of-theory type of experiment encountered in the literature for validating the proposed adsorption models. Each of these models eventually is then applied to more complex systems (e.g., adsorption of electrolytes and use of clays or whole soils), and to alternative experiments (e.g., electrokinetics, kinetics, and spectroscopy) if these have not yet been done.

III. In Search of a Model: The Tools for Soil Chemists

A. The Freundlich Equation

Freundlich was a German-born chemist well known for his work on colloid and capillary chemistry (e.g., Freundlich, 1922, gave a treatise on the subject). The Freundlich equation was an empirical equation used extensively by Freundlich but not developed by him. The parabolic equation was used to describe adsorption behavior of soils as early as 1859 by Boedeker.

The Freundlich equation was originally applied empirically. Henry (1922) finally derived the equation by assuming that the free surface energy (F) is linearly related to the fraction of surface coverage (Θ)

$$F = F_0(1 - \Theta) + F_1\Theta \tag{7}$$

or

$$F = F_0 - (F_0 - F_1)\Theta, \tag{8}$$

where F_0 = free surface energy while in contact with the pure solvent (or vacuum if adsorbing a gas), F_1 = free surface energy while covered with a monolayer of solute (or gas), x = weight of adsorbate, m = weight of adsorbent, $(x/m)_{max}$ = maximum monolayer capacity, and $\Theta = (x/m)/(x/m)_{max}$. The Θ and F values are then correlated to the aqueous concentration through use of the Gibbs and Gibbs–Duhem equations. The exact form of the Gibbs (1961, original work 1876–1878) equation is:

$$-\mathrm{d}\gamma = \sum_i \Gamma_i \,\mathrm{d}\mu_i \tag{9}$$

where γ is the surface tension of the solution with chemical potential μ_i, and surface concentration Γ_i for each component i. For a two-component mixture, let $i = 1$ for the solvent and $i = 2$ for the adsorbate, so that

$$-\mathrm{d}\gamma = \Gamma_1 \,\mathrm{d}\mu_1 + \Gamma_2 \,\mathrm{d}\mu_2. \tag{10}$$

The Gibbs–Duhem equation then correlates μ_i to the mole fraction of adsorbate (f_i) on the solid phase:

$$f_1 \, d\mu_1 + f_2 \, d\mu_2 = 0 \tag{11}$$

or

$$d\mu_1 = \frac{-f_2 \, d\mu_2}{f_1}, \qquad d\mu_2 = \frac{-f_1 \, d\mu_1}{f_2}. \tag{12}$$

Combining Eqs. (10) and (12), and rearranging:

$$\Gamma_2^{(N)} = \frac{f_2 \, d\gamma}{d\mu_1} = \Gamma_2 f_1 - \Gamma_1 f_2 \tag{13}$$

or

$$\Gamma_2^{(N)} = \frac{-f_1 \, d\gamma}{d\mu_2} = \Gamma_2 f_1 - \Gamma_1 f_2. \tag{14}$$

The term $\Gamma_i^{(N)}$ is surface *excess* (in moles), whereas Γ_i is surface *concentration* of component *i* (Kipling, 1965). Surface excess is the extent by which the surface concentration exceeds the aqueous concentration. Guggenheim and Adam (1933) summarize several other surface excess terms expressed with alternate units. Because $d\mu_2 = RT \, d(\ln a_2)$, Eq. (14) becomes

$$\Gamma_2^{(N)} = \frac{-f_1 \, d\gamma}{RT \, d(\ln a_2)}, \tag{15}$$

where a_2 is the activity of adsorbate in the liquid phase. For dilute solutions, Eq. (14) can be simplified because $f_1 \approx 1.0$ and $f_2 \approx 0.0$, resulting in

$$\Gamma_2^{(N)} \approx \Gamma_2 \approx \frac{-d\gamma}{d\mu_2} = \frac{-d\gamma}{RT \, d(\ln a_2)} \tag{16}$$

or

$$\Gamma = \frac{-c \, d\gamma}{RT \, dc}, \tag{17}$$

where the aqueous concentration *c* can be used instead of the activity *a* because the activity coefficient is also near unity in dilute solutions (Kipling, 1965). The Γ value used here is often referred to as surface excess or surface concentration, thus causing some confusion particularly at the higher concentrations.

For dilute solutions, the Gibbs surface excess can be approximated by x/m; thus, Henry's (1922) derivation of Freundlich's equation continues as follows (γ and F are numerically identical):

$$x/m = \frac{-c \, dF}{RT \, dc}. \tag{18}$$

From Eq. (8), we obtain

$$dF = -(F_0 - F_1)d\Theta = \frac{-(F_0 - F_1)d(x/m)}{(x/m)_{max}}. \tag{19}$$

Substituting Eq. (19) into Eq. (18)

$$x/m = \frac{c}{RT} \cdot \frac{F_0 - F_1}{(x/m)_{max}} \cdot \frac{d(x/m)}{dc}. \tag{20}$$

Integrating and rearranging,

$$\ln(x/m) = \frac{RT(x/m)_{max}}{F_0 - F_1} \cdot \ln(c) + constant \tag{21}$$

or

$$x/m = kc^{1/n} \tag{22}$$

where $1/n = RT(x/m)_{max}/(F_0 - F_1)$. Equation (22) is known as the Freundlich equation. The exponent $1/n$ = osmotic work in adsorption (RT) ÷ total work in adsorption (RT + nonosmotic work). Henry (1922) pointed out that the larger the molecules of adsorbate, the larger the value of n. One should note that this derivation is only applicable to dilute solutions, where surface excess $\Gamma^{(N)} \approx x/m$. Alternate derivations are discussed by Adamson (1982) and Trapnell (1955) who concluded that if data fit the Freundlich isotherm, then the solid surface is most probably heterogenous (i.e., the heat of adsorption varies on the solid phase). There remains, however, no proof for this equation; for example, Henry's (1922) linear correlation of free surface energy (F) to surface coverage (Θ) [Eq. (7)] is not proven.

As is illustrated in Figure 2, the major limitation of the Freundlich equation

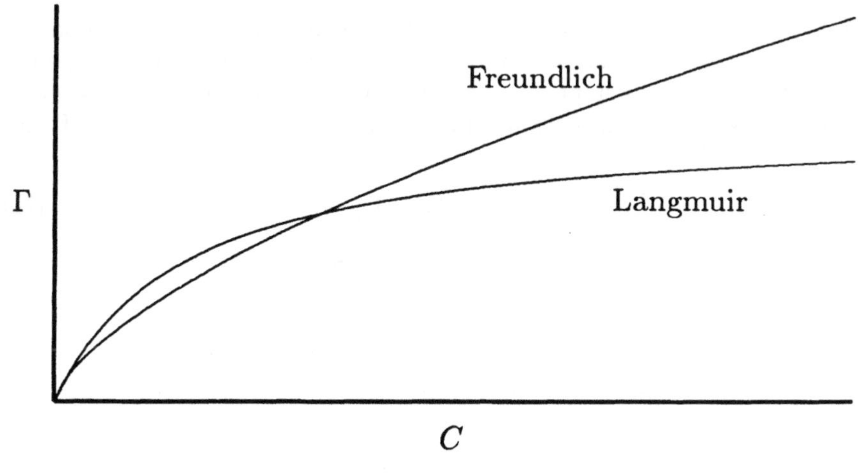

Figure 2. General shape of the well-known Freundlich and Langmuir isotherms. Note the lack of an adsorption maximum (Γ_{max}) with the Freundlich isotherm as the aqueous concentration of the adsorbate (c) increases.

is that no maximum adsorption is predicted. This equation is easily applied for predicting adsorption behavior at low aqueous concentrations but generally overestimates the adsorption capacity of the particles at high aqueous concentrations of the adsorbates. This is because the total number of monolayer sites on the surface of a particle should be finite. An additional problem with using Eq. (22) is that it does not elucidate any physically tangible adsorption mechanism. Consequently, force fitting Freundlich isotherm curves to adsorption data does not really improve one's understanding of the adsorption phenomena.

B. The Langmuir Equation

The Langmuir equation is perhaps the best known equation due to its adherence to mass action laws and to its derivation being relatively easy and straight forward. Langmuir (1918) considered separately the rates of evaporation (desorption) and condensation (adsorption). Thus, for a surface adsorption reaction

$$S + A \underset{k_{-1}}{\overset{k_1}{\rightleftharpoons}} SA \tag{23}$$

the law of mass action derives

$$\frac{\mathrm{d}\{SA\}}{\mathrm{d}t} = k_1\{S\}(A) - k_{-1}\{SA\} = 0 \quad \text{at equilibrium} \tag{24}$$

and

$$\frac{k_1}{k_{-1}} = K = \frac{\{SA\}}{\{S\}(A)}, \tag{25}$$

where $\{SA\}$ = concentration of sites adsorbing component A, $\{S\}$ = concentration of sites lacking component A, (A) = activity of A in the aqueous phase, k_1 = forward rate constant, k_{-1} = backward rate constant, and K = equilibrium constant. From mass balance principles, the total number of sites (Γ_{max}) is defined as

$$\Gamma_{max} = \Gamma_0 + \Gamma_A, \tag{26}$$

where $\Gamma_0 = \{S\}$, and $\Gamma_A = \{SA\}$. Combining Eqs. (25) and (26),

$$\Gamma_{max} = \frac{\Gamma_A}{K(A)} + \Gamma_A = \Gamma_A\left[\frac{1 + K(A)}{K(A)}\right] \tag{27}$$

and solving for the amount of A adsorbed,

$$\Gamma_A = \frac{\Gamma_{max}Kc}{1 + Kc} \tag{28}$$

where $c = (A)$. Equation (28) is known as the Langmuir equation. The Γ_{max} and K values are easily determined by plotting c/Γ_A versus c; the intercept equals $1/K\Gamma_{max}$ and the slope equals $1/\Gamma_{max}$. Figure 2 illustrates the general shape followed by the Langmuir equation. Based on the Langmuir equation, the plateau often observed in adsorption isotherm assays are generally interpreted as the maximum adsorption capacity of the solid phase (e.g., Olsen and Watanabe, 1957).

One should keep in mind that the Langmuir equation is derived from well-understood mass action laws. Conversely, the Freundlich equation is derived in reverse, that is, from an initial linear assumption of the surface energy relation with surface coverage [Eq. (7)]. An empirical equation can be derived that is intermediate between the Langmuir and Freundlich equations, known as the Temkin equation (Trapnell, 1955; Adamson, 1982; Brunauer et al., 1967). As with the Freundlich equation, however, the Temkin equation has no proof for the assumed correlation between the free surface energy and surface coverage.

Equation (28) predicts that adsorption will occur on a surface based on a single chemical reaction such as the one described by Eq. (23). It does not confirm, however, that the reaction described by Eq. (23) is in fact the reaction that is occurring in the (macroscopic) adsorption isotherm assays. This can be confusing particularly if precipitation or coprecipitation phenomena are possible in the system studied (Veith and Sposito, 1977). From the Kelvin equation (Hiemenz, 1977) it can be argued that if seed formation and subsequent crystal growth require greater energy than does adsorption, then adsorption would be the preferred chemical reaction. Nevertheless, the distinction between these processes is best resolved through spectroscopic research (Sposito, 1986). This problem, however, is not unique to Eq. (28) but rather is shared by *all* equations that are derived from mass action laws and mass balance principles, and that are applied to (macroscopic) adsorption isotherm data. This is because the precipitation and adsorption reactions both involve crystal growth and, consequently, are basically very similar chemical reactions.

In spite of these comments, the Langmuir equation should not be considered an empirical equation because both the Γ_{max} and K values can be determined independently (e.g., through crystallography and kinetics). The coincidence that the Langmuir equation can predict various different phenomena (e.g., adsorption and precipitation) does not justify any retraction on the soundness of its derivation.

Irving Langmuir (1881–1957), of Brooklyn, NY, U.S.A., was awarded the Nobel Prize in chemistry in 1932 for his investigations in surface chemistry. Working with the adsorption of gases by solids, Langmuir (1918) argued that the adsorption of compounds by solids involved strong chemical interactions rather than physical phenomena. The measured energy of adsorption did not depend on the amount adsorbed, which he attributed to the adsorption of neutral salt-like substances or of two components having opposite polarity (Langmuir, 1932). Langmuir (1918) described the adsorbed molecules as those occupying definite fixed positions on the solid surface. If heat is supplied, however, then the adsorbed molecules can move ("hop") from one site to another acting as a two-dimensional gas (Langmuir, 1933). He also described a "normal" surface distribution of adsorbed molecules that, if disturbed, is reestablished through a surface diffusion (or surface mobility) mechanism at a rate characterized by a relaxation time, τ_0 (Langmuir, 1933).

The Langmuir equation assumes that the heat of adsorption remains constant during the entire adsorption process, and is independent of the effects of neighboring adsorbed molecules; that is, K is constant. This condition was experimentally shown by Langmuir (1932) with gaseous adsorption behavior. Brunauer et al. (1967) state that statistical and thermodynamic derivations (rather than Langmuir's kinetic derivation) consider this assumption valid only for energetically uniform surfaces with no lateral interaction. Lateral interactions and surface heterogeneity exert opposing effects on the heat of adsorption; thus, these compensating factors may maintain the heat of adsorption approximately constant for an adsorption isotherm to obey the Langmuir equation (Brunauer, 1961; Brunauer et al., 1967; Hiemenz, 1977).

The presumed heterogeneity of a solid surface does not need to imply that the surface varies only with respect to the nature of the adsorption sites (e.g., presence of cavities or degree of crystallinity). This would suggest that the surface is free of impurities that could change the character of the adsorption site. Instead, the concept of heterogeneity should also include the presence of chemically different adsorption sites (e.g., Al, Fe, or Si related sites). This additional option, which is often overlooked, will also result in rather complex adsorption isotherms that are obviously not well predicted by rather simple surface adsorption models [e.g., Eq. (23) describes adsorption as consisting of *only one* chemical reaction]. Consequently, some caution should be exercised in prematurely concluding that lack of adherence to the Langmuir equation is due to a single type of surface with nonuniform adsorption energies. Alternate models (some are discussed in this chapter) are derived from mass action laws that also assume that each type of solid surface is energetically uniform and that various chemical reactions (and types of surfaces) are involved in the adsorption phenomena. The correct research emphasis should be to describe accurately the various chemical reactions occurring at the solid-aqueous interface through the use of equations that have well-defined (both physically and chemically) mathematical parameters.

C. Ion Exchange Models

Other equations can be derived by the same principles of mass action and mass balance as those used for deriving the Langmuir equation. The differences lie on how the surface reaction and the corresponding equilibrium constant should be defined. For the surface reaction

$$SB + A \rightleftharpoons SA + B \tag{29}$$

known as an ion exchange reaction, the equilibrium constant can be defined as

$$K = \frac{\{SA\}(B)}{\{SB\}(A)}, \tag{30}$$

which predicts adsorption of compound A to be

$$\Gamma_A = \frac{\Gamma_{max} K(A)}{(B) + K(A)}, \tag{31}$$

where $\Gamma_{max} = \Gamma_A + \Gamma_B$. If $B = OH^-$ anions or H^+ cations, then Eq. (31) reduces to the adsorption model described by Langmuir [Eq. (28)] after modifying the equilibrium constant K to $K/(OH^-)$ or $K/(H^+)$. The $K/(OH^-)$ or $K/(H^+)$ value is constant if the pH is maintained constant throughout the adsorption isotherm analysis.

Gans (1913) proposed Eq. (30) defining A and B in concentration units and setting $\{SB\} = 1.0$. Kerr (1928a) emphasized that the soil exchange complex is a solid in solution; he proposed Eq. (30) defining $\{SA\}$ and $\{SB\}$ in units of "active mass," which he then described as equivalent fractions. Vanselow (1932), who had studied the concept of aqueous activity rather than concentrations under G. Lewis, defined all the terms in Eq. (30) in activity units; he described the solid phase activity as mole fractions.

For monovalent–divalent ion exchange, the proposed reactions vary significantly. Suggested ion exchange reactions have included

$$SB + 2A^+ \rightleftharpoons SA_2 + B^{2+}, \tag{32}$$

$$S_2B + 2A^+ \rightleftharpoons 2SA + B^{2+}, \tag{33}$$

and

$$SB_{1/2} + A^+ \rightleftharpoons SA + \tfrac{1}{2}B^{2+}. \tag{34}$$

Kerr (1928b) proposed that the equilibrium constant (K) should be based on mass action laws. For Ca^{2+}/H^+ ion exchange, Kerr (1928a) suggested Eq. (32) and

$$K_K' = \frac{\{SA_2\}(B^{2+})}{\{SB\}(A^+)^2} \tag{35}$$

but for Ca^{2+}/NH_4^+ exchange Kerr (1928b) suggested Eq. (33) and

$$K_K = \frac{\{SA\}^2(B^{2+})}{\{S_2B\}(A^+)^2}. \tag{36}$$

Gapon (1933) reduced Eq. (33) to Eq. (34) and, using activity units, defined K around the adsorption site rather than the adsorbing ion:

$$K_G = \frac{\{SA\}(B^{2+})^{1/2}}{\{SB_{1/2}\}(A^+)}. \tag{37}$$

Gapon's equation has been used extensively and is generally more constant than the other equilibrium equations discussed here.

Concern over how the solid phase activity should be expressed has led to several interesting variations. Vanselow (1932) suggested the formation of mixed crystals, and represented the adsorption reaction of Eq. (33) as

$$K_V = \frac{K_K}{[SA] + [S_2B]}, \tag{38}$$

where the solid activity values are expressed as mole fractions

$$\{SA\} = \frac{[SA]}{[SA] + [S_2B]}. \tag{39}$$

Krishnamoorthy and Overstreet (1949), using statistical thermodynamics, criticized Vanselow's concepts for not restricting a divalent ion from occupying two sites that are far apart. They suggested instead for Eq. (33)

$$K_{KO} = \frac{K_K}{[SA] + 1.5[S_2B]}. \tag{40}$$

Gaines and Thomas (1953) replaced mole fractions with equivalent fractions, and suggested for Eq. (33) (as expressed by Thomas, 1974)

$$K_{GT} = \frac{K_K}{2([SA] + [S_2B])}. \tag{41}$$

Schofield's (1947) "ratio law," which can be derived from the Gapon equation [Eq. (37)], and the quantity–intensity (Q/I) technique presented by Matthews and Beckett (1962) have been used mostly to describe K^+ cation availability in soils (Beckett, 1964; Moss and Beckett, 1971). The quantity of exchangeable K^+ in the soil (ΔK) is plotted as a function of the intensity factor, or activity ratio (AR^K), where $AR^K = a_K/\sqrt{a_{Ca} + a_{Mg}}$, and a_i = activity of cation *i*. These techniques further illustrate the complexity involved in obtaining "workable" expressions of the ion exchange reactions.

Equations (36) and (38)–(41) are empirical in their assumptions as to how the solid phase activity should be defined. Note that Gapon's equation [Eq. (37)] remains essentially unchanged no matter how the solid phase activity is defined; because the definition of K centers around the adsorption site, the denominators of the mole (or equivalent) fractions [e.g., Eq. (39)] will always cancel out.

Two basic assumptions, therefore, determine how ion exchange will be predicted. First, a model that describes the chemical reaction must be proposed, such as those described by Eqs. (29) and (32)–(34). Second, after applying the law of mass action to obtain K, the aqueous and solid phase activity values are not uniquely defined and must be estimated (Sparks, 1984; Sposito, 1984a; Lane, 1983; Argersinger et al., 1950; Gaines and Thomas, 1953). The Gapon equation is exceptional in that the ratio of $\{SA\}/\{SB_{1/2}\}$ does not vary regardless of how they are defined. The solid phase activity coefficient, γ_s, if one could be independently defined, would presumably cancel out in Eq. (37). Additionally, when comparing Eqs. (32)–(34), it makes sense to describe the ion exchange mechanisms around the phase with the least freedom of motion. Similarly, for monovalent ion exchange [Eqs. (29) and (30)], the ratio $\{SA\}/\{SB\}$ does not vary regardless of how they are defined; the γ_s values presumably would also cancel out.

Elucidating the validity of one ion exchange model over another is a difficult task, particularly due to the lack of a definitive expression for the solid phase activity. For example, Ogwada and Sparks (1986a) showed that both the Vanselow equation and the Gaines and Thomas equation will result in equivalent thermodynamic parameters (e.g., standard free energies of exchange, ΔG°). Perhaps due to the frustration of not being able to attain a model with a reliable (constant) K value, Thomas (1977) observed that the need for an equilibrium constant may have been misplaced, and perhaps an equilibrium coefficient is more useful (i.e., affinities with a specific applicable concentration range). If one merely wishes to know how much is adsorbed, then a simple isothermal experiment and equilibrium "coefficient" will suffice. This concept suggests either that the models are complete but the mass action laws do not yield true equilibrium constants, or that our pursuit for a more accurate model is not worth so much trouble. Our understanding of the adsorption mechanism, however, relies heavily on these models.

Modeling and predicting ion exchange, which was first elucidated by Way (1850, 1852), is not completely described by Eqs. (29) or (32)–(34) and the mass action laws. This is apparent from the continuous drive to obtain more accurate predictions. For example, as will be discussed in the following sections, the suggestion of multiple adsorption sites and pH-dependent reactions have recently been proposed. That is, equilibrium constants derived from mass action laws are not disputed; instead, the completeness of the equations describing the surface adsorption reactions are questioned.

D. Potential-Determining Ion (PDI) Models—Basic Concepts

Ion exchange behavior is pH-dependent. The pH-dependent adsorption envelopes of anions (Hingston et al., 1967) and cations (Belloni et al., 1959; Grebenshchikova and Davydov, 1961a,b) suggest that a model based on a single ion exchange reaction may be oversimplified. Consequently, the aqueous phase and solid phase speciations are commonly incorporated in model formulations. (Speciation describes the various forms, or species, in which a compound may exist, and it usually describes the degree of protonation, or degree of complexation, of a compound). Interest in describing the solid phase interactions as a function of pH led to correlating ion exchange with several other pH-dependent phenomena—more specifically, with electrokinetics.

Electrokinetics, the study of the mobility of particles under an applied or induced electric field, has played a major role in our interpretation of the solid phase speciation. Helmholtz (1879) developed a quantitative theory for electrosmosis and electrophoresis, initiating the concept of the electrical double layer at the interface of the surface and the liquid. Through various historical developments, numerous interpretations exist today on how the double layer theories should be applied in describing the solid phase speciation. Basically, the solid surface has an electrical charge that is neutralized by counterions in the liquid phase. The charge exhibits a potential that influences the potential at the plane of shear (the plane where the solid and liquid phase shear and travel in opposite directions under the applied electric field). The potential at the plane of shear is measured through electrokinetics; note, however, that the exact location of the plane of shear, or distance from the solid surface, cannot be unequivocally determined. Helmholtz assumed that the double layer was very thin. From electrostatic studies, the surface potential, Ψ_0, varies linearly with surface charge, σ_0

$$\sigma_0 = \mathrm{C}\Psi_0 \tag{42}$$

where C = capacitance density. The C parameter is empirically adjusted in ion adsorption modeling.

An important theoretical development for explaining electrokinetics was the diffuse layer theory advanced predominantly by Gouy (1910) and Chapman (1913) (often referred to as the Gouy–Chapman theory), and it is commonly expressed as

$$\sigma_0 = \sqrt{8\,\varepsilon\,\epsilon_0 RTI}\,\sinh(F\Psi_0/2RT) \tag{43}$$

where ε = dielectric constant of the aqueous medium, ϵ_0 = permittivity of free space, F = Faraday's constant, RT = energy terms, and I = ionic strength; that is, the particle travels under an applied electric field because it has a surface charge, σ_0, with a surface potential, Ψ_0. The counter ion atmosphere decreases gradually in concentration (thus the term *diffuse*), and its thickness varies with the ionic strength.

The Stern (1924) (diffuse double-layer, DDL) and triple-layer ("extended" Stern) models include an inner layer(s) adhering to Eq. (42) and an outer diffuse layer expressed by Eq. (43) (Westall and Hohl, 1980). These models also require empirical estimations of the inner-layer capacitance density (C) in ion adsorption modeling. Note that all the terms are defined in Eq. (43). The dielectric constant, however, is probably considerably less in the Stern layer than in the bulk medium due to the high electric field assumed present in this region. This results in a dielectric saturation effect [the ε value in Eq. (43) is affected] that cannot be evaluated experimentally for the Stern layer (Hiemenz, 1977).

Electrophoresis, which measures the mobility of a suspended particle under an applied electric field, is perhaps the most commonly used procedure in electrokinetic studies. As is illustrated in Figure 3 for an aluminum (Al) oxide, the results generally show that the particles travel toward the anode at low pH (due to a positive charge on the particle surface) and toward the cathode at high pH (due to a negative charge on the particle surface) under an applied electric field. The isoelectric point (IEP) is the pH where the particle does not

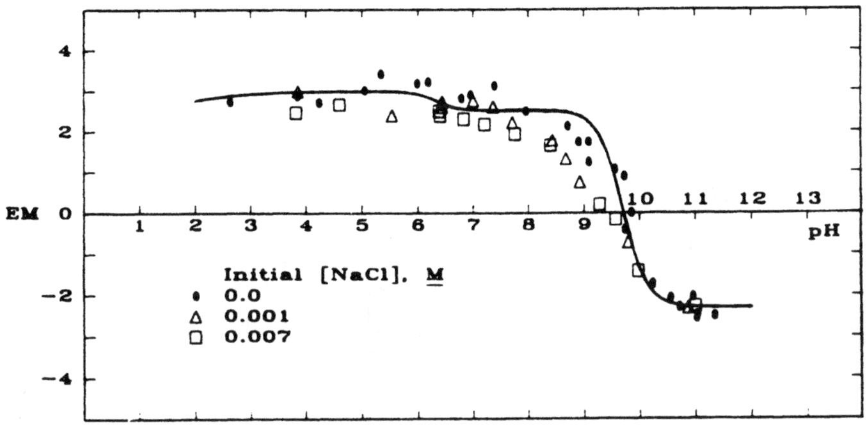

Figure 3. Electrophoretic mobility (EM; μm cm V^{-1} s^{-1}) as a function of aqueous pH for an Al oxide. Notice that the mobility, whose magnitude and sign are determined by the surface charge, reaches a maximum value at low and high pH values relative to the IEP. Solid line is predicted EM values. (*Source:* Schulthess and Sparks, 1988).

move. Consequently, the surface speciation is generally described as

$$SOH \rightleftharpoons SO^- + H^+ \tag{44}$$

$$SOH + H^+ \rightleftharpoons SOH_2^+, \tag{45}$$

where the negative surface, SO^-, forms at high pH, and the positive surface, SOH_2^+, forms at low pH.

It follows from Eqs. (44) and (45) that the adsorption or desorption of protons should be measurable. Schulthess and Sparks (1986) describe this procedure through potentiometric titration of solid suspensions. Theoretically, a mass-balanced equation is used in which the acid (or base) concentration added is known, and the equilibrium pH of the supernatant liquid (remaining H^+ or OH^- concentrations) is also known. Schulthess and Sparks (1986) criticized this procedure, which, unfortunately, has been used extensively in colloid chemistry since circa 1920. The error is that this mass-balanced procedure does not account for the variable solubility of the solid phase, which also acts as a source or sink for the aqueous protons. They suggested instead a back-titration technique: at a specific pH value, the amount of protons adsorbed is equal to a known acid (or base) addition subtracted by the equilibrium acidity (or alkalinity) remaining in the liquid phase (as determined by *titratable* acidity or alkalinity). Figure 4 illustrates the dramatically different estimates of the adsorption of protons (often considered synonymous with the surface charge) by an Al oxide as a function of pH based on these two methods.

Regardless of this solubility oversight, the proton adsorption behavior (Fig. 4a) is described by initially modifying the activity of the proton in the vicinity

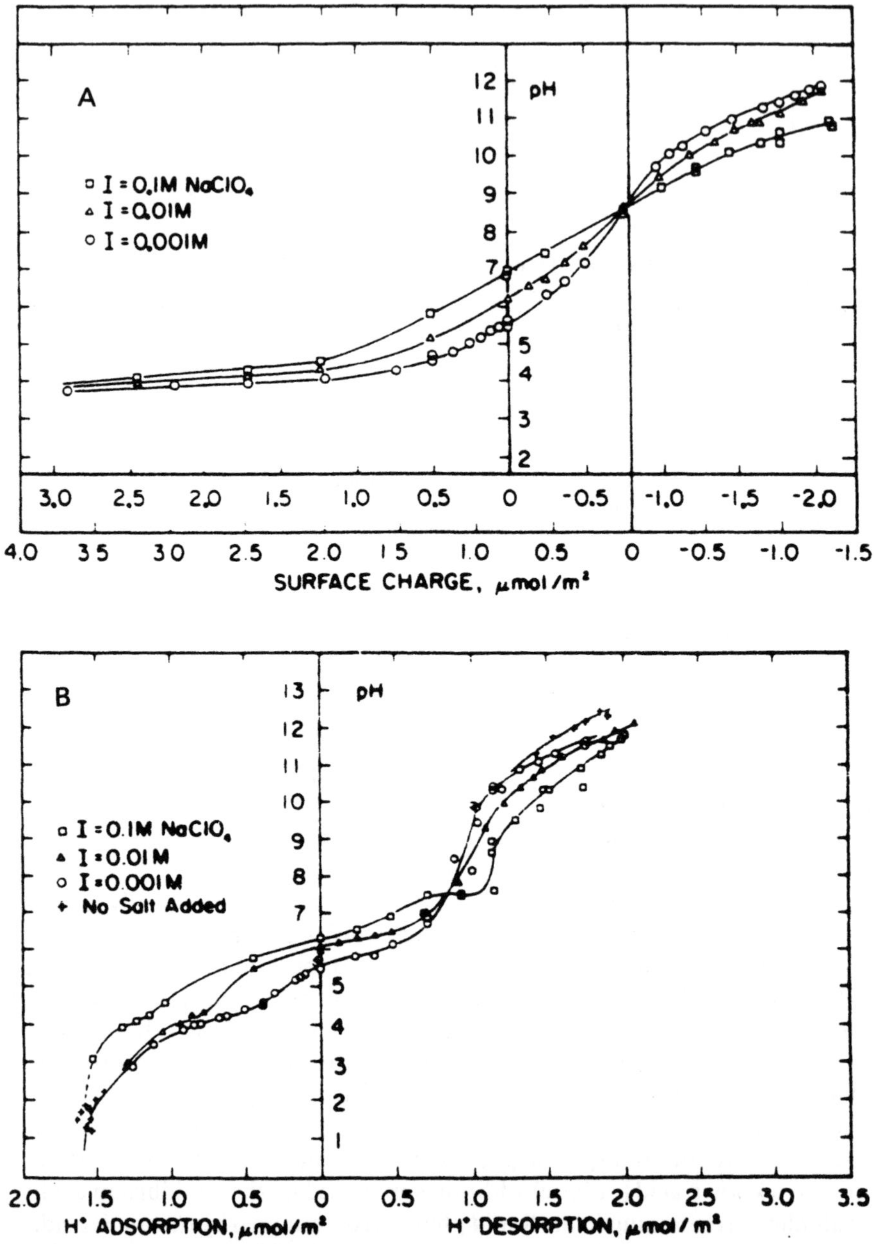

Figure 4. Proton adsorption isotherms on an Al oxide based on mass-balanced potentiometric titration methods. A, Traditional method: assumes no interfering solubility effects, and measures the acidity remaining in solution through a simple pH measurement; B, Backtitration technique: assumes pH-dependent Al solubility effects, and measures the acidity remaining in solution through a *titratable* acidity (or alkalinity) analysis. (*Source:* Schulthess and Sparks, 1986).

of the solid surface as a function of the potential on the surface of the particle. This is done by applying Boltzmann's equation

$$\frac{(H^+)_{SURFACE}}{(H^+)_{BULK}} = \exp\left(\frac{-w}{kT}\right) = \exp\left(\frac{-ze\Psi_0}{kT}\right), \tag{46}$$

where w = work required to bring the proton from the bulk solution to the surface with a potential Ψ_0, z = valence of the ion, e = unit charge, and kT = energy terms. The amount of work, w, required to ionize the surface is assumed to be related to the concentration ratio of the potential-determining ion (PDI, i.e., H^+) (Hiemenz, 1977). Consequently, the proton activity in the vicinity of the surface should be $(H^+)_{BULK}\exp(-e\Psi_0/kT)$. Substituting this activity expression into the mass action laws for the reactions expressed by Eqs. (44) and (45) one obtains

$$K_1 = \frac{\{SO^-\}(H^+)\exp(-e\Psi_0/kT)}{\{SOH\}} \tag{47}$$

and

$$K_2 = \frac{\{SOH_2^+\}}{\{SOH\}(H^+)\exp(-e\Psi_0/kT)}, \tag{48}$$

where $(H^+) = (H^+)_{BULK}$ = activity of the proton in the bulk solution.

The constant capacitance model or DDL theory may be used to correlate Ψ_0 to surface charge [Eqs. (42) and (43)], where

$$\sigma_0 = \{SOH_2^+\} - \{SO^-\}. \tag{49}$$

From mass balance principles,

$$\Gamma_{max} = \{SOH_2^+\} + \{SOH\} + \{SO^-\}. \tag{50}$$

Solving for $\{SOH\}$ using Eqs. (47) and (48)

$$\Gamma_{max} = \{SOH\}K_2(H^+)\exp(-e\Psi_0/kT) + \{SOH\} + \frac{\{SOH\}K_1}{(H^+)\exp(-e\Psi_0/kT)} \tag{51}$$

and

$$\alpha_0 = \frac{\{SOH\}}{\Gamma_{max}} = \frac{1}{D}, \tag{52}$$

where $D = K_2(H^+)\exp(-e\Psi_0/kT) + 1 + K_1(H^+)^{-1}\exp(e\Psi_0/kT)$. Similarly, solving for $\{SOH_2^+\}$ and $\{SO^-\}$ one obtains

$$\alpha_+ = \frac{\{SOH_2^+\}}{\Gamma_{max}} = \frac{K_2(H^+)\exp(-e\Psi_0/kT)}{D} \tag{53}$$

and

$$\alpha_- = \frac{\{SO^-\}}{\Gamma_{max}} = \frac{K_1}{(H^+)\exp(-e\Psi_0/kT)D}. \tag{54}$$

Note that $\sum \alpha_i \equiv 1.0$. Finally, the potentiometric titration data (Fig. 4a) are predicted from Eq. (49) using

$$\sigma_0 = \Gamma_{max}(\alpha_+ - \alpha_-) \tag{55}$$

with the values Γ_{max}, K_1, K_2, and Ψ_0 needing to be determined. Huang (1981) discusses techniques that one may use to solve for Γ_{max}, K_1, and K_2. The pH of zero point of charge (ZPC) is

$$ZPC = \frac{pK_1 + pK_2}{2}, \tag{56}$$

where $pK_i = -\log K_i$. The ZPC is equivalent to the IEP as defined by electrokinetics. Potentiometric titration techniques result in intersecting surface charge curves as the ionic strength is varied. The pH at the point of intersection is called the *point of zero salt effect* (PZSE) (Sposito, 1984b). The PZSE and IEP (or ZPC) are not numerically identical (Schulthess and Sparks, 1988); nevertheless, the ZPC value is often taken to be equivalent to the PZSE.

Because the Ψ_0 value is a function of σ_0, Eq. (55) will require a few iterations prior to solving. The choice of a model that correlates Ψ_0 to σ_0 may depend on the complexity of the data being fitted (i.e., shape of the isotherms). Westall (1986) observed that all these models have a great deal of uncertainty, and, for this reason, he recommended use of the simpler models [e.g., constant capacitance model based on Eq. (42)]. Additionally, Koopal et al. (1987) emphasized that in the triple-layer model no unique solution is possible due to the large number of parameters that need to be optimized as well as the presence of some experimental and model inaccuracies. Westall and Hohl (1980) were able to fit the surface charge data (from potentiometric titration studies) of an Al oxide equally well using several models that correlate Ψ_0 to σ_0 (Fig. 5). Consequently, the validity of one model over another remains unresolved (e.g., constant capacitance versus the DDL model).

Sposito (1984b) noted that the exponential correction term used on equations describing proton adsorption [e.g., Eqs. (47) and (48) together with the constant capacitance model] is not significant when describing divalent metal adsorption behavior. This is not surprising because, in experiments measuring the mass balance of cations other than protons, there are usually no unknown sources or sinks (e.g., solid phase pH-dependent solubility) that can interfere with the aqueous measurement of the cation concentration remaining in solution. Unlike the potentiometric measurements of aqueous proton concentrations, the measurements of aqueous cation concentrations are not generally affected by the degree of metal speciation or complexation (e.g., by using atomic absorption spectroscopy).

One should also keep in mind that proton adsorption data by traditional

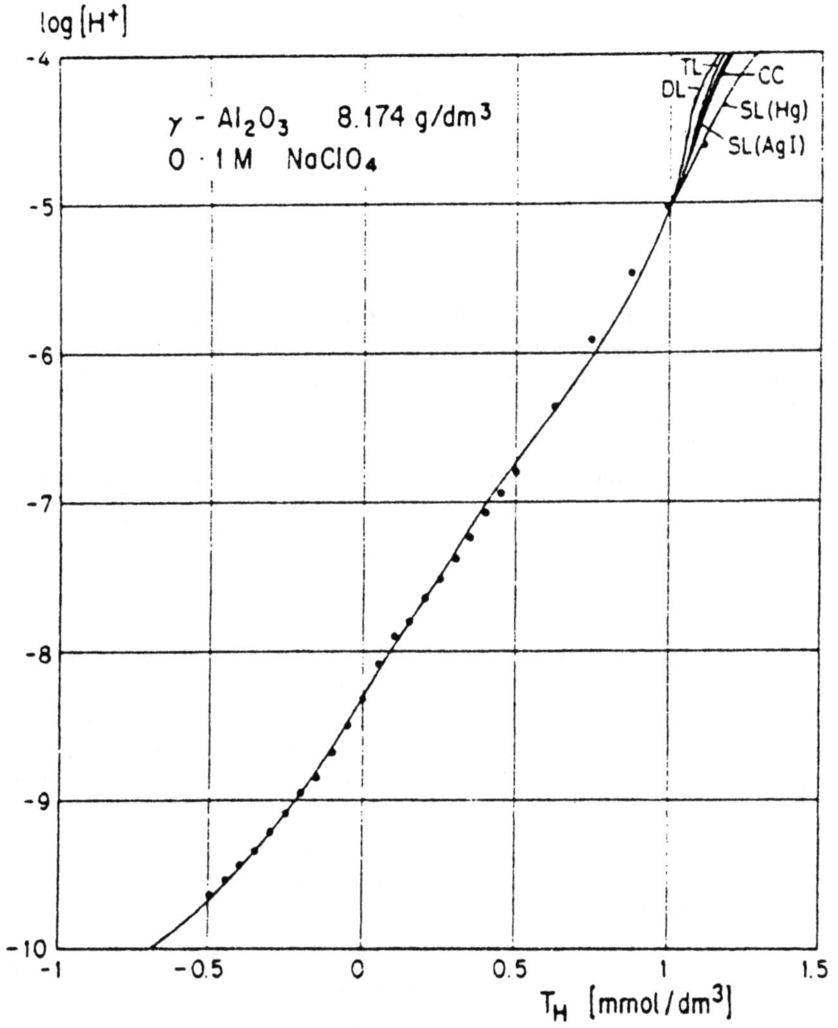

Figure 5. Predictions from five models of a proton adsorption isotherm on an Al oxide (traditional titration methods as described by Fig. 4a). CC, constant capacitance; DL, diffuse layer; SL, Stern layer; TL, triple layer. (*Source:* Westall and Hohl, 1980).

potentiometric titration methods (e.g., Fig. 4a and 5) have not considered the experimental error introduced by the solid phase solubility (Fig. 4b) (Schulthess and Sparks, 1986) and, therefore, the applicability of all these models to an erroneous set of data is questionable (Schulthess and Sparks, 1988). From a mathematical perspective, the exponential terms used in Eqs. (47) and (48) (and others discussed in the following section) are most probably merely tracking the pH-dependent solubility effects.

E. PDI Models—Additional Theories

As a logical follow-up to the modeling of surface protonation or deprotonation [Eqs. (44) and (45)], the adsorption of charged compounds can be hypothesized as

$$SO^- + M^+(xH_2O) \rightleftharpoons SO^-M^+(xH_2O) \tag{57}$$

or

$$SOH_2^+ + L^- \rightleftharpoons SOH_2^+L^-, \tag{58}$$

where the cation, $M^+(xH_2O)$, and ligand (or anion), L^-, are electrostatically attracted to the charged surface. The products in Eqs. (57) and (58) have at least one solvent molecule between the surface and the adsorbing ion (Sposito, 1984b), and are generally referred to as outer-sphere complexes or nonspecific adsorption.

Inner-sphere complexes, or specific adsorption, are often hypothesized to occur as a ligand exchange mechanism

$$SOH + M^+ \rightleftharpoons SOM + H^+ \tag{59}$$

$$SOH_2^+ + L^{l-} \rightleftharpoons SL^{1-l} + H_2O. \tag{60}$$

Stumm (1986) states that compounds adsorbed as inner-sphere complexes should be included in mass action equations and rate laws. Though this statement should be extended to include all outer-sphere complexes that are potentially competitive in the adsorption phenomena (Schulthess and Sparks, 1989), the outer-sphere complexes are often ignored (i.e., they are not included in the charge and mass-balanced equations).

From the mass action laws, the equilibrium constants used in ligand exchange models include Eqs. (47) and (48) corresponding to the surface charging reactions described by Eqs. (44) and (45), and

$$K_3 = \frac{\{SL^{1-l}\}(H_2O)}{\{SOH_2^+\}(L^{l-})} \tag{61}$$

for Eq. (60). Additional equations and equilibrium constants can be assumed that correspond to the various pH-dependent speciation reactions that the aqueous ligand may undergo. The surface charge is expressed as

$$\sigma_0 = \{SOH_2^+\} - \{SO^-\} - (1 - l)\{SL^{1-l}\}, \tag{62}$$

where the ligand adsorption decreases the surface charge (or makes it more negative). A corollary for this reasoning is that the drop in the IEP in the presence of electrolytes is attributed to the adsorption of anions. [Note that this corollary was questioned by Schulthess and Sparks (1988) who suggested instead that the drop in the IEP in the presence of electrolytes is due to the adsorption (or coadsorption) of cations, which are easily sheared off the surface–aqueous interface in the presence of an applied electric field]. From mass balance, the total number of sites is

$$\Gamma_{max} = \{SOH_2^+\} + \{SOH\} + \{SO^-\} + \{SL^{1-l}\} \tag{63}$$

from which one obtains

$$\alpha_+ = \{SOH_2^+\}/\Gamma_{max} = \frac{K_2(H^+)\exp(-e\Psi_0/kT)}{D} \tag{64}$$

$$\alpha_0 = \{SOH\}/\Gamma_{max} = \frac{1}{D} \tag{65}$$

$$\alpha_- = \{SO^-\}/\Gamma_{max} = \frac{K_1}{(H^+)\exp(-e\Psi_0/kT)D} \tag{66}$$

$$\alpha_L = \{SL^{1-l}\}/\Gamma_{max} = \frac{K_2K_3(L^{l-})(H^+)\exp(-e\Psi_0/kT)}{(H_2O)D}, \tag{67}$$

where

$$D = K_2(H^+)\exp(-e\Psi_0/kT) + 1 + K_1(H^+)^{-1}\exp(e\Psi_0/kT) + K_2K_3(L^{l-})(H^+)\exp(-e\Psi_0/kT)(H_2O)^{-1}.$$

These models can get much more comprehensive by including $\{SO^-M^+\}$ terms (e.g., James, 1981) or several ligand adsorption species (e.g., Goldberg, 1985) in Eqs. (62) and (63). The adsorption predictions are then computed based on the assumed correlation between Ψ_0 and σ_0 (e.g., constant capacitance, DDL models, and several others). Because Ψ_0 is a function of σ_0, solving Eq. (62) will require a few iterations.

Solid surfaces are known to exhibit a charge in an applied electric field, and are assumed to remain charged in absence of the electric field. Schulthess and Sparks (1988) critically questioned this assumption. They suggested instead that the particles are always neutral but will exhibit a charge under an applied electric field because the electric field force is shearing the ions off the solid–aqueous interface into the bulk solution. An electrical imbalance is induced on the particle resulting in the observed particle mobility. Schulthess and Sparks (1988) further questioned, therefore, the validity of the Boltzmann equation [Eq. (43)] and its applicability in surface adsorption studies [e.g., Eqs. (51)–(54) and (64)–(67)]; that is, if the surface is neutral or closely associated with counterions, then the surface charge (σ_0) and surface potential (Ψ_0) should be zero.

F. Multisite Models

Because soils are complex mixtures, Syers et al. (1973) proposed extending the Langmuir model to two surfaces. The surface adsorption reactions are then described as

$$S' + A \rightleftharpoons S'A \tag{68}$$

$$S'' + A \rightleftharpoons S''A, \tag{69}$$

and the equilibrium constants are

$$K_1 = \frac{\{S'A\}}{\{S'\}(A)} \tag{70}$$

$$K_2 = \frac{\{S''A\}}{\{S''\}(A)}. \tag{71}$$

The predicted adsorption of component A by the solid phase consisting of surfaces S' and S'' is

$$\Gamma_A = \frac{\Gamma'_{max} K_1(A)}{K_1(A) + 1} + \frac{\Gamma''_{max} K_2(A)}{K_2(A) + 1}, \tag{72}$$

where $\Gamma'_{max} = \Gamma'_0 + \Gamma'_A$, and $\Gamma''_{max} = \Gamma''_0 + \Gamma''_A$. Equation (72) is reasonable if the equilibrium constants K_1 and K_2 describe two independent reactions. The S' and S'' surfaces do not interact with each other; for example, an S' surface cannot move to an $S''A$ surface and displace the A component to form an $S'A$ surface. The mixture is a heterogenous surface that is split into two independently homogenous surfaces. Langmuir (1918) also suggested a similar equation for adsorption occurring on several different kinds of surfaces. Equation (72) is solved by plotting $\Gamma_A/(A)$ versus Γ_A (also called a Scatchard plot) (Sposito, 1982; Klotz, 1982).

This concept suggests that the S' and S'' surfaces can be uniquely identified. Early views of multiple sites were suggested by Mattson (1931) who argued that ion exchange and adsorption reactions could be described in terms of the relative ratios of the oxide components forming the soil colloids. Holford et al. (1974, 1975) and Shuman (1975) showed the applicability of the two-surface Langmuir equation, and recognized the existence of different ion adsorption energies by different surfaces. Posner and Bowden (1980) criticized this conclusion, however, and attributed the phenomena to the variable energy of adsorption of the aqueous ions by the solid phase (refer to the PDI models previously discussed). One should keep in mind that the applicability of Boltzmann's equation (as was used by Posner and Bowden, 1980) was questioned by Schulthess and Sparks (1988). Posner and Bowden (1980) further stated that splitting an adsorption isotherm into a series of traditional Langmuir equations [e.g., Eq. (72)] is erroneous "unless there is some *a priori* reasoning for doing so." However, the various solid components known to be present in soils overwhelmingly suggest that different adsorbing surfaces should be considered for soils. More recently, Schulthess and Huang (1990, 1991) have proposed that adsorption by clay minerals of heavy metals and natural organic acids occur on specific Si and Al sites, and that the pH-dependent adsorption behavior of the adsorbates are expressed independently by each of these sites.

G. Competitive Ion Adsorption Models

If at least two different ions are able to adsorb on a given solid phase, then a competitive adsorption model can be described as

$$S + A \rightleftharpoons SA \tag{73}$$

$$S + B \rightleftharpoons SB \tag{74}$$

with equilibrium constants

$$K_1 = \frac{\{SA\}}{\{S\}(A)} \tag{75}$$

$$K_2 = \frac{\{SB\}}{\{S\}(B)}. \tag{76}$$

Huang and Smith (1981) applied a competitive model for Cd^{2+} and H^+ adsorption on activated carbon. Equations (73) and (74) were illustrated as

$$\begin{array}{lcl} \mathrm{C} & + \mathrm{Cd}^{2+} \overset{K_1}{\rightleftharpoons} & \mathrm{CCd} \\ +\mathrm{H}^+ & & \\ \updownarrow K_2 & & \\ \mathrm{CH} & & \end{array} \tag{77}$$

which, written this way, emphasizes the competitive nature of the compounds for the surface. Once again, from mass balance principles

$$\Gamma_{max} = \{\mathrm{C}\} + \{\mathrm{CCd}\} + \{\mathrm{CH}\}, \tag{78}$$

and solving for {CCd}

$$\Gamma_{max} = \frac{\{\mathrm{CCd}\}}{K_1(\mathrm{Cd}^{2+})} + \{\mathrm{CCd}\} + \frac{K_2(\mathrm{H}^+)\{\mathrm{CCd}\}}{K_1(\mathrm{Cd}^{2+})}, \tag{79}$$

$$\Gamma_{\mathrm{Cd}} = \{\mathrm{CCd}\} = \frac{\Gamma_{max} K_1(\mathrm{Cd}^{2+})}{1 + K_1(\mathrm{Cd}^{2+}) + K_2(\mathrm{H}^+)}, \tag{80}$$

or

$$\frac{1}{\Gamma_{\mathrm{Cd}}} = \frac{1 + K_2(\mathrm{H}^+)}{\Gamma_{max} K_1(\mathrm{Cd}^{2+})} + \frac{1}{\Gamma_{max}}. \tag{81}$$

If one plots $1/\Gamma_{Cd}$ versus $1/(Cd^{2+})$, then

$$intercept = \frac{1}{\Gamma_{max}} \tag{82}$$

and

$$slope = \frac{1 + K_2(\mathrm{H}^+)}{\Gamma_{max} K_1}. \tag{83}$$

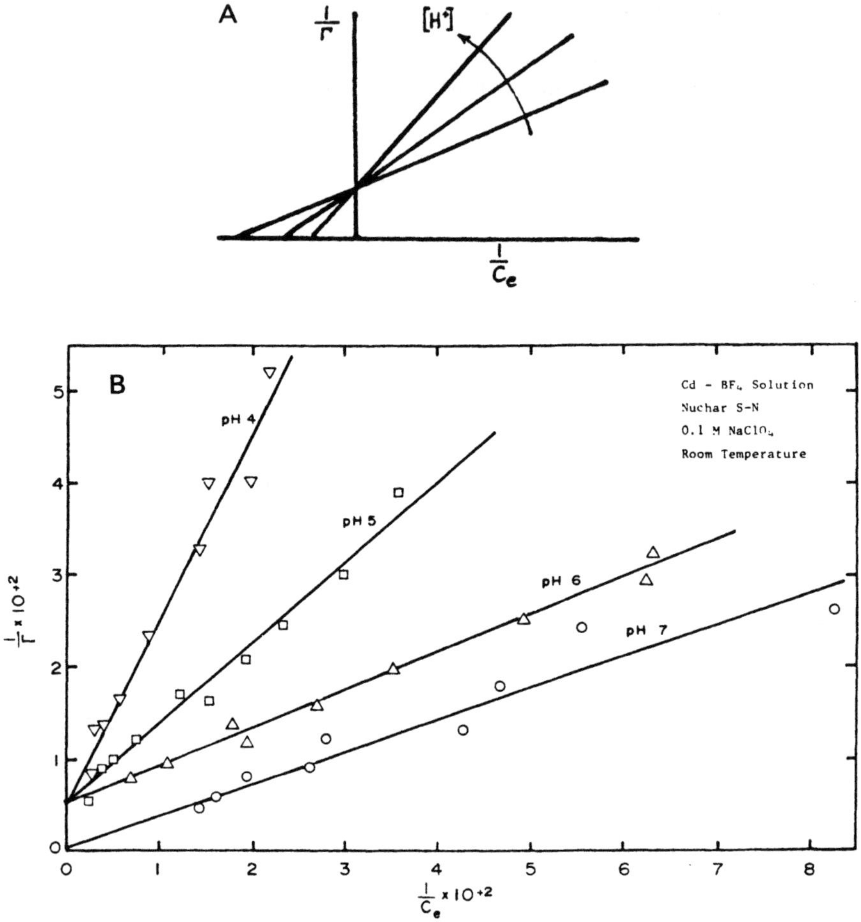

Figure 6. A, Effect from varying one of the components (e.g., pH) in a competitive ion adsorption model for the reactions represented by Eqs. (73) and (74); B, Linearized data for competitive Cd^{2+}/H^+ adsorption on activated C based on Eq. (77). (*Source:* Huang and Smith, 1981).

The K_1 and K_2 values are solved if isotherms are obtained for at least two pH conditions (Fig. 6). Alternate pH effects were discussed by Huang and Smith (1981), which they termed *uncompetitive* and *noncompetitive models.* For an excellent review on this subject see Segel (1975). Murali and Aylmore (1983a) also discussed competitive adsorption, including the use of Freundlich-like isotherms, and later concluded that competitive adsorption occurs on a wide variety of soils (Murali and Aylmore, 1983b). Boyd et al. (1947) and Harter and Baker (1977) correctly emphasized the need for accurate models [e.g., Eq. (80)] to evaluate adsorption dynamics. Note that competitive ion *adsorption* equations [e.g., Eq. (80)] should not be confused with competitive ion *exchange* equations, which are discussed below.

H. Competitive Ion Exchange Models

Ion exchange models view the solid surface as having exchangeable ions on it. If the exchange involves H^+ or OH^- ions on the surface with other cations or anions in the aqueous phase, then the adsorption isotherm will be pH-dependent. Additionally, if the isotherm is obtained at a fixed pH value (i.e., the aqueous H^+ concentration is constant), then the resulting isotherm would probably obey the Langmuir equation. The concept of competitive ion exchange extends these ideas to incorporate all the ions in the aqueous phase as being potentially competitive for adsorption by the surface.

In an aqueous medium, the competitive ions must include H^+ and OH^-, especially if adsorption of a given compound is observed to be pH-dependent. All ions introduced as conjugate acids or bases (e.g., during pH adjustments) should be considered potentially competitive. Finally, other compounds pre-existing on the surface that are exchangeable should be considered in the mass balance equations and rate laws; these include bicarbonates, particularly in subsurface environments where CO_2 partial pressure may reach two orders of magnitude higher than under normal atmospheric conditions (Butler, 1982; Schulthess and McCarthy, 1990). All compounds that adsorb are considered potentially competitive. Competitiveness depends on the aqueous concentration present and the affinity of the compound for the surface.

Assume a given surface, S, is saturated with compound A and is then exposed to a mixture of compounds B and C. The compounds A, B, and C vary in concentration and affinity for the surface. Schematically, the possible reactions are

$$\begin{array}{l} SA + B \overset{K_1}{\rightleftharpoons} SB + A \\ +\, C \\ \quad \upharpoonleft\!\downharpoonright K_2 \\ SC + A \end{array} \tag{84}$$

The equilibrium constants are then defined based on mass action laws as

$$K_1 = \frac{\{SB\}(A)}{\{SA\}(B)} \tag{85}$$

$$K_2 = \frac{\{SC\}(A)}{\{SA\}(C)}, \tag{86}$$

and the total number of surface sites is

$$\Gamma_{max} = \{SA\} + \{SB\} + \{SC\}. \tag{87}$$

Solving for each surface species, and rearranging, the surface speciation ratios are defined by

$$\alpha_A = \frac{\{SA\}}{\Gamma_{max}} = \frac{(A)}{D}, \tag{88}$$

$$\alpha_B = \frac{\{SB\}}{\Gamma_{max}} = \frac{K_1(B)}{D}, \text{ and} \tag{89}$$

$$\alpha_C = \frac{\{SC\}}{\Gamma_{max}} = \frac{K_2(C)}{D}, \tag{90}$$

where $D = (A) + K_1(B) + K_2(C)$, and $\Gamma_i = \alpha_i\Gamma_{max}$. Thus,

$$\Gamma_B = \frac{\Gamma_{max}K_1(B)}{(A) + K_1(B) + K_2(C)}. \tag{91}$$

Solving for K_1, K_2, and Γ_{max} by linearized methods becomes increasingly difficult (e.g., by plotting the reciprocal values $1/(B)$ versus $1/\Gamma_B$) as the number of parameters increases. If numerous reactions are being considered, then the use of computer-assisted graphics and iteration procedures may prove to be easier in evaluating the magnitude of the equilibrium constants.

Schulthess and Sparks (1987) applied a competitive ion exchange model (Fig. 7) to describe the proton adsorption isotherms of an Al oxide (Fig. 8). The competitive compounds were H^+, Na^+, OH^-, Cl^-, and HCO_3^-. The aqueous HCO_3^- concentration was assumed constant and, consequently, simplified the equations; thus, the authors acknowledge that modifications on their competitive model for Al proton adsorption may be forthcoming. It is important to emphasize, however, that this single model predicted not only proton adsorption, but also Cl^-, Na^+, and HCO_3^- adsorption. Additionally, the competitive ion exchange model developed from interpretation of titration data also predicted the electrophoretic mobility data of the Al oxide (Fig. 3) (Schulthess and Sparks, 1988).

The theories advanced with competitive ion exchange models are not restricted to the adsorption of ions. As was discussed by Schulthess and McCarthy (1990), macroscopic observations on the adsorption of a protonated neutral compound (held perhaps by H-bonding forces) is not discernable from the adsorption of the same compound as an unprotonated and charged ion (held perhaps by direct chemical bonding). The adsorption of both unprotonated charged ions and protonated neutral compounds can be described by competitive ion exchange models because the release of exchanged OH^- anions is mathematically the same as the coadsorption of H^+ cations. The term *ion* in *competitive ion exchange models* thus can refer to a combined cation and anion (or salt-like) exchange. This, obviously, also includes competitive exchange with the water (H–OH) molecules on the solid–aqueous inter-

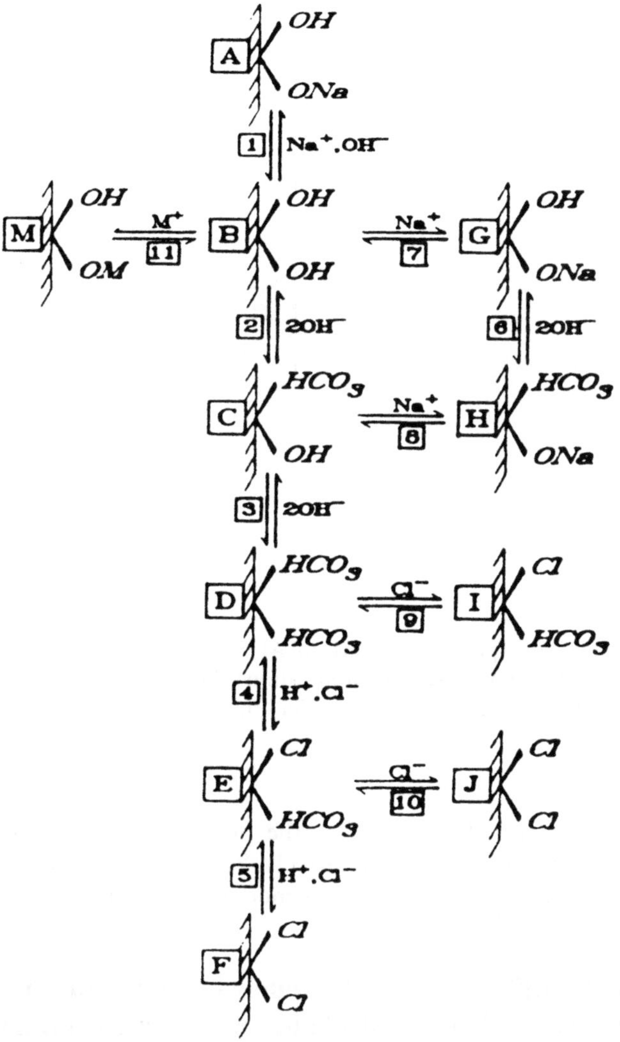

Figure 7. Proposed competitive ion exchange model for Al oxide. The pH- and salt-dependent reactions are illustrated vertically, whereas the salt-dependent reactions are illustrated horizontally. (*Source:* Schulthess and Sparks, 1987).

face. The underlying theme, however, remains the same; that is, the surface must be neutral at all times based on the *principle of electroneutrality*.

For samples more complex than oxides, one may wish to consider a multisite competitive ion exchange model. The emphasis on such models should be placed on the relative surface area of each type of adsorption site, rather than on the weight distribution of the minerals present (Schulthess and Huang,

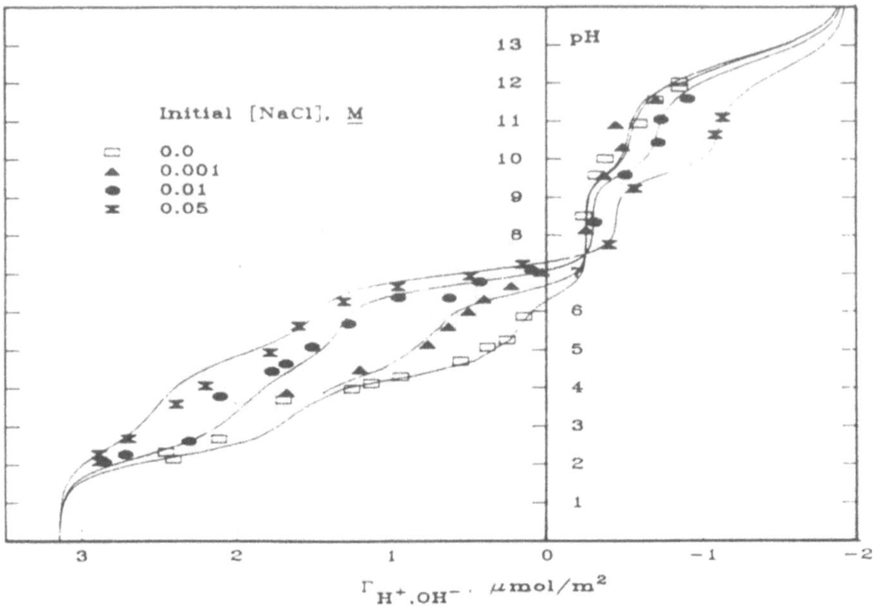

Figure 8. Proton adsorption isotherms on Al oxide obtained by the backtitration technique for several initial NaCl concentrations. The lines drawn are the predicted values of the competitive ion exchange model (Fig. 7). (*Source:* Schulthess and Sparks, 1987).

1990, 1991; Kuo and Yen, 1988). At the present time, however, modeling ion adsorption (or exchange) by relatively "simple" oxides is not completely resolved as witnessed by the numerous models still being discussed today.

IV. Use of Equilibrium-Based Models in Combination with Other Approaches for Modeling Sorption Reactions

The previous models are strictly equilibrium-based and due to their macroscopic nature provide no definitive mechanistic information. They certainly provide no reaction rate information, which is the purpose of kinetic studies. A combination of equilibrium-based models and kinetic studies is a fine way to study sorption phenomena and to assist in elucidating reaction mechanisms. Another excellent way to gain direct evidence about sorption mechanisms on soils and soil constituents is to make spectroscopic measurements. Such measurements can be used to definitively ascertain molecular information about sorption phenomena.

In this section of the chapter we shall discuss briefly if kinetic and spectroscopic techniques substantiate the assumptions made in the macroscopic research of the solid–aqueous interface.

A. Spectroscopic Approaches

Johnston and Sposito (1987) have clearly pointed out that surface complexation models are molecular models that are characterized by different molecular hypotheses concerning an interfacial region. Yet, even if these hypotheses are incorrect, the models may still successfully describe laboratory data that are not microscopic. For example, adsorption data are macroscopic and are not sensitive to molecular processes over a narrow range of adsorptive concentrations (Johnston and Sposito, 1987). Consequently, as Johnston and Sposito (1987) have noted, "It is inevitable that methodologies not equipped to explore molecular structure will produce ambiguous results in the study of surface speciation." To evaluate molecular structure best, one should employ spectroscopic measurements (Davis and Hayes, 1986). Surface spectroscopy can be used to validate the molecular hypotheses in surface speciation models.

There have been a number of attempts at using spectroscopic techniques for gaining insights into surface chemical phenomena. Invasive spectroscopic techniques, such as electron microscopy and microprobe analysis, X-ray photoelectron (XPS), infrared (IR)-transmission, IR-ellipsometric, inelastic electron tunneling (IETS), and electron energy loss spectroscopies (EELS), have been used to delineate interactions on surfaces better (Bell and Hair, 1980; Johnston and Sposito, 1987). For example, Martin and Smart (1987) used XPS to study SO_4 adsorption on goethite. Their results are consistent with the IR studies observed by Parfitt and Smart (1978), who showed that an adsorbed SO_4 ion replaces two A-type OH groups (singly coordinated to Fe^{3+} ions) on (100) and (010) goethite surface sites.

However, the above invasive methods require sample desiccation and high-vacuum techniques, which may cause changes in the surface species. Consequently, the spectroscopic information that is obtained from such methods may not in any way resemble actual mechanisms occurring in a hydrated setting (Johnston and Sposito, 1987).

The best spectroscopic techniques for studying interactions of colloids on aqueous systems are in situ methods. A number of these, including electron spin resonance (ESR), nuclear magnetic resonance (NMR), Fourier transform infrared (FT-IR), and Raman spectroscopy, have been used for colloidal–aqueous systems and are described in Johnston and Sposito (1987).

A beautiful example of using in situ spectroscopic techniques for describing sorption behavior on soil components can be found in the study of Hayes et al. (1987). They used X-ray absorption spectroscopy (XAS) to determine mechanisms for adsorption of the selenium oxyanion on α-FeOOH. They definitively showed that selenate forms a weakly bonded, outer-sphere complex, and that selenite forms a much stronger bonded inner-sphere complex. It was also found that the selenite ion was bonded directly to the goethite surface in a bidentate manner with two iron atoms 3.38 Å from the selenium atom. Adsorbed selenate had no iron atom in the second coordination shell of selenium, which showed that its hydration spheres were retained with sorption.

A limitation of spectroscopic methods is that they cannot elucidate whether an aqueous compound will be preferentially adsorbed over another. However, if the aqueous conditions are varied (e.g., various aqueous concentrations are used), then perhaps the relative surface affinity can be analyzed through spectroscopic methods. This latter procedure is really a combination of spectroscopic and macroscopic methods. In principle, the bonding strength may be elucidated from the knowledge of the physicochemical space occupied by an adsorbate, but this would be possible only after correlating such data with macroscopically determined affinity data.

Spectroscopic research is essential to surface modeling because it offers clear evidence of the actual characteristics of the adsorbed species. This knowledge, in turn, aids in the development of more accurate surface complexation models. A great deal of attention must be paid to the contributions made by spectroscopic research.

B. Kinetic Approaches

Kinetics of soil chemical processes is a complex area and many aspects of the topic are not well understood (Sparks, 1985, 1987, 1989a,b, 1991). Many of the methods that are employed in making kinetic measurements are plagued with problems. In some methods, both transport and chemical kinetics phenomena are being measured simultaneously, and the rate parameters one measures are only "apparent." Consequently, it is difficult for one to obtain mechanistic information. Additionally, many kinetic methods do not allow one to measure reaction rates with half-lives ≤ 30 s. Many sorption reactions on soils are more rapid than this and occur on millisecond and microsecond time scales (Sparks, 1989b). The problems of apparent kinetic measurements and inability to measure fast reaction rates can be eliminated by using relaxation methods, where rapid reaction rates on millisecond and microsecond time scales can be measured.

With these methods, mechanistic rate laws are being determined (i.e., only chemical kinetics phenomena are being studied). Thus, one is studying the rate of chemical reactions and of the molecular processes by which reactions occur where transport is not limiting (Gardiner, 1969). Evidence for this is shown in Table 1 from the work of Zhang and Sparks (1989, 1990) on molybdate and sulfate adsorption–desorption on goethite. The consistency between intrinsic equilibrium constants calculated from equilibrium studies (K_{eq}^{int}) and from the ratio of k_1^{int}/k_{-1}^{int} to yield K_{kin}^{int} is excellent for both systems. This indicates that chemical kinetics is being measured with pressure (p)-jump relaxation because the relationship $k_1^{int}/k_{-1}^{int} = K_{kin}^{int} = K_{eq}^{int}$ will not hold if mass transfer and/or transport processes are occurring and limiting the reaction rates (Ogwada and Sparks, 1986b).

These relaxation methods operate on the principle that the equilibrium of a reaction mixture is perturbed by a rapid change of some external factor such as temperature, pressure, electric field strength, or concentration. One then ob-

Table 1. Intrinsic equilibrium constants determined from equilibrium and kinetics studies[a]

Molybdate Adsorption/Desorption[b,d]		Sulfate Adsorption/Desorption[c]	
K_{eq}^{int}	K_{kin}^{int}	$\log K_{eq}^{int}$	$\log K_{kin}^{int}$
0.50	0.46	9.60	9.14

[a]Kinetic studies were conducted using pressure-jump relaxation.

[b]*Source:* Zhang and Sparks (1989).

[c]*Source:* Zhang and Sparks (1990).

[d]l/mol.

tains kinetic information by measuring the "relaxation time" of the approach to a new equilibrium. The perturbation is small such that measurements are made close to the final equilibrium throughout the relaxation period. Because of the small perturbation, all rate expressions are reduced to first-order equations regardless of molecularity or reaction order. Thus, the rate equations are linearized, which simplifies elucidation of complex reaction mechanisms (Bernasconi, 1986; Sparks, 1989b, 1991). The linearization of the rate equation results in the relationship given below:

$$\tau^{-1} = k_1(\bar{c}_A + \bar{c}_B) + k_{-1} \tag{92}$$

where τ^{-1} is the reciprocal relaxation, k_1 and k_{-1} are the forward and backward rate constants, respectively, and $\bar{c}_A$ and $\bar{c}_B$ are equilibrium aqueous concentrations of reactants A and B. From a linear plot of τ^{-1} versus $(\bar{c}_A + \bar{c}_B)$ in Eq. (86) one can calculate k_1 and k_{-1} from the slope and intercept, respectively.

The pressure-jump method is one relaxation technique that has been used to model adsorption–desorption reactions on soil constituents. It is based on the fact that chemical equilibria are dependent on pressure (Bernasconi, 1976):

$$\left[\frac{\delta(\ln K_{eq})}{\delta p}\right]_T = \frac{-\Delta V}{R} \tag{93}$$

or

$$\frac{\Delta K_{eq}}{K_{eq}} = \frac{-\Delta V}{RT}\Delta p, \tag{94}$$

where ΔV is the standard molar volume change of the reaction (liters), p is pressure (MPa), and R is the molar gas constant.

Pressure-jump measurements can be detected using either optical or conductivity detection. Conductivity detection is usually preferred because the equilibrium displacement after the p-jump is quite small (Bernasconi, 1976).

It has been the detection technique that has been used by all investigators studying the kinetics of reactions on soil constituents because of the high sensitivity obtained and because suspensions are being studied. A more in depth discussion on theoretical and experimental aspects of *p*-jump relaxation can be found in Bernasconi (1986) and Sparks (1989b).

Pressure-jump measurements in combination with various chemical adsorption models have been successfully used to describe sorption of metal ions on clay minerals and oxides. The combination of these approaches offers a unique way to glean mechanistic information about sorption phenomena that cannot be obtained by only using equilibrium-based models. Such studies have included the hydrolysis of hydroxyl groups on Zeolite surfaces (Ikeda et al., 1982a), adsorption–desorption kinetics of protons at the TiO_2–H_2O interface (Ashida et al., 1978), kinetics of IO_3 on TiO_2 (Hachiya et al., 1980), proton adsorption–desorption mechanisms on Fe oxides (Astumian et al., 1981), adsorption–desorption dynamics of acetic acid on silica–alumina particles (Ikeda et al., 1982b), and NH_4^+ adsorption on zeolites (Ikeda et al., 1984). A number of other adsorption and/or desorption kinetic studies, all on Al oxide, have included interactions with Pb (Hachiya et al., 1979; Hayes and Leckie, 1986), phosphate (Mikami et al., 1983a), chromate (Mikami et al., 1983b), and other metals (Hachiya et al., 1984a,b).

Recently, Zhang and Sparks (1989, 1990) were the first soil scientists to employ relaxation techniques to soil constituent reactions. They employed the *p*-jump technique for ascertaining adsorption–desorption mechanisms of molybdate and sulfate on goethite. Molybdate adsorption on goethite was described as an inner-sphere complex with a 1:1 stoichiometry between the surface groups and the adsorbed molybdate. The molybdate anions appear to react primarily with the protonated sites, SOH_2^+, and not with the neutral sites, SOH.

Combining the information from the *p*-jump relaxation studies and from the overall equilibrium partitioning and the reaction stoichiometry, Zhang and Sparks (1989) postulated a two-step reaction to describe the mechanism of molybdate adsorption on goethite:

$$SOH_2^+ + MoO_4^{2-} \underset{\text{Step 1}}{\rightleftharpoons} SOH_2^+MoO_4^{2-} \underset{\text{Step 2}}{\rightleftharpoons} SMoO_4^- + H_2O, \tag{95}$$

where MoO_4 quickly adsorbs electrostatically followed by a slower dehydration step.

In all these kinetic models discussed, the law of mass action and the law of chemical equilibrium are central to the kinetic data interpretation. In general, kinetic research has been interested in determining the rate of those reactions that are also typically assumed to exist in equilibrium-based studies. Consequently, these kinetic models are subject to a similar set of uncertainties as those encountered in equilibrium-based models. These include uncertainties on the degree of surface heterogeneity, evaluating the solid phase activity coefficients, elucidating the presence or absence of a surface charge, and the applicability of exponential terms in the evaluation of the equilibrium constants (from k_1/k_{-1} rate ratios). In truth, kinetic models cannot independently resolve these questions. Instead, the combined efforts of equilibrium, kinetic, and spectroscopic techniques should be pursued if the elusive surface–aqueous phase reactions are to be properly resolved.

References

Adamson, A.W. 1982. *Physical Chemistry of Surfaces*, 4th ed. John Wiley & Sons, New York.

Argersinger, W.J., Jr., A.W. Davidson, and O.D. Bonner. 1950. Thermodynamics and ion exchange phenomena. *Trans. Kansas Acad. Sci.* 53:404–410.

Ashida, M., M. Sasaki, H. Kan, T. Yasunaga, K. Hachiya, and T. Inoue. 1978. Kinetics of proton adsorption–desorption at TiO_2–H_2O interface by means of pressure-jump technique. *J. Colloid Interface Sci.* 67:219–225.

Astumian, R.D., M. Sasaki, T. Yasunaga, and Z.A. Schelly. 1981. Proton adsorption–desorption kinetics on iron oxides in aqueous suspensions, using the pressure-jump method. *J. Phys. Chem.* 85:3832–3835.

Beckett, P.H.T. 1964. Studies on soil potassium. I. Confirmation of the ratio law: Measurement of potassium potential. *J. Soil Sci.* 15:1–8.

Bell, A.T., and M.L. Hair. 1980. Vibrational spectroscopies for adsorbed species. *ACS Symp. Ser. 137*. Am. Chem. Soc., Washington, D.C.

Belloni, J., M. Haissinsky, and H.N. Salama. 1959. On the adsorption of some fission products on various surfaces. *J. Phys. Chem.* 63:881–887.

Bernasconi, C.F. 1976. *Relaxation Kinetics*. Academic Press, NY.

Bernasconi, C.F. (Ed.). 1986. *Investigations of Rates and Mechanisms of Reactions*, 4th ed. John Wiley & Sons, Inc. New York.

Boedeker, C. 1859. Über das Verhältnis zwischen Masse und Wirkung beim Kontakt ammoniakalscher Flüssigkeiten mit Ackererde und mit kohlensaurem Kalk. *Z. Acker- u. Pflanzenbau* 7:48–58.

Boyd, G.E., J. Schubert, and A.W. Adamson. 1947. The exchange adsorption of ions from aqueous solutions by organic zeolites. I. Ion-exchange equilibria. *J. Am. Chem. Soc.* 69:2818–2829.

Boyd, S.A., S. Shaobai, J.F. Lee, and M.M. Mortland. 1988a. Pentachlorophenol sorption by organo-clays. *Clays Clay Minerals* 36:125–130.

Boyd, S.A., M.M. Mortland, and C.T. Chiou. 1988b. Sorption characteristics of organic compounds on hexadecyltrimethylammonium-smectite. *Soil Sci. Soc. Am. J.* 52:652–657.

Boyd, S.A., J.F. Lee, and M.M. Mortland. 1988c. Attenuating organic contaminant mobility by soil modification. *Nature* 333:345–347.

Brunauer, S. 1961. Solid surfaces and the solid–gas interface. *In* R.F. Gould (Ed.). *Advances in Chemistry Series 33*. Am. Chem. Soc., Washington, D.C., pp. 5–17.

Brunauer, S., L.E. Copeland, and D.L. Kantro. 1967. The Langmuir and BET theories. *In* E.A. Flood (Ed.). *The Solid–Gas Interface*, Vol. 1. Marcel Dekker, NY, pp. 77–103.

Butler, J.N. 1982. *Carbon Dioxide Equilibria and their Applications*. Addison–Wesley Publ. Co., Reading, MA.

Chapman, D.L. 1913. A contribution to the theory of electrocapillarity. *Phil. Magazine* (series 6) 25:475–481.

Chiou, C.T., and T.D. Shoup. 1985. Soil sorption of organic vapors and effects of humidity on sorptive mechanism and capacity. *Environ. Sci. Technol.* 19:1196–1200.

Davis, J.A., and K.F. Hayes (Eds.). 1986. *Geochemical Processes at Mineral Surfaces. ACS Symp. Ser. 323*, Am. Chem. Soc., Washington, D.C., pp. 1–18.

Elliott, H.A., and D.L. Sparks. 1981. Electrokinetic behavior of a paleudult profile in relation to mineralogical composition. *Soil Sci.* 132:402–409.

Freundlich, H. 1922. *Colloid and Capillary Chemistry*, 3rd. ed. Translated by H.S. Hatfield (1926), Methuen & Co., Ltd., London.

Gaines, G.L., Jr., and H.C. Thomas. 1953. Adsorption studies on clay minerals. II. A formulation of the thermodynamics of exchange adsorption. *J. Chem. Phys.* 21:714–718.

Gans, R. 1913. Ueber die chemische oder physikalische Natur der kolloidalen wasserhaltigen Tonerdesilikate. *N. Jahrb. Mineralogie, Geologie u. Palaontalogie* 699–712, 728–741.

Gapon, E.N. 1933. Theory of exchange adsorption in soils. *J. Gen. Chem.* (USSR) 3:144–152. (Abstract in *Chem. Abs.* 28:4149–4150, 1934).

Gardiner, W.C. 1969. *Rates and Mechanisms of Chemical Reactions.* Benjamin, NY.

Gibbs, J.W. 1961. On the equilibrium of heterogeneous substances. *In* H.A. Bumstead and R.G. van Name (Eds.). *The Scientific Papers of J. Willard Gibbs*, Vol. I: *Thermodynamics.* Reprint, Dover Publ., Inc., New York, pp. 55–371.

Goldberg, S. 1985. Chemical modeling of anion competition on goethite using the constant capacitance model. *Soil Sci. Soc. Am. J.* 49:851–856.

Gouy, G. 1910. Sur la constitution de la charge électrique à la surface d'un électolyte. *Ann. Phys. (Paris)* (series 4) 9:457–468.

Grebenshchikova, V.I., and Yu.P. Davydov. 1961a. Research on the state of Pu(IV) in dilute solutions of nitric acid. *U.S. Atomic Energy Commission* (Engl. transl. no. 4830), *Radiokhimiya* 3:155–164.

Grebenshchikova, V.I., and Yu.P. Davydov. 1961b. Adsorption of Pu(IV) on the surface of glass. *Soviet Radiochemistry* 3:177–184 (Engl. transl.), *Radiokhimiya* 3:165–172.

Greenland, D.J. 1971. Interactions between humic and fulvic acids and clays. *Soil Sci.* 111:34–41.

Guggenheim, E.A., and N.K. Adam. 1933. The thermodynamics of adsorption at the surface of solutions. *Proc. R. Soc. Lond. [Math. & Phys.]* 139A:218–236.

Hachiya, K., M. Ashida, M. Sasaki, H. Kan, T. Inoue, and T. Yasunaga. 1979. Study of the kinetics of adsorption–desorption of Pb^{2+} on a γ-Al_2O_3 surface by means of relaxation techniques. *J. Phys. Chem.* 83:1866–1871.

Hachiya, K., M. Ashida, M. Sasaki, M. Karasuda, and T. Yasunaga. 1980. Study of the adsorption-desorption of IO_3^- on a TiO_2 surface by means of relaxation techniques. *J. Phys. Chem.* 84:2292–2296.

Hachiya, K., M. Sasaki, Y. Saruta, N. Mikami, and T. Yasunaga. 1984a. Static and kinetic studies of adsorption–desorption of metal ions on a γ-Al_2O_3 surface. 1. Static study of adsorption–desorption. *J. Phys. Chem.* 88:23–27.

Hachiya, K., M. Sasaki, T. Ikeda, N. Mikami, and T. Yasunaga. 1984b. Static and kinetic studies of adsorption–desorption of metal ions on a γ-Al_2O_3 surface. 2. Kinetic study by means of pressure-jump technique. *J. Phys. Chem.* 88:27–31.

Harter, R.D., and D.E. Baker. 1977. Applications and misapplications of the Langmuir equation to soil adsorption phenomena. *Soil Sci. Soc. Am. J.* 41:1077–1080. See also letters to the editor *SSSAJ* 42:986–988 (1978).

Hayes, K.F., and J.O. Leckie. 1986. Mechanism of lead ion adsorption at the goethite–water interface. *In* J.A. Davis and K.F. Hayes (Eds.). *Geochemical Processes at Mineral Surfaces. ACS Symp. Ser. 323*, Am. Chem. Soc., Washington, D.C., pp. 114–141.

Hayes, K.F, A.L. Roe, G.E. Brown, Jr., K.O. Hodgson, J.O. Leckie, and G.A. Parks. 1987. In situ X-ray adsorption study of surface complexes: Selenium oxyanions on α-FeOOH. *Science* 238:783–786.

Helmholtz, H. 1879. Studien über elektrische Grenzschichten. *Ann. Physik u. Chemie* (Leipzig) 7:337–382.

Henry, D.C. 1922. A kinetic theory of adsorption. *Phil. Mag.* (series 6) 44:689–705.

Hiemenz, P.C. 1977. *Principles of Colloid and Surface Chemistry.* Marcel Dekker, New York.

Hingston, F.J., R.J. Atkinson, A.M. Posner, and J.P. Quirk. 1967. Specific adsorption of anions. *Nature* 215:1459–1461.

Ho, C.H., and N.H. Miller. 1985. Effect of humic acid on uranium uptake by hematite particles. *J. Colloid Interface Sci.* 106:281–288.

Holford, I.C.R., R.W.M. Wedderburn, and G.E.G. Mattingly. 1974. A Langmuir two-surface equation as a model for phosphate adsorption by soils. *J. Soil Sci.* 25:242–255.

Holford, I.C.R., and G.E.G. Mattingly. 1975. The high- and low-energy phosphate adsorbing surfaces in calcareous soils. *J. Soil Sci.* 26:407–417.

Huang, C.P. 1981. The surface acidity of hydrous solids. *In* M.A. Anderson and A.J. Rubin (Eds.). *Adsorption of Inorganics at Solid–Liquid Interfaces.* Ann Arbor Sci. Publ., Ann Arbor, MI, pp. 183–217.

Huang, C.P., and E.H. Smith. 1981. Removal of Cd(II) from plating waste water by an activated carbon process. *In* W.J. Cooper (Ed.). *Chemistry in Water Reuse*, Vol. 2. Ann Arbor Science, Ann Arbor, MI, pp. 355–400.

Ikeda, T., M. Sasaki, and T. Yasunaga. 1982a. Kinetics of the hydrolysis of hydroxyl groups on zeolite surfaces using the pressure-jump relaxation method. *J. Phys. Chem.* 86:1678–1680.

Ikeda, T., M. Sasaki, K. Hachiya, R.D. Astumian, T. Yasunaga, and Z.A. Schelly. 1982b. Adsorption–desorption kinetics of acetic acid on silica–alumina particles in aqueous suspensions, using the pressure-jump relaxation method. *J. Phys. Chem.* 86:3861–3866.

Ikeda, T., M. Sasaki, and T. Yasunaga. 1984. Kinetic studies of ion exchange of the ammonium ion for H^+ in zeolite H-ZSM-5 by the chemical relaxation method. *J. Colloid Interface Sci.* 98:192–195.

James, R.O. 1981. Surface ionization and complexation at the colloid/aqueous electrolyte interface. *In* M.A. Anderson and A.J. Rubin (Eds.). *Adsorption of Inorganics at Solid–Liquid Interfaces.* Ann Arbor Sci. Publ., Ann Arbor, MI. pp. 219–261.

James, R.O., and T.W. Healy. 1972. Adsorption of hydrolyzable metal ions at the oxide–water interface: II. Charge reversal of SiO_2 and TiO_2 colloids by adsorbed Co(II), La(III), and Th(IV) as model systems. *J. Colloid Interface Sci.* 40:53–64.

Johnston, C.T., and G. Sposito. 1987. Disorder and early sorrow: Progress in the chemical speciation of soil surfaces. *In* L.L. Boersma et al. (Eds.). *Future Developments in Soil Science Research.* Soil Sci. Soc. Am. Madison, WI, pp. 89–99.

Kauffman, G.B. 1972. Cato Maximilian Guldberg. *In* C.C. Gillispie (Ed.). *Dictionary of Scientific Biography*, Vol. V. Charles Scribner's Sons, NY, pp. 586–587.

Kauffman, G.B. 1976. Peter Waage. *In* C.C. Gillispie (Ed.). *Dictionary of Scientific Biography*, Vol. XIV. Charles Scribner's Sons, NY, pp. 108–109.

Kerr, H.W. 1928a. The nature of base exchange and soil acidity. *J. Am. Soc. Agr.* 20:309–335.

Kerr, H.W. 1928b. The identification and composition of the soil alumino–silicate active in base exchange and soil acidity. *Soil Sci.* 26:385–398.

Kipling, J.J. 1965. *Adsorption from Solutions of Non-Electrolytes.* Academic Press, New York.

Klotz, I.M. 1982. Number of receptor sites from Scatchard graphs: Facts and fantasies. *Science* 217:1247–1249.

Koopal, L.K., W.H. van Riemsdijk, and M.G. Roffey. 1987. Surface ionization and complexation models: A comparison of methods for determining model parameters. *J. Colloid Interface Sci.* 118:117–136.

Krishnamoorthy, C., and R. Overstreet. 1949. Theory of ion-exchange relationships. *Soil Sci.* 68:307–315.

Kuo, J.F., and T.F. Yen. 1988. Some aspects in predicting the point of zero charge of a composite oxide system. *J. Colloid Interface Sci.* 121:220–225.

Lagaly, G., R. Witter, and H. Sander. 1983. Water on hydrophobic surfaces. *In* R.H. Ottewill, C.H. Rochester, and A.L. Smith (Eds.). *Adsorption from Solution.* Academic Press, NY, pp. 65–77.

Lane, J.E. 1983. Surface activity coefficients. *In* R.H. Ottewill, C.H. Rochester, and A.L. Smith (Eds.). *Adsorption from Solution.* Academic Press, NY, pp. 51–64.

Langmuir, I. 1918. The adsorption of gases on plane surfaces of glass, mica and platinum. *J. Am. Chem. Soc.* 40:1361–1403.

Langmuir, I. 1932. Vapor pressures, evaporation, condensation and adsorption. *J. Am. Chem. Soc.* 54:2798–2832.

Langmuir, I. 1933. An extension of the phase rule for adsorption under equilibrium and non-equilibrium conditions. *J. Chem. Phys.* 1:3–12.

Leenheer, J.A., and E.W.D. Huffman, Jr. 1979. Analytical method for dissolved-organic carbon fractionation. *U.S. Geological Survey, Water-Resources Investigations* 79–4.

Martin, R.R., and R.St.C. Smart. 1987. X-ray photoelectron studies of anion adsorption on goethite. *Soil Sci. Soc. Am. J.* 51:54–56.

Matthews, B.C., and P.H.T. Beckett. 1962. A new procedure for studying the release and fixation of potassium ions in soil. *J. Agric. Sci.* 58:59–64.

Mattson, S. 1931. The laws of soil colloidal behavior: V. Ion adsorption and exchange. *Soil Sci.* 31:311–331.

Mikami, N., M. Sasaki, K. Hachiya, R.D. Astumian, T. Ikeda, and T. Yasunaga. 1983a. Kinetics of the adsorption–desorption of phosphate on the γ-Al_2O_3 surface using the pressure-jump technique. *J. Phys. Chem.* 87:1454–1458.

Mikami, N., M. Sasaki, T. Kikuchi, and T. Yasunaga. 1983b. Kinetics of adsorption–desorption of chromate on γ-Al_2O_3 surfaces using the pressure-jump technique. *J. Phys. Chem.* 87:5245–5248.

Mortland, M.M. 1970. Clay–organic complexes and interactions. *Adv. Agron.* 22:75–117.

Moss, P., and P.H.T. Beckett. 1971. Sources of error in the determination of soil potassium activity ratios by the *Q/I* procedure. *J. Soil Sci.* 22:514–536.

Murali, V., and L.A.G. Aylmore. 1983a. Competitive adsorption during solute transport in soils: 1. Mathematical models. *Soil Sci.* 135:143–150.

Murali, V., and L.A.G. Aylmore. 1983b. Competitive adsorption during solute transport in soils: 3. A review of experimental evidence of competitive adsorption and an evaluation of simple competition models. *Soil Sci.* 136:279–290.

Nabzar, L., A. Carroy, and E. Pefferkorn. 1986. Formation and properties of the kaolinite–polyacrylamide complex in aqueous media. *Soil Sci.* 141:113–119.

Ogwada, R.A., and D.L. Sparks. 1986a. Use of mole or equivalent fractions in determining thermodynamic parameters for potassium exchange in soils. *Soil Sci.* 141:268–273.

Ogwada, R.A., and D.L. Sparks. 1986b. A critical evaluation on the use of kinetics for determining thermodynamics of ion exchange in soils. *Soil Sci. Soc. Am. J.* 50:300–305.

Olsen, S.R., and F.S. Watanabe. 1957. A method to determine a phosphorus adsorption maximum of soils as measured by the Langmuir isotherm. *Soil Sci. Soc. Am. Proc.* 21:144–149.

Parfitt, R.L., and R.St.C. Smart. 1978. The mechanism of sulfate adsorption on iron oxides. *Soil Sci. Soc. Am. J.* 42:48–50.

Podoll, R.T., K.C. Irwin, and S. Brendlinger. 1987. Sorption of water-soluble oligomers on sediments. *Environ. Sci. Technol.* 21:562–568.

Posner, A.M., and J.W. Bowden. 1980. Adsorption isotherms: Should they be split? *J. Soil Sci.* 31:1–10.

Schmuck, M.N., M.P. Nowlan, and K.M. Gooding. 1986. Effects of mobile phase and ligand arm on protein retention in hydrophobic interaction chromatography. *J. Chromatogr.* 371:55–62.

Schofield, R.K. 1947. A ratio law governing the equilibrium of cations in the soil solution. *Proc. 11th Int. Congr. Pure Appl. Chem.* 3:257–261.

Schulthess, C.P., and C.P. Huang. 1990. Adsorption of heavy metals by silicon and aluminum oxide surfaces on clay minerals. *Soil Sci. Soc. Am. J.* 54:679–688.

Schulthess, C.P., and C.P. Huang. 1991. Humic and fulvic acid adsorption by silicon and aluminum oxide surfaces on clay minerals. *Soil Sci. Soc. Am. J.* 54:34–42.

Schulthess, C.P., and J.F. McCarthy. 1990. Competitive adsorption of aqueous carbonic and acetic acids by an aluminum oxide. *Soil Sci. Soc. Am. J.* 54:688–694.

Schulthess, C.P., and D.L. Sparks. 1986. Backtitration technique for proton isotherm modeling of oxide surfaces. *Soil Sci. Soc. Am. J.* 50:1406–1411.

Schulthess, C.P., and D.L. Sparks. 1987. Two-site model for aluminum oxide with mass balanced competitive pH + salt/salt dependent reactions. *Soil Sci. Soc. Am. J.* 51:1136–1144.

Schulthess, C.P., and D.L. Sparks. 1988. A critical assessment of surface adsorption models. *Soil Sci. Soc. Am. J.* 52:92–97.

Schulthess, C.P., and D.L. Sparks. 1989. Competitive ion exchange behavior on oxides. *Soil Sci. Soc. Am. J.* 53:366–373.

Segel, I.H. 1975. *Enzyme Kinetics.* John Wiley & Sons, New York.

Shuman, L.M. 1975. The effect of soil properties on zinc adsorption by soils. *Soil Sci. Soc. Am. Proc.* 39:454–458.

Silbermann, W.E. 1961. Law of mass action. *In* J. Thewlis (Ed.). *Encyclopaedic Dictionary of Physics, Vol. 4.* Pergamon Press, NY, pp. 505–506.

Sparks, D.L. 1984. Ion activities: An historical and theoretical overview. *Soil Sci. Soc. Am. J.* 48:514–518.

Sparks, D.L. 1985. Kinetics of ionic reactions in clay minerals and soils. *Adv. Agron.* 38:231–266.

Sparks, D.L. 1987. Kinetics of soil chemical processes: Past progress and future needs. *In* L.L. Boersma et al. (Eds.). *Future Developments in Soil Science Research.* Soil Sci. Soc. Am., Madison, WI, pp. 61–73.

Sparks, D.L. 1989a. Soil chemistry: Kinetics and mechanisms. *McGraw-Hill Yearbook of Science and Engineering*. McGraw-Hill, NY, pp. 362–365.

Sparks, D.L. 1989b. *Kinetics of Soil Chemical Processes.* Academic Press, San Diego, CA.

Sparks, D.L. 1991. Chemical kinetics and mass transfer processes in soils and soil constituents. *In* Y. Corapcioglu and J. Bear (Eds.). *Transport Processes in Porous Media.* Kluwer Publ., Dordrecht, The Netherlands. (*in press.*)

Sposito, G. 1982. On the use of the Langmuir equation in the interpretation of "adsorption" phenomena: II. The "two-surface" Langmuir equation. *Soil Sci. Soc. Am. J.* 46:1147–1152.

Sposito, G. 1984a. The future of an illusion: Ion activities in soil solutions. *Soil Sci. Soc. Am. J.* 48:531–536.

Sposito, G. 1984b. *The Surface Chemistry of Soils.* Oxford Univ. Press, NY.

Sposito, G. 1986. Distinguishing adsorption from surface precipitation. *In* J.A. Davis and K.F. Hayes (Eds.). *Geochemical Processes at Mineral Surfaces. ACS Symp. Ser. 323* Washington, D.C., pp. 217–228.

Stern, O. 1924. Zur Theorie der elektrolytischen Doppelschicht. *Z. Elektrochemie* 30:508–516.

Stumm, W. 1986. Coordinative interactions between soil solids and water—an aquatic chemist's point of view. *Geoderma* 38:19–30.

Syers, J.K., M.G. Browman, G.W. Smillie, and R.B. Corey. 1973. Phosphate sorption by soils evaluated by the Langmuir adsorption equation. *Soil Sci. Soc. Am. Proc.* 37:358–363.

Thomas, G.W. 1974. Chemical reactions controlling soil solution electrolyte concentration. *In* E.W. Carson (Ed.). *The Plant Root and Its Environment.* University Press of Virginia, Charlottesville, VA, pp. 483–506.

Thomas, G.W. 1977. Historical developments in soil chemistry: Ion exchange. *Soil Sci. Soc. Am. J.* 41:230–238.

Tipping, E. 1981a. The adsorption of aquatic humic substances by iron oxides. *Geochem. Cosmochim. Acta* 45:191–199.

Tipping, E. 1981b. Adsorption by goethite (α-FeOOH) of humic substances from three different lakes. *Chem. Geology* 33:81–89.

Trapnell, B.M.W. 1955. *Chemisorption.* Academic Press, New York.

Vanselow, A.P. 1932. Equilibria of the base-exchange reactions of bentonites, permutites, soil colloids, and zeolites. *Soil Sci.* 33:95–113.

Veith, J.A., and G. Sposito. 1977. On the use of the Langmuir equation in the interpretation of "adsorption" phenomena. *Soil Sci. Soc. Am. J.* 41:697–702.

Way, J.T. 1850. On the power of soils to absorb manure. *J. R. Agric. Soc. Engl.* 11:313–379.

Way, J.T. 1852. On the power of soils to absorb manure. *J.R. Agric. Soc. Engl.* 13:123–143.

Weiß, A. 1966. Modellversuche zur Hydrophobierung hydrophiler Grenzflächen an Schichtsilicaten. *Kolloid Z.Z. Polymere* 211:94–97.

Westall, J., and H. Hohl. 1980. A comparison of electrostatic models for the oxide/solution interface. *Adv. Colloid Interface Sci.* 12:265–294.

Westall, J.C. 1986. Reactions at the oxide–solution interface: Chemical and electrostatic models. *In* J.A. Davis and K.F. Hayes (Eds.). *Geochemical Processes at Mineral Surfaces. ACS Symp. Ser. 323.* Am. Chem. Soc., Washington, D.C., pp. 54–78.

Zachara, J.M., C.C. Ainsworth, C.E. Cowan, and B.L. Thomas. 1987. Sorption of binary mixtures of aromatic nitrogen heterocyclic compounds on subsurface materials. *Environ. Sci. Technol.* 21:397–402.

Zhang, P.C., and D.L. Sparks. 1989. Kinetics and mechanisms of molybdate adsorption/desorption at the goethite/water interface using pressure-jump relaxation. *Soil Sci. Soc. Am. J.* 53:1028–1034.

Zhang, P.C., and D.L. Sparks. 1990. Kinetics and mechanisms of sulfate adsorption/desorption on goethite using pressure-jump relaxation. *Soil Sci. Soc. Am. J.* 54:1266–1273.

[illegible]

Determination of Nutrient Availability and Elemental Toxicity by AB-DTPA Soil Test and ICPS

P.N. Soltanpour

I. Introduction

The ammonium bicarbonate-diethylenetriaminepentaacetic acid (DTPA) (AB-DTPA) soil test was developed by Soltanpour and Schwab (1977) to extract simultaneously labile NO_3–N, P, K, Zn, Fe, Mn, and Cu from neutral and calcareous soils. Later, it was used for extraction of potentially toxic elements from soils and mine spoils and soils treated with sewage sludge. The list of elements whose bioavailability can be determined with AB-DTPA also include Pb, Cd, Ni, Se, As, B, Mo, and S. To use the AB-DTPA test more efficiently one can use an inductively coupled plasma (ICP) spectrometer. An

Advances in Soil Science, Volume 16

ICP spectrometer is capable of simultaneous or sequential determination of the above elements in the AB-DTPA extract accurately and rapidly. The use of AB-DTPA soil test in conjunction with an ICP spectrometer enabled the discovery of Cd, Pb, and Zn contamination in garden soil samples from some Colorado mountain communities during a routine soil fertility test. The metal contamination originated from old mine spoils. As a result of this discovery and further testing by the U.S. Environmental Protection Agency (EPA) and Colorado Department of Public Health these communities were earmarked for cleanup by the U.S. Government.

In this chapter, I elucidate the ability of the AB-DTPA–ICP system for determining soil fertility indexes and for screening soils contaminated with sewage sludge, mine spoil, and other materials for their potential elemental toxicity to plants, animals, and humans.

II. Theoretical Principles of the Ammonium Bicarbonate-DTPA (AB-DTPA) Soil Test

The AB-DTPA solution is 1 *M* NH_4HCO_3 and 0.005 *M* DTPA. It has an initial pH of 7.6. Ten grams of soil are mixed with 20 ml of AB-DTPA solution and shaken for 15 min in an open flask (Tables 1 and 2). The NH_4^+ replaces K^+ on the exchange complex and thus gives a measure of exchangeable plus soluble K^+. One molar NH_4OAc is the most commonly used reagent to

Table 1. Preparation of NH_4HCO_3–DTPA solution

1. Dissolve 1.97 g of DTPA in 800 ml of pure water.
2. Add about 2 ml of 1:1 NH_4OH to facilitate dissolution.
3. After DTPA is in solution, add 79.06 g of NH_4HCO_3, and stir gently until dissolved.
4. Adjust pH to 7.6 using either NH_4OH or HCl.
5. Bring to a final volume of 1 L.
6. The extracting solution is unstable with regard to pH and should be made immediately before using or kept under mineral oil.

Table 2. Extracting procedure

1. Place 10 g of soil in 250-ml Erlenmeyer flask.
2. Add 20 ml of extracting solution and shake on a reciprocating shaker at 180 cycles per minute for 15 min. The flasks are kept open.
3. Filter through a medium-speed filter paper.
4. Place 0.25 ml of concentrated HNO_3 in a 10-ml beaker.[a]
5. Carefully add 2.5 ml of AB-DTPA extract or standard solution to the beaker and mix on a rotary shaker for 15 min to eliminate carbonate species.[a]
6. Analyze on ICP.

[a] Eliminate 4 and 5 and analyze on ICP with a Legere or other high-salt rebulizers.

extract plant-available (soluble plus exchangeable) K^+. The AB-DTPA test extracts similar amounts of K^+ (Soltanpour and Schwab, 1977).

The theoretical development of DTPA as a micronutrient soil test has been well documented (Lindsay and Norvell, 1969, 1978; Lindsay et al., 1967; Lindsay, 1974; 1979; Norvell, 1972). Plant uptake of micronutrients is related to the metal activity in soil solution and to the ability of the solid phases to resupply the solution with those cations. Once added to the soil solution, DTPA forms stable soluble chelates with the various metal cations. The binding strength of a chelate for a metal cation depends on the stability constant of the metal chelate complex. The higher the stability constant, the stronger the metal–chelate complex is. As free metal cations are removed from solution through this complexation (or through plant root uptake), labile solid phases dissolve or surface cations desorb to replenish those in solution. Rule and Graham (1976) used labeled ^{59}Fe, ^{54}Mn, and ^{65}Zn to calculate soil-labile pools of these nutrients and obtained similar results between plant uptake and soil equilibration with DTPA for Fe, Mn, and Zn. Lauer (1971) also obtained similar results with Zn.

The purpose of the HCO_3^- anion in the extracting solution is to extract phosphates and other anions such as sulfates, arsenates, selenates, and molybdates. Olsen et al. (1954) demonstrated that the $NaHCO_3$ soil test is a good measure of plant-available P. As CO_2 (g) evolves during the extraction with NH_4HCO_3–DTPA, the pH of the soil-extractant mixture rises from its initial value of 7.6 to 8.5. At this pH with 1.0 *M* carbonate–bicarbonate, $CaCO_3$ precipitates, allowing the labile phases of calcium phosphates to dissolve,

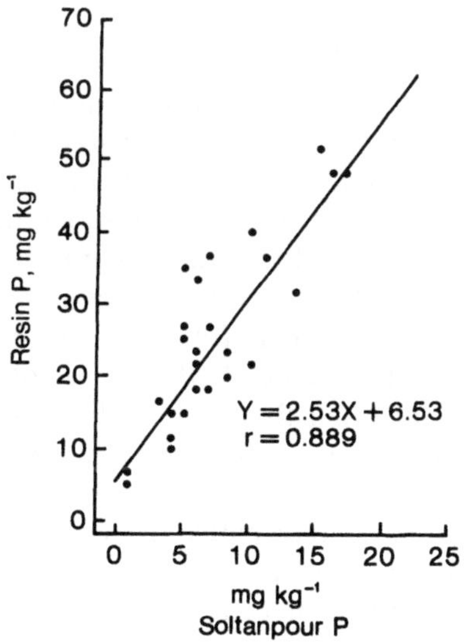

Figure 1. The relationship of Soltanpour AB-DTPA extractable P to resin-extractable P for 30 alkaline soils. (*Source:* Nesse et al., 1988, reprinted from *Commun. Soil Sci. Plant. Anal.* 19:675–689 by courtesy of Marcel Dekker, Inc.).

which increases solution P. At the same time, HCO_3^- ions desorb phosphates from the surface of clay-size minerals. Iron and Al also precipitate as hydroxides and thus labile Fe and Al phosphates dissolve. The same principles (desorption and dissolution) explain the extraction of other labile anions such as arsenates, molybdates, selenates, and sulfates. Previous studies (Soltanpour and Schwab, 1977) have shown that AB-DTPA extracts approximately one-half the P, compared with $NaHCO_3$, from medium-texture Colorado soils. In heavy Moroccan soils the relationship between $NaHCO_3$P (Y) and AB-DTPA P (X) was determined to be $Y = 5.20 + 3.11X$ with a r^2 of 0.96 (Azzaoui, et al., 1989). Because the intercept is high, this regression equation should not be used at low levels of P. This difference between the extracted levels is mainly a result of a 20:1 solution/soil ratio for the $NaHCO_3$ test and a 2:1 ratio for the NH_4HCO_3–DTPA soil test. Nesse et al. (1988) showed a good correlation between resin-extractable P and AB-DTPA P (Fig. 1). Initially, Soltanpour and Schwab (1977) added activated charcoal to the soil during the AB-DTPA nutrient extraction to remove the organic color from the extract. Activated charcoal use was discontinued because it adsorbed metal-DTPA complexes (Soltanpour and Workman, 1979).

III. Theoretical Principles of Inductively Coupled Plasma Optical Emission Spectrometry (ICPS)

For a detailed discussion on ICP spectrometry refer to Soltanpour et al. (1982a). In this paper, a brief explanation of how ICP operates and its advantages compared to other common methods of elemental analysis is given.

A. Principles of ICP Generation

Inductively coupled plasma is a highly ionized and hot Ar gas used to excite atoms and ions simultaneously in order to produce their specific radiation lines whose intensities are related to their atomic concentrations in the sample. The ICP is produced by passing initially ionized Ar gas through a quartz torch located inside a Cu coil connected to a radio frequency (RF) generator. The RF generator provides up to 3 kW forward power (in most commercial units) at a frequency of 27.1 MHz. The high-frequency currents flowing in the Cu coil generate oscillating magnetic fields whose lines of force are axially oriented inside the quartz tube and that follow elliptical closed paths outside the coil as shown schematically (Fig. 2) (Fassel and Kniseley, 1974). Electrons and ions passing through the oscillating electromagnetic field flow at high acceleration rates in closed annular paths inside the quartz tube space. The induced magnetic fields are time variable in their direction and strength, resulting in electron acceleration on each half cycle. Collisions between accelerated electrons (and ions) and ensuing unionized Ar gas cause further ionization. The collisions cause ohmic heating and give temperatures ranging from 6000 to 10,000°K (Fig. 3) (Fassel, 1977). The quartz torch has three concentric channels. The outer channel conducts Ar gas at about 17 L/min to the plasma

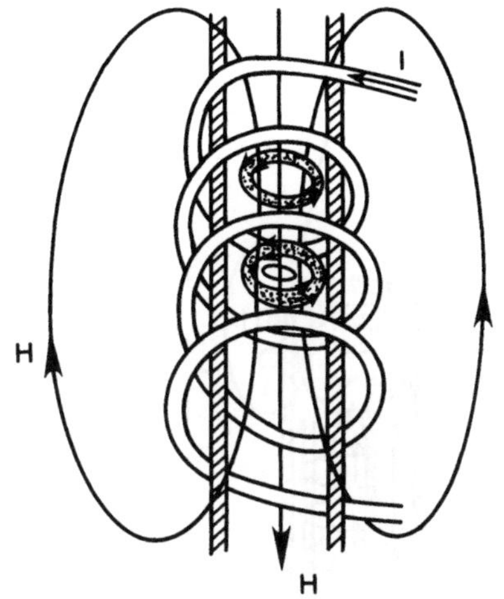

Figure 2. Magnetic fields (H) and eddy currents (shaded) generated by high-frequency currents (I) flowing through coil. (*Source:* Fassel and Kniseley, Inductively coupled plasmas. Anal. Chem. 46:1155A–1164A. © 1974 American Chemical Society.)

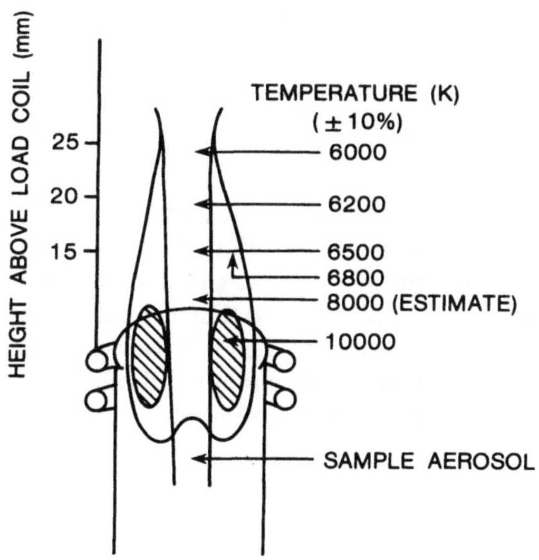

Figure 3. Temperatures in the plasma as measured by the spectroscopic slope method. (*Source:* Fassel, 1977, Pure and Appl. 49:1533–1545, Pergamon Press Ltd., Oxford.)

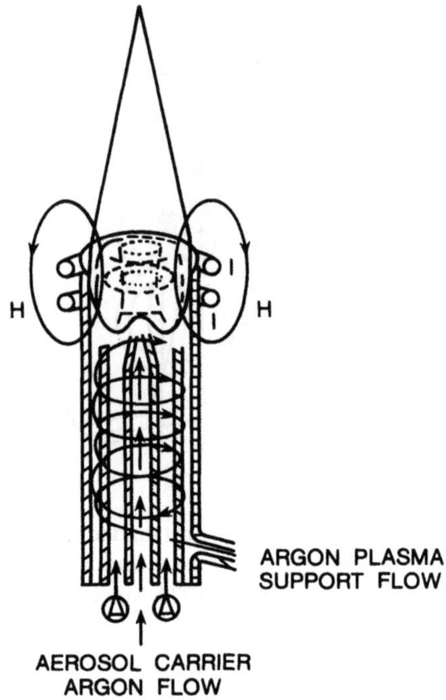

Figure 4. Typical quartz torch and inductively coupled plasma configuration. Flow A is auxilliary flow used for organic samples. (*Source:* Fassel and Kniseley, Inductively coupled plasmas. Anal. Chem. 46:1155A–1164A. © 1974 American Chemical Society.)

to sustain the plasma and to isolate the quartz tube from high temperatures. The innermost channel is for introduction of sample into the plasma. The middle channel conducts the auxiliary Ar gas at about 1 L/min to the plasma and is used with organic samples (Fig. 4) (Fassel and Knisely, 1974). The ICP has an annular or doughnut shape when it is viewed from above. The hole has a lower temperature than the doughnut and offers less resistance to the sample injection. The sample is injected into the plasma using Ar carrier gas at a rate of about 1 L/min.

B. Properties of ICP

The ICP generated, as discussed above, has unique physical properties that make it an excellent source for vaporization, atomization, and excitation of elements.

In the ICP, the analyte is vaporized, ionized, and atomized in the high temperature region of the plasma in the vicinity of the Cu coil (Fig. 4). The analyte emission is observed at a 15- to 20-mm height above the coil in an environment relatively low in background emission. The background consists of Ar lines and some weak band emissions from OH, NO, and CN molecules present in the plasma (Ward, 1978). By the time the decomposition products of the sample reach observation height, they have had a residence time of about 2 ms at temperatures ranging from about 8000 to 5000°K (Fassel, 1977). The residence time and temperature experienced by samples introduced into the plasma are about twice as large as those in the hottest flames, for example, $N_2O{-}C_2H_2$. The high temperature and residence time combination leads to complete sample vaporization and atomization in contrast to flames that require releasing agents for refractory compounds (Larson et al., 1975). Once the free compounds, atoms, and ions are formed, they are in a chemically inert environment in contrast to highly reactive combustion flame environments. Ionization interferences are generally negligible in an ICP. Self-absorption (a phenomenon responsible for the flattening of the standard curve at high analyte concentrations) is practically absent, which leads to a wide linear dynamic analytical range of three to five decades. No electrodes are used in ICP, and therefore contamination from them is absent.

C. Sample Introduction Systems*

1. Nebulizers

Nebulizers are devices used for the injection of the sample into the plasma. There are two general types of nebulizers, namely pneumatic and ultrasonic. The former uses the Venturi effect to draw sample solutions into the spray chamber. Larger solution droplets settle out in the spray chamber and drain off, and the finer aerosol droplets are transported into the plasma. In ultrasonic nebulizers, transducers are used to produce the sample aerosol. Ultrasonic nebulizers improve the detection limit of ICP spectrometers by one or two orders of magnitudes compared to pneumatic nebulizers (Olson et al., 1977). This is due to the production of finer sprays and improved sample delivery. The ultrasonic nebulizers are more expensive than pneumatic nebulizers and have a longer memory (effect of the previous sample on the readings of the sample being analyzed) than the latter.

The "cross-flow" type of pneumatic nebulizer was previously used in our laboratory. It has two capillary tubes positioned at right angles to each other and a small spacing between the two. One capillary carries Ar gas at about 1 L/min and the other sample solution. The "cross-flow" nebulizer clogs with high-salt solutions. Soltanpour et al. (1979b) diluted 1 M NH_4HCO_3–0.005 M DTPA soil extracts with 15 M HNO_3 to overcome clogging. However, presently a Legere teflon nebulizer**, consisting of a vertical V groove into which the AB-DTPA extract is pumped (Legere and Burgener, 1985) is used.

*This section was adapted from Soltanpour et al. (1982a) and was originally written by S.M. Workman.

**Distributed by Burtec Instrument Corporation, P.O.B. 235, Delmar, New York 12054, U.S.A.

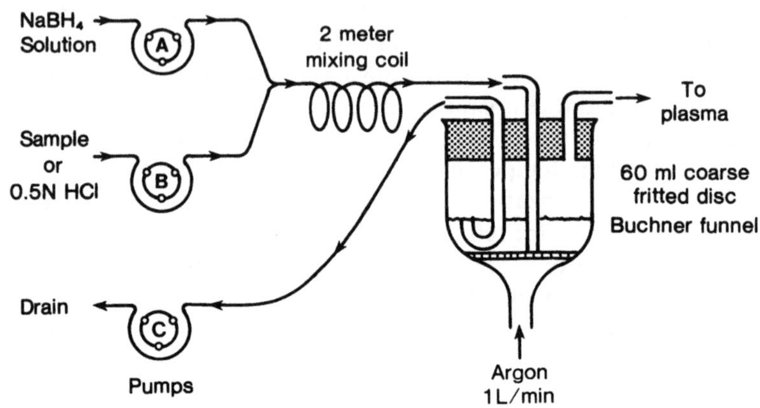

Figure 5. Workman hydride generator. (*Source:* Workman and Soltanpour, 1980, with permission from Soil Science Soc. Am. J.)

The solution is ruptured by Ar gas coming from a small hole in the groove. The use of Legere nebulizer eliminates the need for HNO_3 pretreatment.

2. Hydride and Mercury Vapor Generator

Certain elements when reduced by $NaBH_4$, form gases that can be directly introduced into the plasma. Arsenic, Sb, Bi, Se, and Te can be reduced to form hydrides. Also, Hg can be reduced to Hg vapor. This method of sample introduction greatly improves the detection limits of these elements compared with pneumatic nebulization due to an increase in sample concentration injected into plasma and a decrease in matrix effect. Thompson et al. (1978a,b) simultaneously determined As, Sb, Bi, Se, and Te by use of ICP spectrometry and a hydride generator. Workman and Soltanpour (1980) showed that by reducing As and Se to their hydrides and Hg to its vapor form using the hydride generator shown in Figure 5, and introducing these gases into the ICP, they could be quantitatively detected at 1.0, 0.5, and 0.5 μg/l, respectively.

D. Spectrometers*

Atoms of elements in a sample when excited emit light of characteristic wavelength with an intensity directly proportional to the element concentration. The light is focused on the entrance slit of the spectrometer to illuminate the diffraction grating, which separates the light into its component wavelengths or lines (spectrum). The spectral line of an analyte passes through the aperture of an exit slit and strikes a photomultiplier tube. Photomultiplier

*This section was adapted from Soltanpour et al. (1982a) and was originally written by S.M. Workman.

tubes produce signals directly proportional to the intensity of the spectral line. The signal is fed to the readout system, which is computer controlled, and displays either intensities or concentrations. The computer stores the intensities of standards and uses these data to calculate the concentrations of unknowns.

Two types of spectrometers are commonly used, direct-reading polychromators (direct readers) and scanning monochromators (Slavin, 1971). Some systems are equipped with both spectrometers.

Direct readers have a fixed concave grating. Fixed exit slits and photomultiplier tubes are located on the focal curve of the grating at specific wavelengths. Direct readers must be carefully adjusted so that all the spectral lines are centered in their respective exit slits. The fine adjustments are made using refractor plates located just before the exit slits. Once these refractor plates have been set, there is usually no need to readjust them. All the analytical lines are positioned in reference to a nonanalytical line (often produced by a Hg lamp). The assumption is made that if the reference line is properly aligned, then the other lines are in proper alignment, too. The reference spectrum is kept correctly positioned by adjusting another refractor plate located at the entrance slit.

Direct readers are designed to reduce the possibility of unwanted light reaching the photomultiplier tubes. The refractor plates used for fine alignment of the spectral lines also are filters that exclude stray light. The photomultiplier tube–exit slit assemblies are protected by light shields and the internal surfaces of the spectrometer are blackened to reduce reflections.

Scanning or sequential monochromators use a movable grating and a single photomultiplier tube to measure the intensities of the special lines. This allows the analyst to select any spectral line. The scanning is computer controlled and is fast and accurate.

Direct readers have the advantage of being faster and requiring a smaller sample volume compared with scanning monochromators. The disadvantage of direct readers is their fixed wavelengths. For laboratories engaged in both research and routine activities, a spectrometer with both a scanning monochromator and a polychromator is the best system to use.

The manufacturers of the spectrometers usually provide the required software (computer program) for the operation of the spectrometer. These programs enable the computer to do many tasks automatically. Through computer program commands, modern spectrometers are able to perform standardization, normalization of the standard solution readings, correction of the interelemental spectral interferences, printing of the data output, and so on. When a spectrometer is purchased such factors as computer size, available software, speed of the printer, automatic interelemental spectral interference corrections, and other computer-related factors should be considered in addition to the factors mentioned with regard to optical systems.

In summary, ICP spectrometers can analyze the AB-DTPA extracts simultaneously (or sequentially), accurately, and rapidly for many elements, thus allowing the scientists to evaluate soil fertility and potential elemental toxicity.

IV. Elements Extracted with Ammonium Bicarbonate-DTPA

A. Plant Nutrients

Presently, AB-DTPA is used to extract NO_3^-, P, K, B, Zn, Fe, Cu, and Mn from Colorado soils. Theoretically SO_4^{2-} is extracted from soils by dissolution and desorption action of AB-DTPA. We have not conducted research on SO_4^{2-} because its deficiency in Colorado soils is rare.

Calcium and Mg deficiencies in Colorado agricultural soils are also rare. However, exchangeable plus soluble Ca cannot be determined by AB-DTPA because Ca precipitates as $CaCO_3$. Magnesium carbonates may also precipitate. But, R.C. Severson of the U.S. Geological Survey Office in Denver, Colorado (personal communications) found a high correlation between AB-DTPA extractable Mg and 1 *M* ammonium acetate extractable Mg in some mine overburden samples. Hanlon and Johnson (1984) found a high correlation between AB-DTPA Mg and 1 *M* ammonium acetate Mg. It appears Mg minerals do not precipitate during the short shaking period of 15 min.

Boron deficiency in Colorado has not been documented. Gestring and Soltanpour (1984) evaluated the AB-DTPA soil test to determine available B and found that soil pH, organic matter content and clay content should also be considered along with extractable B in determining B uptake ($R^2 = 0.70$). Hot-water-soluble B alone gave an r^2 value of 0.87. However, for determination of B toxicity AB-DTPA test was as effective as the hot-water extraction method (Gestring and Soltanpour, 1987).

As mentioned before AB-DTPA extracts the exchangeable plus soluble K^+ (Soltanpour and Schwab, 1977). Hanlon and Johnson (1984) found a very high correlation between K extracted by AB-DTPA and 1 *M* ammonium acetate.

Nitrate (NO_3^-) is an important analyte in the western U.S. to determine residual fertilizer N in soils (Hergert, 1987). The AB-DTPA extracting solution, like any water-based solution extracts NO_3^- from soils and eliminates the necessity for use of a separate extracting solution for nitrates (Soltanpour and Schwab, 1977). A flow injection autoanalyzer (FIA) is used to determine nitrates in AB-DTPA extracts at a rate of more than 80 samples per hour (Workman et al., 1988).

Labile P is extracted with AB-DTPA. Greenhouse experiments using alfalfa (*Medicago sativa*) as the test plant have shown that the AB-DTPA, 0.5 *M* $NaHCO_3$ (Olsen et al., 1954), ethylenediamilnetetraacetic acid (EDTA) (Ahmed and Islam, 1975; Onken et al., 1980), and 0.01 *M* $CaCl_2$ extracting solutions were equally effective in separating deficient from nondeficient soils (Fig. 6) (Labhsetwar and Soltanpour, 1985). Also field studies in Colorado, using wheat (*Triticum aestivum*) as the test plant have shown that these same

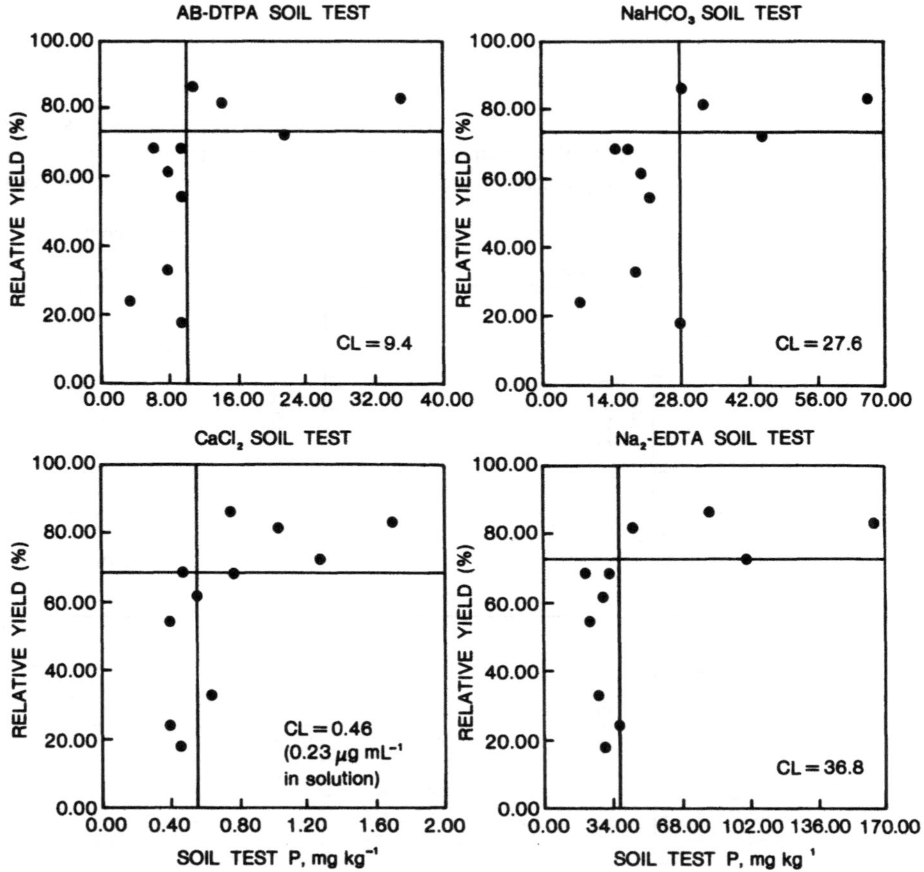

Figure 6. Graphical method of Cate and Nelson as modified by the analysis of variance method of Nelson and Anderson for critical levels (CL) in soils. (*Source:* Labhsetwar and Soltanpour, 1985, with permission from Soil Science Soc. Am. J.)

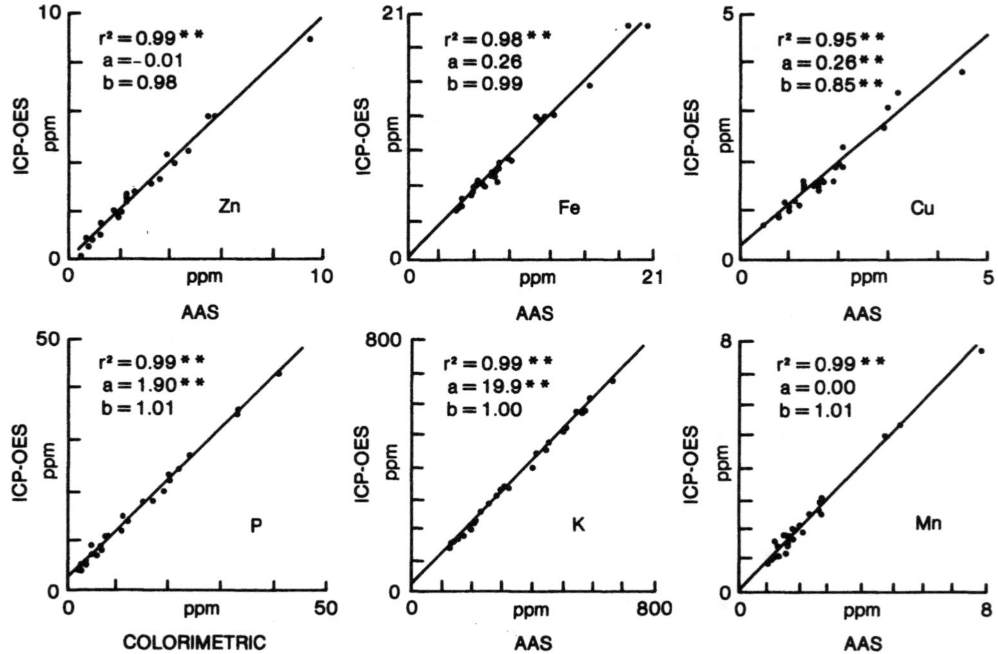

Figure 7. Regressions of ICPS values on atomic absorption spectrophotomety values for cations and on colorimetric P results. **Indicates significance at the 1% level of probability. (*Source:* Soltanpour et al., 1979b, with permission from Soil Sci. Soc. Am. J.)

extracting solutions were equally effective in separating deficient from nondeficient soils (Labhsetwar, 1984). A critical soil P level (point below which response to fertilizer is expected) of 6.7 mg/kg extracted by AB-DTPA was found by the Cate–Nelson method (Cate and Nelson, 1971) for dryland wheat grown in Eastern Colorado. Rodriquez et al. (1989) calibrated the AB-DTPA soil test for dryland millet (*Panicum miliaceum* L.) grown in Eastern Colorado and found a critical level of 6 mg P/kg (0–15-cm sample depth). In calcareous soils of Morocco, 0.5 *M* $NaHCO_3$, 0.01 *M* $CaCl_2$, and AB-DTPA tests were equally effective in separating P-deficient from nondeficient soils (Azzaoui et al., 1989). Hanlon and Johnson (1984) found a good correlation between AB-DTPA P, and Bray and Kurtz (0.015 *M* HCl plus 0.03 *M* NH_4F) P in acid and calcareous Oklahoma soils. The ICP spectrometer determines both inorganic and organic P in AB-DTPA extracts (Soltanpour et al., 1979b; Labhsetwar and Soltanpour, 1985). The amount of organic P in AB-DTPA extracts was about 1.8 ppm in Colorado soils. The relationship between ICP and colorimetric (inorganic) P is shown in Figure 7 (Soltanpour et al., 1979b). There is a need for research in high-organic soils to determine the importance of AB-DTPA extractable organic P and its availability to plants.

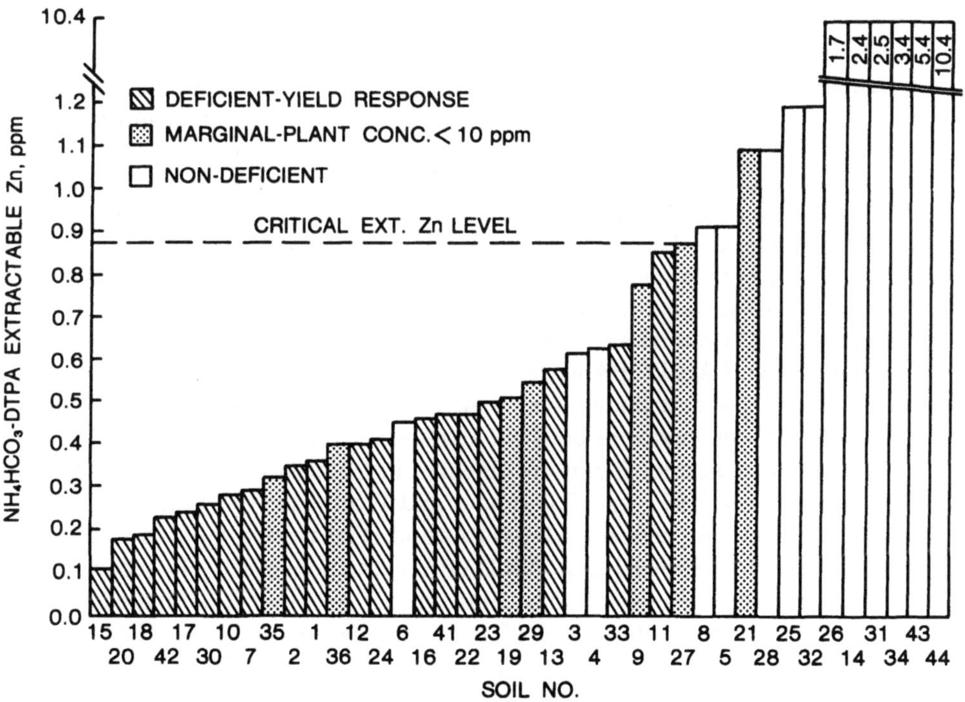

Figure 8. Corn response of 40 Colorado soils to Zn as a function of NH_4HCO_3–DTPA soil test levels. (*Source:* Havlin and Soltanpour, 1981, with permission from Soil Sci. Soc. Am. J.)

Zinc levels in AB-DTPA extracts separated deficient from nondeficient soils for greenhouse-grown corn in 37 out of 40 cases (Fig. 8) (Havlin and Soltanpour, 1981). The critical Zn level was determined to be 0.9 mg/kg of soil. Barbarick and Workman (1987) evaluated the AB-DTPA soil test for determining Zn availability in sludge-treated soils and obtained r values >0.90 between Zn concentrations in Swiss chard (*Beta vulgaris* L.) leaves and soil extractable Zn for both a calcareous soil and an acid soil (Fig. 9). Rohman and Cox (1988) found a high correlation (0.89) between AB-DTPA soil test values and Zn uptake by corn plants. Havlin and Soltanpour (1984) showed that AB-DTPA extraction measures the residual effect of $ZnSO_4$. Boon (1985) showed that total Zn level and AB-DTPA Zn levels were highly correlated ($r = 0.89$) in contaminated garden soils.

Havlin and Soltanpour (1981, 1982) separated Fe-deficient from nondeficient soils for sorghum (*Sorghum bicolor*) using the AB-DTPA soil test. A critical level of 4.8 ppm was determined for AB-DTPA Fe (Fig. 10). Residual levels of iron chelate (FeEDDHA) fertilizer was effectively measured with AB-DTPA (Havlin and Soltanpour, 1984).

Copper deficiency has not been documented in Colorado. Soltanpour and Schwab (1977) and Soltanpour and Workman (1979) found a high degree of correlation between DTPA- and AB-DTPA-extractable Cu. Tills and Alloway (1983) found a low correlation between Cu concentration of wheat plants and AB-DTPA Cu. However, the R^2 changed from 0.14 to 0.47 when other soil parameters (Mn-oxide, Fe-oxide, clay, pH, and loss on ignition) were included. Cox (1987) in a review paper showed that soil properties other than the extractable level of micronutrients should be considered when soil test results are interpreted. Makarim and Cox (1983) found an R^2 value of 0.47 for AB-DTPA Cu and percent maximum yield using the Cate–Nelson analysis of variance method. This value was lower than the value for the Mehlic 1 test ($R^2 = 0.70$). Rohman and Cox (1988) found a relatively low correlation (0.47) between AB-DTPA-extractable Cu and wheat Cu uptake. Barbarick and Workman (1987) found an r value of 0.92 between Cu concentrations in Swiss chard leaves and AB-DTPA extracts of a sludge-treated calcareous soil but an r value of 0.64 for a sludge-treated acid soil (Fig. 9). For Cu, it seems that the AB-DTPA test performs better in alkaline soils than in acid soils.

Manganese deficiency is not documented in Colorado. However, Soltanpour and Schwab (1977) and Soltanpour and Workman (1979) found a high degree of correlation between AB-DTPA- and DTPA-extractable Mn. Schuman et al. (1980) showed that AB-DTPA extraction effectively separated Mn-deficient from -sufficient soils (Figs. 11, 12), and that it compared well with other extracting solutions, DTPA, NH_4Cl–NH_4F, and HCl–H_2SO_4. A critical Mn level of 0.4 ppm was determined for the AB-DTPA soil test. Rohman and Cox (1988) found a high correlation (0.81) between Mn extracted with AB-DTPA and Mn uptake by soybean plants. Shepard and Bates (1982) determined the relationship between Mn uptake and other parameters of 70

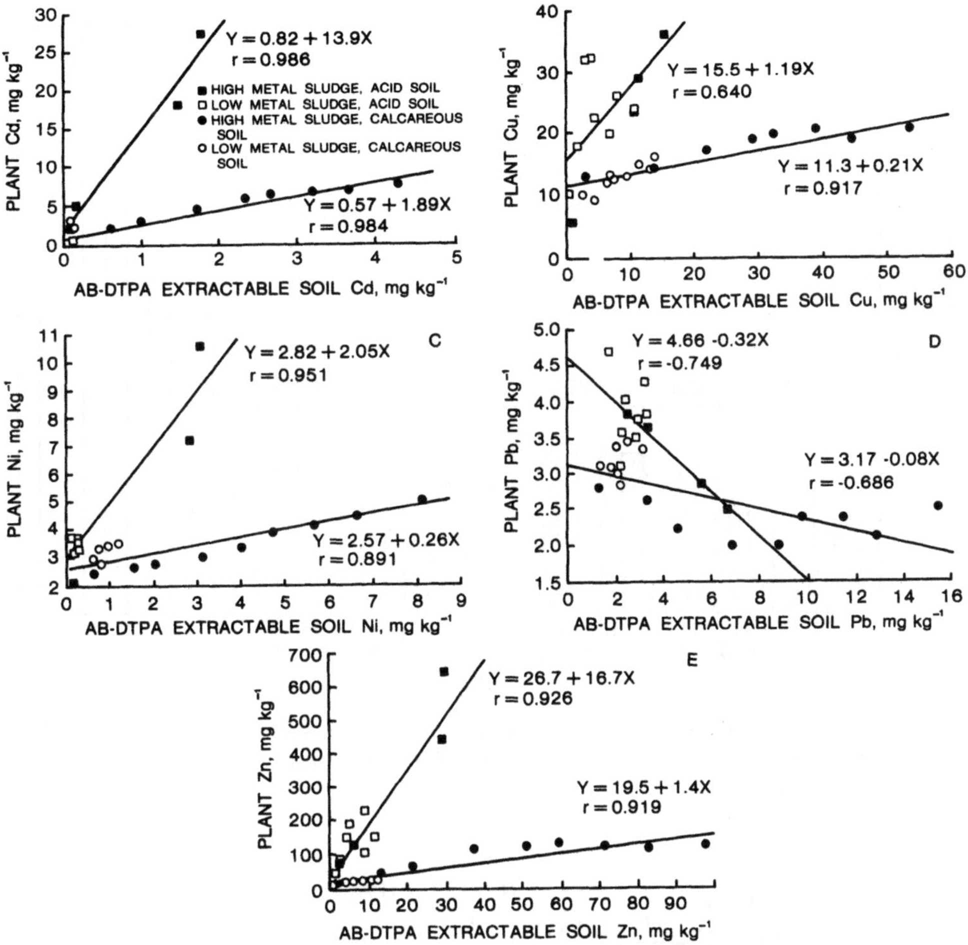

Figure 9. Metal concentration in leaves of Swiss chard versus AB-DTPA levels. (*Source:* Barbarick and Workman, 1987, with permission from J. Envir. Quality.)

Canadian soils. Manganese extracted by AB-DTPA, soil pH, their interaction, and yield entered significantly into the interpretation. The AB-DTPA test compared well with other tests for Mn.

Molybdenum deficiency has not been documented in Colorado. However, Mo toxicity to animals (molybdenosis) is of concern in areas contaminated with Mo due to mining activities. Boon and Soltanpour (1983) and Boon (1984) showed that there is a good correlation between Mo in plants with their respective AB-DTPA extraction levels in soil. Pierzynski and Jacobs (1986) found that the AB-DTPA extractant was more effective in predicting Mo concentrations in plant tissue than Tamm's acid ammonium oxalate with

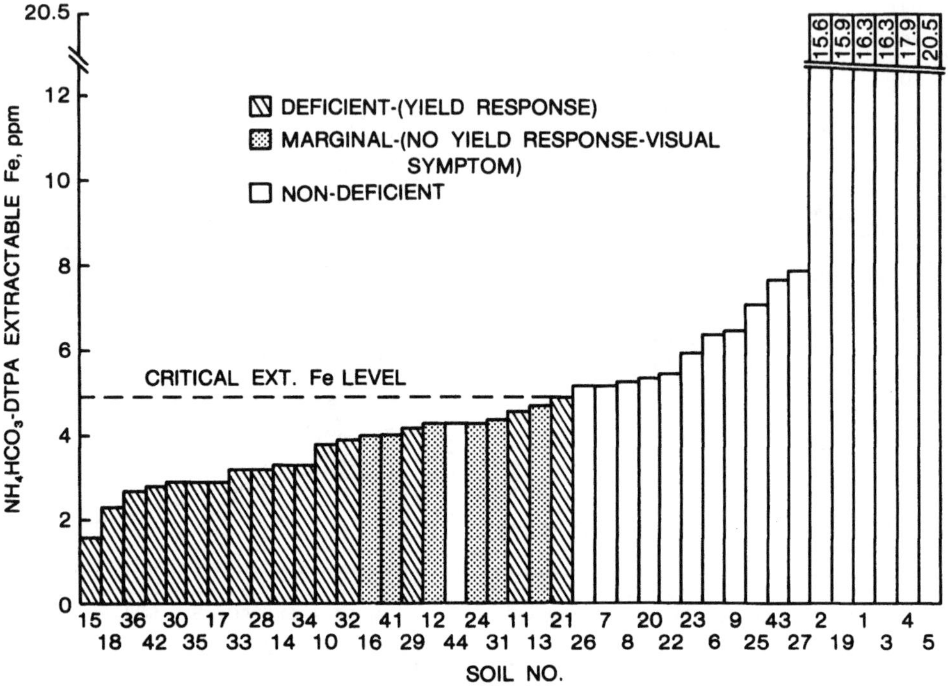

Figure 10. Sorghum response of 40 Colorado soils to Fe as a function of NH_4HCO_3–DTPA soil test levels. (*Source:* Havlin and Soltanpour, 1981, with permission from Soil Sci. Soc. Am. J.)

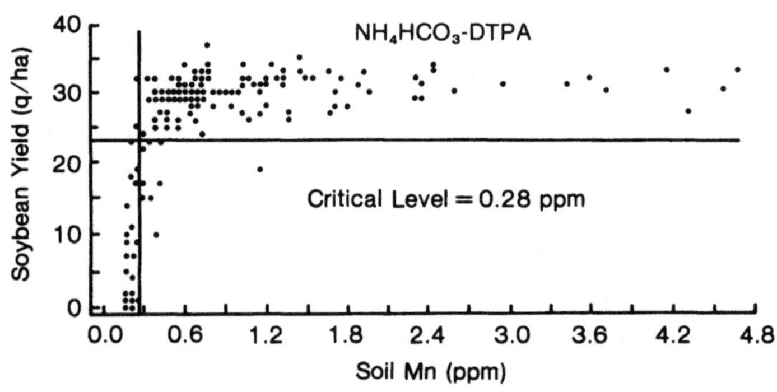

Figure 11. Cate–Nelson plot for soybean yield versus AB-DTPA-extractable soil Mn for 1977 with the two experiments combined. (*Source:* Shuman et al., 1980, with permission from Soil Sci. Soc. Am. J.)

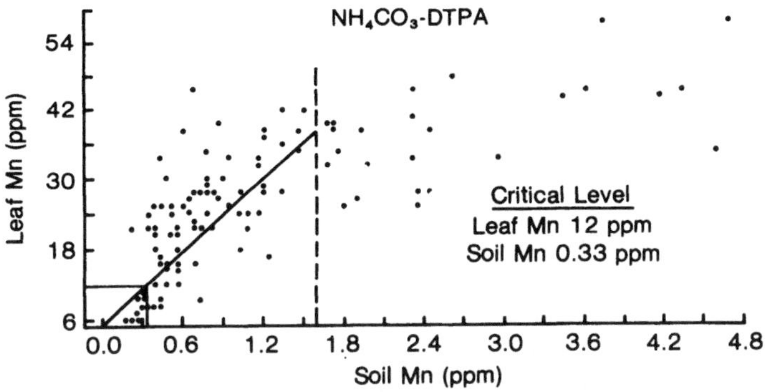

Figure 12. Soybean leaf Mn versus AB-DTPA-extractable soil Mn for 1977 to determine the soil critical deficiency level from the leaf critical level. (*Source:* Shuman et al., 1980, with permission from Soil Sci. Soc. Am. J.)

sludge Mo additions and is more effective with the sludge source Mo as compared to the inorganic source Mo. These authors showed that soil pH should be considered along with AB-DTPA Mo for determining Mo availability to plants.

From the literature cited above, it seems that AB-DTPA soil test determines availability indexes for NO_3–N, P, K, Zn, Fe, Cu, Mn, and Mo. It can determine availability of B if soil pH and organic matter content are used to help interpret the results. For Mo, inclusion of pH improves interpretation of results. For Cu, the AB-DTPA test did perform better in alkaline soils than it did in acid soils. In acid soils, AB-DTPA can be used to determine an availability index for Mg. The use of AB-DTPA test for SO_4^{2-} extraction should be evaluated.

B. Other Elements

Exchangeable plus soluble Na can be determined with AB-DTPA. Therefore, sodic soils can be diagnosed with this test if the cation exchange capacity of soil and the amount of soluble Na are known. Boon (1985) showed that total Pb and Cd and AB-DTPA-extractable levels were highly correlated ($r = 0.92$) in mine-spoil-contaminated garden soils. Boon and Soltanpour (1983) and Boon (1985) showed that AB-DTPA-extractable levels of Pb and Cd in soils contaminated with silver mine-spoil material were highly correlated with their respective plant concentrations in spinach and lettuce leaves (Fig. 13). Their results showed that broccoli and cabbage (*Brassica oleracea*) leaves did not accumulate Pb even when grown in a soil that was highly contaminated with Pb. They showed that the AB-DTPA test is a good indicator of soil contamination with Pb, Cd. Zn, and Mo. Barbarick and Workman (1987) showed a high correlation between AB-DTPA-extractable Ni and Cd levels, and

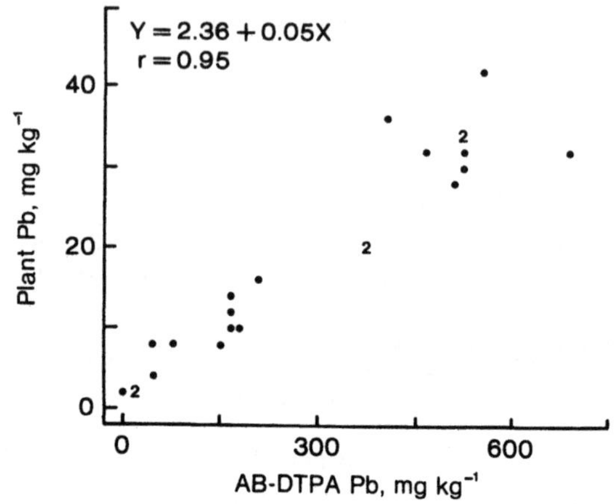

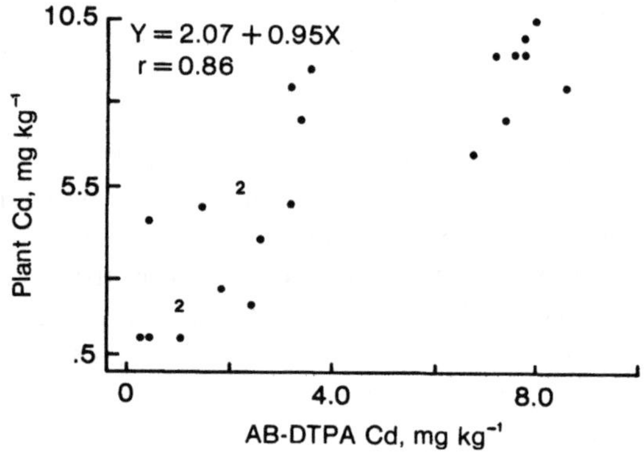

Figure 13. Regression of AB-DTPA extracts on plant concentrations of Pb and Cd. (*Source:* Boon, 1985.)

concentration of these metals in leaves of Swiss chard grown in sludge-treated soils (Fig. 9). The latter authors showed a negative relationship between extractable Pb and plant Pb concentration (Fig. 9). The negative relationship is probably be due to the positive effect of sewage sludge on dry matter yield (dilution effect).

Selenium-contaminated soils pose a hazard for wildlife and humans. Therefore, the Se availability should be determined in suspected areas. A greenhouse study of Soltanpour and Workman (1980) showed a high degree of correlation between Se uptake by alfalfa plants (*Medicago sativa*) and AB-DTPA-extractable Se levels. Soltanpour et al. (1982b) showed a high degree of cor-

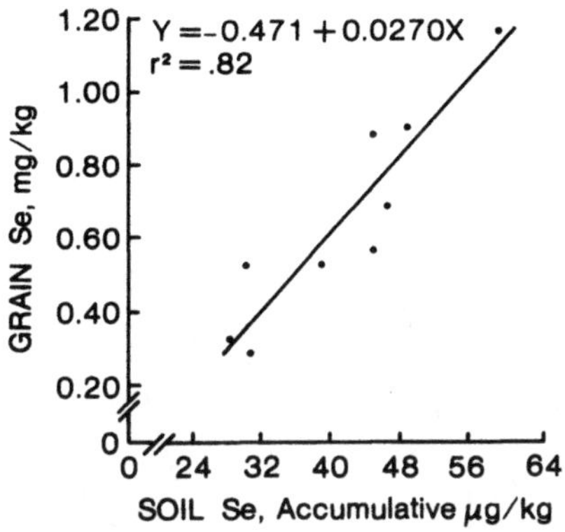

Figure 14. Grain selenium as a function of soil NH_4HCO_3–DTPA-extractable Se, 0- to 90-cm depth. (*Source:* Soltanpour et al., 1982b, with permission from Soil Sci. Soc. Am. J.)

relation ($r = 0.90$) between Se level in wheat (*Triticum aestivum*) grain grown in eastern Colorado and soil-AB-DTPA-extractable Se if deep samples (0–90 cm) were taken (Fig. 14).

The relationship between As extracted by AB-DTPA and plant As has not been evaluated. However, theoretically HCO_3^- should desorb or solubilize arsenates and arsenites. Woolson et al. (1971) showed that 0.5 *M* $NaHCO_3$ extracted plant available As.

Future research is needed to evaluate the AB-DTPA test for determining the bioavailability of Cr, Hg, Sb, Be, Co, F, Ag, Ti, V and other elements of interest to agronomists, soil scientists, botanists, and ecologists.

V. Analysis of Ammonium Bicarbonate-DTPA Extracts Using Inductively Coupled Plasma Spectrometry

Soil samples should not be kept at field-moisture level and at temperatures that change the levels of elements of interest. For example, air drying of soil is recommended to avoid changes in NO_3–N (Westfall et al., 1978). Soil samples should not be oven dried as this results in an increase in the extractability of Fe, Zn, Mn, and Cu (Khan and Soltanpour, 1978; Leggett and Argyle, 1983). The degree of grinding should be standardized as it affects the extractability of Fe, Zn, and probably other metals (Fig. 15) (Soltanpour et al., 1976, 1979a). The vigor of shaking also affects the amount of extractable elements (Soltanpour et al., 1976). Therefore, all of the above factors should be standardized to allow comparison of data and use of similar critical levels. It is important to use grinders that do not contaminate the soil with elements

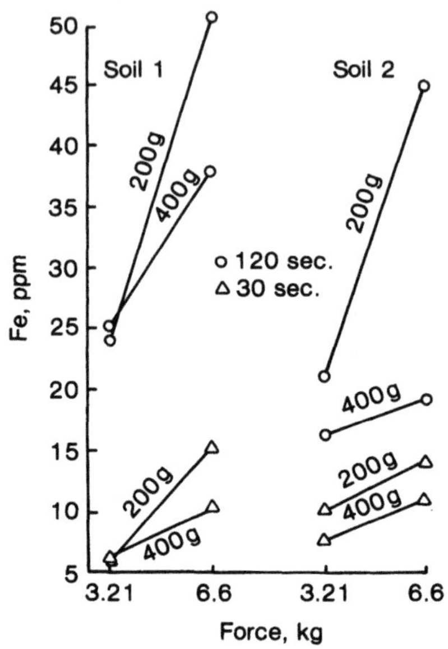

Figure 15. Effect of force and time of grinding and size of sample on extractable Fe. (*Source:* Soltanpour et al., 1976, reprinted from Commun. Soil. Sci. Plant Anal. 7:797–821 by courtesy of Marcel Dekker, Inc.).

of interest. For practical purposes, soils are air dried and ground before analysis. However, air drying and grinding are not recommended for Cr (Reisenauer, 1982). Moist soils can be analyzed, but interpretation of results cannot be based on data obtained with air-dry soils. For example, Khan and Soltanpour (1978) and Khan and Banwart (1979) found that the level of extractable Fe, Zn, Cu, and Mn decreased with air drying. The latter authors indicated that these changes were nonmicrobial in nature.

The AB-DTPA extracting solution should be made daily (Table 1) or alternatively should be covered with mineral oil to prevent CO_2 evolution and change in pH.

Ten grams of air-dry soil are placed in a 125-ml polyethylene conical flask and 20 ml of the AB-DTPA solution is introduced into the flask. The flasks are placed on a reciprocal shaker and shaken at 180 oscillations per minute for 15 min (Table 2). After shaking, the extracts are filtered and analyzed directly by an ICP spectrometer. Standards are made in AB-DTPA solution as shown in Table 3 (Workman et al., 1988). Ultratrace levels of As, Se, and Hg are determined by a hydride generator (Workman and Soltanpour, 1980).

Interelemental spectral interferences are usually absent in AB-DTPA extracts, however, if one analyzes highly contaminated soils, then one should determine if there is any interference present and correct for it (Soltanpour et al., 1982a).

If the ICP spectrometer is equipped with a high-salt nebulizer, such as

Table 3. ICP standards

Elements	Soil µg/g	Concentradtion in AB-DTPA high standard µg/ml	ml stock solution per 500 ml of AB-DTPA high standard	Stock solution µg/ml
P	20	10	5	1000[a]
K	200	100	25	2000[b]
Zn	10	5	2.5	1000[c]
Fe	10	5	2.5	1000[c]
Mn	10	5	2.5	1000[c]
Cu	10	5	2.5	1000[c]
Pb	2	1	0.5	1000[c]
Cd	2	1	0.5	1000[c]
Ni	2	1	0.5	1000[c]
Mo	2	1	0.5	1000[c]
B	2	1	0.5	1000[c]
As	2	1	0.5	1000[c]
Se	2	1	0.5	1000[c]
Ca	200	100	25	2000[b]
Mg	200	100	25	2000[b]
Na	200	100	25	2000[b]
Cr	2	1	0.5	1000[c]
Sr	2	1	0.5	1000[c]
Ba	2	1	0.5	1000[c]
Al	2	1	0.5	1000[c]
Ti	2	1	0.5	1000[c]
Hg	2	1	0.5	1000[c]

Note: Low standard = AB-DTPA blank. High standard for P, K, Zn, Fe, Mn, Cu, Pb, Cd, Ni, Mo, B, As, Se, Ca, Ma, Na.

[a] P: 100 µg per ml = 1.85 g $NH_4H_2PO_4$ per 500 ml.

[b] Ca, Mg, Na, K: 2000 µg per ml = 2.500 g $CaCO_3$, 1.000 g Mg metal, 2.5413 g NaCl, and 1.9066 g KCl dissolved in 500 ml. Dissolve the Mg metal and $CaCO_3$ first in 5 ml concentrated HCl, then dissolve NaCl and KCl, and bring to volume.

[c] Commercially avaiable 1000 µg per ml atomic absorption standards.

Legere nebulizer, then no pretreatment of AB-DTPA extract is necessary. But, an acid pretreatment is necessary to eliminate carbonate species that will precipitate at the tip of cross-flow nebulizers and cause erratic readings.

VI. Interpretation of Results

Interpretation for AB-DTPA-extractable plant nutrients and some other elements in soils are shown in Tables 4 and 5. In using these tables one should be aware that these interpretations should be regarded as general guidelines and should be verified for different crops in different regions.

Table 4. Interpretation for AB-DTPA-extractable plant nutrients in soils (0–30 cm)

Element	Low	Medium	High	Very high
	mg/kg			
No_3–N	0–10	11–20	21–30	> 30
P (sandy, loamy sand and sandy loam soils)	0–5	6–10	11–15	>15
P (loam soils)	0–2.5	3–5	5.5–8	> 8
P (clay soils)	0–1.5	2–3	3.5–5	>5
K	0–60	61–120	121–180	> 180
Zn	0–0.9	1.0–1.5	> 1.5	—
Fe	0–3.0	3.1–5.0	> 5.0	—
Mn	0–0.5	0.6–1.0	> 1.0	—
Cu	0–0.2	0.3–0.5	>0.5	—

Table 5. Toxic levels for AB-DTPA-extractable B, Pb, Cd, Se, and As in soils

Element	Potentially toxic range[a]
B	> 5[b]
Pb	> 100[c]
Pb	> 20[d]
Cd	> 1.5[e]
Se	> 0.1[f]
As	> 2.5[g]
Mo	> 1.7[h]
Zn	> 70[i]

[a] Tentative.

[b] For susceptible plants growing on the suspect soil. (*Source:* Gestring and Soltanpour, 1987.)

[c] For human ingestion of soil. (*Source:* Boon, 1985.)

[d] For susceptible plants. (*Source:* calculated from Dang et al., 1990 and Boon, 1985.)

[e] For susceptible plants. (*Source:* calculated from Dang et al., 1990 and Boon, 1985.)

[f] For animals grazing on the plants grown on the suspect soil. (*Source:* Soltanpour and Workman, 1980.)

[g] For susceptible plants growing on the suspect soil. (*Source:* Calculated from Woolson et al. 1971, assuming the AB-DTPA tests extracts 0.5 as much As from soils as $NaHCO_3$ test.)

[h] For ruminants grazing on the plants grown on the suspect soil (*Source:* Vlek, 1975, assuming AB-DTPA extracts the same amount of Mo as 1 *M* $(NH_4)_2CO_3$.)

[i] For susceptible plants. (*Source:* Calculated from Dang et al., 1990, Boon, 1985, and Soltanpour and Workman, 1979.)

VII. Conclusions

Literature data and our experience since 1977 show that AB-DTPA soil test in conjunction with ICP optical emission spectrometry is a very efficient tool for determination of bioavailability indices of essential and nonessential elements. By use of the AB-DTPA–ICP it was possible to detect Zn, Pb, and Cd contamination in garden soils of old mining towns in western Colorado during a routine soil fertility evaluation. As a result, these towns were earmarked for cleanup by the U.S. EPA. The AB-DTPA–ICP method is recommended for soils, mine overburden, and mine spoils to evaluate them for soil fertility and potential elemental toxicity to plants and their consumers and to animals and humans who may inhale or injest the contaminated material. Total elemental analysis of soils can determine contamination, but AB-DTPA test is more rapid and more economical. The AB-DTPA test needs evaluation for determination of bioavailability of soil S, Cr, Hg, Sb, Be, Co, F, Ag, Ti, V and possibly other elements of interest to soil scientists, agronomists, and environmental scientists.

References

Ahmed, B., and A. Islam. 1975. The use of sodium EDTA as an extractant for determining available phosphate in soil. *Geoderma* 14:261–265.

Azzaoui, A., R.G. Hanson and P.N. Soltanpour. 1989. Wheat P requirements on calcareous Moroccan soils: I. A comparison of Olsen, Soltanpour and Schwab and $CaCl_2$ soil tests. *Commun. Soil Sci. Plant Anal.* 20:869–891.

Barbarick, K.A. and S.M. Workman. 1987. Ammonium bicarbonate-DTPA and DTPA extractions of sludge-amended soils. *J. Environ Quality*. 16:125–130.

Boon, D.Y. 1984. The ammonium bicarbonate-DTPA soil test for determination of plant-available Pb, Cd, Ni, and Mo in mine tailings and contaminated soils. *Sixth High Altitude Revegetation Workshop, Colorado State Univ.* Information Series No. 53.

Boon, D.Y. 1985. Lead, cadmium and zinc: Evaluation of contamination using the NH_4HCO_3–DTPA soil test. Colorado State Univ., Fort Collins (M.Sc. Thesis).

Boon, D.Y. and P.N. Soltanpour. 1983. The ammoniumbicarbonate-DTPA soil test for determination of plant available lead, cadmium and molybdenum in mine tailings and contaminated soils. *In Agronomy Abstracts*, ASA, Madison, WI, pp. 29.

Cate, R.B., Jr. and L.A. Nelson. 1971. A simple statistical procedure for partitioning soil test correlation data into two classes. *Soil Sci. Soc. Am. Proc.* 35:658–660.

Cox, F.R. 1987. Micronutrient soil tests: Correlation and calibration. *In* J.R. Brown (Ed.). *Soil Testing: Sampling, Correlation, Calibration, and Interpretation.* SSSA Spec. Publ. 21, SSSA, Madison, WI, pp. 97–118.

Dang, Y.P., R. Chhabra, and K.S. Verma. 1990. Effect of Cd, Ni, Pb, and Zn on growth and chemical composition of onions and fenugreek. *Commun. Soil Sci. Plant Anal.* 21:717–735.

Fassel, V.A. 1977. Current and potential applications of inductively coupled plasma (ICP) atomic emission spectroscopy (AES) in the exploration, mining, and processing of materials. *Pure Appl. Chem.* 49:1533–1545.

Fassel, V.A. and R.N. Kniseley. 1974. Inductively coupled plasmas. *Anal. Chem.* 46:1155A–1164A.

Gestring, W.D. and P.N. Soltanpour. 1984. Evaluation of the ammonium bicarbonate-DTPA soil test for assessing boron availability to alfalfa. *Soil Sci. Soc. Am. J.* 48:96–100.

Gestring, W.D. and P.N. Soltanpour. 1987. Comparison of soil tests for assessing boron toxicity to alfalfa. *Soil Sci. Soc. Am. J.* 51:1214–1219.

Hanlon, E.A. and G.V. Johnson. 1984. Bray/Kurtz, Mehlich III, AB/DTPA and ammonium acetate extraction of P, K, and Mg in four Oklahoma soils *Commun. Soil Sci. Plant Anal.* 15:277–294.

Havlin, J.L. and P.N. Soltanpour. 1981. Evaluation of the NH_4HCO_3–DTPA soil test for iron and zinc. *Soil Sci. Soc. Am. J.* 45:70–75.

Havlin, J.L. and P.N. Soltanpour. 1982. Greenhouse and field evaluation of the NH_4HCO_3–DTPA soil test for Fe. *J. Plant Nutr.* 5:769–783.

Havlin, J.L. and P.N. Soltanpour. 1984. Changes in NH_4HCO_3–DTPA-extractable zinc and iron as affected by various soil properties. *Soil Sci.* 137:188–193.

Hergert, G.W. 1987. Status of residual nitrate nitrogen soil tests in the United States of America. *In* J.R. Brown (Ed.). *Soil Testing; Sampling, Correlation, Calibration and Interpretation.* SSSA Spec. Publ. 21, SSSA, Madison, WI, pp. 73–88.

Khan, A. and P.N. Soltanpour. 1978. Effect of wetting and drying on DTPA-extractable Fe, Zn, Mn, and Cu in soils. *Commun. Soil Sci. Plant Anal.* 9:193–202.

Khan, A. and W.L. Banwart. 1979. Effect of incubation and microbial inhibition at field moisture capacity on changes in DTPA-extractable Fe, Zn, and Cu in soils of varying pH. *Commun. Soil Sci. Plant Anal.* 10:613–622.

Labhsetwar, V.K. 1984. Evaluation of the NH_4HCO_3–DTPA soil test for phosphorus. Ph.D. diss. Colorado State Univ., Fort Collins (Diss. Abstr. 84-27859).

Labhsetwar, V.K. and P.N. Soltanpour. 1985. A. comparison of NH_4HCO_3–DTPA. $NaHCO_3$, $CaCl_2$, and Na_2–EDTA soil tests for phosphorus. *Soil Sci. Soc. Am. J.* 49:1437–1440.

Larson, G.F., V.A. Fassel, R.H. Scott and R.N. Kniseley. 1975. Inductively coupled plasma-optical emission analytical spectrometry. A study of some interelemental effects. *Anal. Chem.* 47:238.

Lauer, D.A. 1971. Evaluation of plant available Zn by the DTPA soil test, 0.1 *N* HCI extraction, and labile Zn measurements. Ph.D. Thesis. Colorado State Univ. (Libr. Congr. Card no. Mic. 72-16, 114). Univ. Microfilms, Ft. Collins, Colo. (Diss. Abstr. 31-6157-B).

Legere, G. and P. Burgener. 1985. Elimination of high-salt interference effects caused by lithium metaborate using a new teflon high-salt rebulizer. *ICP Information Newsl.* 11:447–456.

Leggett, G.E. and D.P. Argyle. 1983. The DTPA-extractable iron, manganese, copper, and zinc from neutral and calcareous soils dried under different conditions. *Soil Sci. Soc. Am. J.* 47:518–522.

Lindsay, W.L. 1974. Role of chelation in micronutrient availability. *In* E.W. Carson (Ed.). *The Plant Root and its Environment.* Univ. Press of Virginia, Charlottesville, pp. 507–524.

Lindsay, W.L. 1979. *Chemical Equilibria in Soils.* Wiley-Interscience, New York.

Lindsay, W.L., J.F. Hodgson, and W.A. Norvell. 1967. The physico-chemical equilibrium of metal chelates in soils and their influence on the availability of micronutrient cations. *In* G.V. Jacks (Ed.). *Soil Chemistry and Fertility.* Int. Soc. Soil

Sci., Trans. Comm. II, IV (Aberdeen, Scotland). Univ. of Aberdeen, 1966, pp. 305–316.

Lindsay, W.L. and W.A. Norvell. 1969. Equilibrium relationships of Zn, Fe, Ca, and H with EDTA and DTPA in soils. *Soil Sci. Soc. Am. Proc.* 33:62–68.

Lindsay, W.L. and W.A. Norvell. 1978. Development of a DTPA soil test for zinc, iron, manganese, and copper. *Soil Sci. Soc. Am. J.* 42:421–428.

Makarim, A.K. and F.R. Cox. 1983. Evaluation of the need for copper with several soil extractants. *Agron. J.* 10:903–909.

Nesse. P., J. Grava, and P.R. Bloom. 1988. Correlation of several tests for phosphorus with resin-extractable phosphorus for 30 alkaline soils. *Commun. Soil Sci. Plant Anal.* 19:675–689.

Norvell, W.A. 1972. Equilibria of metal chelates in soil solution. *In* J.J. Mortvedt et al. (Ed.). *Micronutrients in Agriculture.* Am. Soc. of Agron., Inc., Madison, WI, pp. 115–138.

Olson, K.W., W.J. Haas, Jr., and V.A. Fassel. 1977. Multielement detection limits and sample nebulization efficiencies of an improved ultrasonic nebulizer and a conventional pneumatic nebulizer in inductively coupled plasma-atomic emission spectrometry. *Anal. Chem.* 49:632–637.

Olsen, S.R., C.V. Cole, F.S. Watanabe and L.A. Dean. 1954. Estimation of available P in soils by extraction with $NaHCO_3$. USDA Circ. 939. U.S. Government Printing Office, Washington, D.C.

Onken, B.A., R. Matheson and E.J. Williams. 1980. Evaluation of EDTA-extractable phosphorus as a soil test procedure. *Soil Sci. Soc. Am. J.* 44:783–786.

Pierzynski, G.M. and L.W. Jacobs. 1986. Extractability and plant availability of molybdenum from inorganic and sewage sludge sources. *J. Environ. Qual.* 15:323–326.

Reisenauer, H.M. 1982. Chromium. *In* A.L. Page (Ed.). *Methods of Soil Analysis*, Part 2, 2d ed. *Agronomy* 9:337–344.

Rodriguez, J.B., G.A. Peterson, and D.G. Westfall. 1989. Calibration of nitrogen and phosphorus soil tests with yield of proso millet. *Soil Sci. Soc. Am. J.* 53:1737–1741.

Rohman, P.C. and F.R. Cox. 1988. Evaluation of the modified Olsen extracting reagent for copper, zinc, and manganese. *Commun. Soil Sci. Plant Anal.* 19:1859–1870.

Rule, J.H. and E.R. Graham. 1976. Soil-labile pools of Mn, Fe, and Zn as measured by plant uptake and DTPA equilibrium. *Soil Sci. Soc. Am. Proc.* 40:853–857.

Shepard, S.C. and T.E. Bates. 1982. Selection of a soil extraction and a multiple regression model to predict plant available manganese. *Commun. Soil Sci. Plant Anal.* 13:1095–1113.

Shuman, L.M., F.C. Boswell, K. Ohki, M.B. Parker, and D.O. Wilson. 1980. Critical soil manganese deficiency levels for four extractants for soybeans grown in sandy soil. *Soil Sci. Soc. Am. J.* 41:1021–1025.

Slavin, M. 1971. *Emission Spectrochemical Analysis.* Wiley-Interscience, New York.

Soltanpour, P.N., J.B. Jones, Jr. and S.M. Workman. 1982a. Optical emission spectrometry. *In* A.L. Page (Ed.). *Methods of Soil Analysis*, Part 2, 2d ed. *Agronomy* 9:29–65.

Soltanpour, P.N., A. Khan and W.L. Lindsay. 1976. Factors affecting DTPA-extractable Zn, Fe, Mn, and Cu from soils. *Commun. Soil Sci. Plant Anal.* 7:797–821.

Soltanpour, P.N., A. Khan, and A.P. Schwab. 1979a. Effect of grinding variables on the NH_4HCO_3–DTPA soil test values for Fe, Zn, Mn, Cu, P, and K. *Commun. Soil Sci. Plant Anal.* 10:903–909.

Soltanpour, P.N., S.R. Olsen, and R.J. Goos. 1982b. Effect of nitrogen fertilization of dryland wheat on grain selenium concentration. *Soil Sci. Soc. Am. J.* 46:430–433.

Soltanpour, P.N. and A.P. Schwab. 1977. A new soil test for simultaneous extraction of macro- and micro-nutrients in alkaline soils. *Commun. Soil Sci. Plant Anal.* 8:195–207.

Soltanpour, P.N. and S.M. Workman. 1979. Modification of the NH_4HCO_3–DTPA soil test to omit carbon black. *Commun. Soil Sci. Plant Anal.* 10:1411–1420.

Soltanpour, P.N., and S.M. Workman. 1980. Use of NH_4HCO_3–DTPA soil test to assess availability and toxicity of selenium to alfalfa plants. *Commun. Soil Sci. Plant Anal.* 11:1147–1156.

Soltanpour, P.N., S.M. Workman, and A.P. Schwab. 1979b. Use of inductively coupled plasma spectrometry for the simultaneous determination of macro- and micronutrients in NH_4HCO_3–DTPA extracts of soils. *Soil Sci. Soc. AM. J.* 43:75–78.

Thompson, M., B. Pahlavanpour and S.J. Walton. 1978a. Simultaneous determination of trace concentrations of arsenic, antimony, bismuth, selenium, and tellurium in aqueous solution by introduction of the gaseous hydrides into an inductively coupled plasma source for emission spectrometry, Part I. Preliminary studies. *Analyst* (London) 103:568–579.

Thompson, M., B. Pahlavanpour, and S.J. Walton. 1978b. Simultaneous determination of trace concentrations of arsenic, antimony, bismuth, selenium, and tellurium in aqueous solution by introduction of the gaseous hydrides into an inductively coupled plasma source for emission spectrometry, Part II. Interference studies. *Analyst* (London) 103:705–713.

Tills, A.R. and B.J. Alloway. 1983. An appraisal of currently used soil tests for available copper with reference to deficiencies in English soils. *J. Sci. Food Agric.* 34:1190–1196.

Vlek, P.L.G. 1975. The chemistry, availability, and mobility of molybdenum in Colorado soils. Ph.D. diss. Colorado State Univ., Fort Collins (Diss. Abstr. 77-01190).

Ward, A.F. 1978. Inductively coupled argon plasma spectroscopy. Development, technique, and applications. *Am. Lab.* 10:79–87.

Westfall, D.G., M.H. Henson, and E.P. Evans. 1978. The effect of soil sample handling between collection and drying on nitrate concentration. *Commun. Soil Sci. Plant Anal.* 9:169–185.

Woolson, E.A., J.H. Axley and P.C. Kearney. 1971. Correlation between available soil arsenic estimated by six methods and response of corn (*Zea mays* L.). *Soil Sci. Soc. Am. Proc.* 35:101–105.

Workman, S.M. and P.N. Soltanpour. 1980. Importance of prereducing selenium (VI) to selenium (IV) and decomposing organic matter in soil extracts prior to determination of selenium using hydride generation. *Soil Sci. Soc. Am. J.* 44:1331–1333.

Workman, S.M., P.N. Soltanpour and R.H. Follett. 1988. Soil testing methods used at Colorado State Univ. for the evaluation of fertility, salinity, and trace element toxicity. *Colorado Agric. Exp. Stn. Bull.* LTB88-2, pp. 11–14.

Effects of Sewage Sludge Application on Soil Microbial Processes and Soil Fertility

S.R. Smith

I. Introduction

The ability of soil to supply nutrients for crop growth and the maintenance of soil physical conditions to optimize crop yield responses are the most important components of soil fertility that determine the productivity of agricultural systems. Many of the processes that influence soil fertility are mediated by microorganisms in the soil including fungi, bacteria, actinomycetes, and algae. The presence and activity of soil microorganisms are, therefore, of fundamental importance to the productivity of agricultural soils.

The benefits of applying bulky organic materials to soil to improve soil fertility have been recognized throughout the history of agricultural crop

Advances in Soil Science, Volume 16

production (Avnimelech, 1986). The addition of sewage sludge to soil provides a useful source of plant nutrients, particularly N and P, and also organic matter which improves soil physical properties (Pagliai et al., 1981; Chaussod et al., 1985; Wei et al., 1985; Coker and Carlton-Smith, 1986; Coker et al., 1987a,b; Metzger and Yaron, 1987; Smith et al., 1989). In addition, land application of sewage sludge offers a cost-effective method of disposal. However, sewage sludge contains potentially toxic elements (PTEs), particularly heavy metals, that can accumulate in the cultivated zone of the soil. Maximum soil limit values have been developed that ensure PTE accumulations in soil are well below any concentrations that have adverse effects on crop plants, animals and humans that consume them (Department of the Environment, 1989). Therefore, the potential human dietary intake of contaminants is minimized (Carlton-Smith, 1987). In the United Kingdom (U.K.), maximum permissible heavy metal concentrations for sludge treated agricultural soil have been established and are listed in Table 1. Recently, however, there has been increased concern that applications of sludge to land at or below the maximum permitted soil concentrations for heavy metals may have deleterious effects on soil microorganisms, particularly after sludge application has ceased and the organic matter content of the soil has declined (Brookes and McGrath, 1984; Brookes et al., 1986a, b; McGrath et al., 1988; Giller et al., 1989). In contrast, other studies (Stadelmann and Furrer, 1982; Tomati et al., 1985; Aichberger and Öhlinger, 1988; Öhlinger et al., 1990; Smith et al., 1990) show sludge application to improve soil microbial parameters. There is much con-

Table 1. U.K. maximum permissible concentrations for heavy metals in agricultural soils treated with sewage sludge when sampled to a depth of 150 mm

	Maximum permissible concentration of metals in soil (mg/kg dry solids)			
	A. For pH[1]			
PTE	5.0–5.5	5.5–6.0	6.0–7.0	>7.0
Zinc	200	250	300	450
Copper	80	100	135	200
Nickel	50	60	75	110
	B. For pH ≥ 5.0			
Cadmium	3			
Lead	300			
Chromium	400 (provisional)			

Source: Department of the Environment, 1989.

[1] For soils with pH in the ranges 5.0–5.5 and 5.5–6.0 the permitted concentrations of zinc, copper, nickel, and cadmium are provisional and will be reviewed when current research into their effects on certain crops and livestock is completed.

tradictory evidence of the effects of sewage sludge on soil microorganisms. The purpose of this chapter, therefore, is to review current literature relating to sludge effects on soil microorganisms and to consider their agronomic significance. Effects of heavy metals applied in sewage sludge to soil on microbial processes are also discussed in relation to current maximum soil limit concentrations in the U.K.

II. Effects of Sewage Sludge and Heavy Metals on Soil Microbial Processes

A. Microbial Biomass and Activity

Soil microbial biomass is considered to be the agent of breakdown of organic matter in the soil, although the importance of its size in relation to nutrient cycling and decomposition processes is poorly understood. It is well known that biomass increases when organic materials are applied to soil (Jenkinson and Ladd, 1981). However, Brookes and McGrath (1984) showed that microbial biomass in soil from the classical Woburn Market-garden Experiment which had been supplied with anaerobically digested lagoon-dried sewage sludge (at rates of 8.2 and 16.4 t/ha/year of organic matter between 1942–1961) was half that in soil that had received farmyard manure (at rates of 5.2 or 10.4 t/ha/year of organic matter between 1942–1967). Sludge application increased soil metal concentrations up to current limits with the exception of Cd which was three to five times the maximum limit (See references: Brookes et al., 1986b; McGrath et al., 1988 in Table 2). In contrast, the concentrations of metals in soil supplied with farmyard manure were small whereas the organic matter content and pH of both soils were the same. Reduced biomass in the sludged soil was therefore attributed to the toxic effects of metals in sludge on the soil microbial population. These differences were detectable 20 years after the last application of sewage sludge. However, in other studies with the same soils (Brookes et al., 1984) there were no effects of metal concentration on respiration rate [a measure of soil microbial activity (Domsch, 1961)]. Therefore, the respiration rate per unit weight of biomass was considerably greater in the metal-contaminated soil, even though the adenylate charge [a measure of metabolic activity (Eiland, 1985)] was large in both the high- and low-metal soils (Brookes and McGrath, 1987). This response may not necessarily be detrimental, but could be explained because metal-sensitive species have been replaced by other more tolerant groups that respire at a higher rate. Alternatively, the effect could be interpreted as a metal-induced stress response.

Jones et al. (1987) and Stark and Lee (1988) measured microbial biomass in soil 10 years after single, large applications of sewage sludge had been applied to two experiments on the same sandy loam soil. However, the interpretation of results from such experimental sites established to simu-

Table 2. Ranges of total metal concentrations in soils (mg/kg dry solids) and soil properties used in microbial studies

Site/soil description	pH	Organic matter[a] (%)	Total metal concentrations (mg/kg dry solids) Cr	Ni	Cu	Zn	Cd	Pb	Authors
Derelict mine sites	6.5–6.8	2.1–6.6[b]			20–40	71–33,860	1.6–345	61–7580	Rother et al., 1982a
Derelict mine sites	6.3–7.1	2.1–4.5[b]			20–40	71–33,860	1.6–345	61–7580	Rother et al., 1982b
Derelict mine sites	6.3–6.9				20–50	71–20,520	1.4–216	73–4285	Rother et al., 1983
Sludge-treated agricultural	5.1–6.5	8.8–32.6			35–589	109–1900	0.6–120	31–758	Barkay et al., 1985
Woburn Market-garden Experiment									
Farmyard manure-treated	7.0	1.67[b]		4	27	92	2.2		Brookes et al., 1986b
Sludge-treated	6.8	1.97[b]		27	102	289	8.6		
Farmyard manure-treated				16	24	94	1.6		McGrath et al., 1988
Sludge-treated				35	163	469	15.4		
Soil with 50% N_2-fixation				27	99	334	10.0		
Sludge-treated experimental	6.7	10.5	143	2	223	1480	1.29	177	Jones et al., 1987
Sludge-treated experimental									
Control (no sludge applied)	5.4	5.3		35	19	116	0.35[c]		Stark and Lee, 1988
Soil with 15% N_2-fixation	5.9	7.8		40	70	210	2.15[c]		
Soil with 50% N_2-fixation	5.6	7.5		60	110	300	5.38[c]		
Sludge-treated agricultural				66	158	981	4.59	189	
Derelict mine sites	3.8–5.6	0.5–28.7			10–415	70–3800	0.11–10.3	195–30,500	Obbard and Jones, 1989
Sludge-treated experimental									
Control (no sludge applied)				33	20	109	0.35	30	Kwan and Smith, 1990
Low-Cd treatment				39	74	235	2.15	96	
High-Cd treatment				52	95	273	5.38	92	
High-sludge treatment				74	142	334	7.03	124	
Sludge-treated experimental[d]									
Low-metal treatments	5.0–5.7	3.8–5.9	230	63	147	297			Obbard et al., 1990
High-metal treatments	5.0–5.6	3.6–6.7	331	153	235	542			

Sludge-treated agricultural									
Pasture (clover present)	5.7–6.7	6.7–26.8	63–576	30–74	15–808	140–1670	1.2–9.4	28–667	Smith et al., 1990
Arable (clover absent)	6.2–7.1	4.2–11.4	93–379	39–84	19–200	109–643	0.5–13.5	95–321	
Received sludge for 100 years		7.0–32		40–310	50–1250	300–6900	2–48	200–4200	
Sludge-treated experimental									
Sandy loam									
Control (no sludge applied)	5.6	7.0	102	35	64	127	0.5	40	
Sludge-treated	5.1–5.4	9.6–11.6	128–281	47–74	159–170	362–445	4.9–9.5	137–161	
Clay									
Control (no sludge applied)	5.6	12.6	140	52	25	168	0.9	44	
Sludge-treated	4.9–6.0	15.2–16.6	131–392	51–105	96–218	326–644	4.7–17.3	116–185	
Calcareous loam									
Control (no sludge applied)	7.8	8.8	51	23	28	83	0.3	58	
Sludge-treated	7.4–7.6	11.1–12.1	68–181	34–46	81–107	234–274	1.8–3.4	99–122	
Sludge-treated agricultural									
Low-Cd soil	6.6–7.4	4.0–10.3	73–99	28–46	96–380	230–620	1.2–2.2	165–300	Smith, 1991
High-Cd soil	6.6–7.6	2.4–3.6	82–115	22–35	40–85	105–200	2.4–7.6	68–160	

[a] Determined by loss on ignition except [b] where results presented as organic carbon (%).

[c] From Kwan and Smith, 1990.

[d] Concentration for main metal contaminant in soil treated with specific monometallic sludges.

late metal accumulation in sludged soils should be considered with some caution. This is because metal availability may be generally higher than that measured in historically sludged soils that have received repeated, small applications of sludge on an operational basis (Carlton-Smith and Stark, 1987). In these experimental soils, Stark and Lee (1988) showed that biomass generally decreased with increasing soil metal concentration (Table 2). In contrast, Jones et al. (1987) found no effect of sludge application on microbial biomass even when the Zn concentration was five times the soil limit (Table 2). However, inorganic metal salts applied at similar concentrations to those supplied in the sewage sludge significantly reduced microbial biomass.

Sewage sludge application to soil generally raises the activity of soil microorganisms by increasing soil organic matter content (Agbim et al., 1977; Stroo and Jencks, 1985; Seaker and Sopper, 1988; Öhlinger et al., 1990). In contrast, Stark and Lee (1988) showed that sludge application reduced soil respiration. Microbial activity is also significantly reduced by metal salt additions to soil (Bhuiya and Cornfield, 1972; Chaney et al., 1978; Lighthart et al., 1983). However, potentially inhibitory effects of metals on soil respiration are modulated by soil physico-chemical characteristics and decline with increasing soil cation-exchange capacity (Doelman and Haanstra, 1979a) and soil pH (Doelman and Haanstra, 1984) due to lower metal availability.

B. Biological Nitrogen Fixation

Biological N_2-fixation is a sensitive indicator of soil metal pollution (Skujins et al., 1986; Coppola et al., 1988; Stark and Lee, 1988). Nitrogen fixation by free-living heterotrophic microbes (Letunova et al., 1985; Stark and Lee, 1988), growth and fixation by blue–green algae (Brookes et al., 1986b), and effectiveness of symbiotic N_2-fixing *Rhizobium* (McGrath et al., 1988; Stark and Lee, 1988) may decline in soils with elevated metal concentrations. The contribution made by free-living soil microorganisms to the pool of available N for crop growth is relatively small and probably of little agronomic significance. For example, free-living heterotrophic bacteria may fix less than 5 kg N/ha/year, whereas blue green algae may fix up to 10–50 kg N/ha (Henriksson, 1971; Witty et al., 1979). However, symbiotic N_2-fixation provides large N inputs to soil and is of considerable importance to the productivity of agricultural systems. For example, fixation by white clover (*Trifolium repens* L.) can contribute 100–200 kg N/ha/year (Robson et al., 1989).

Laboratory and glasshouse studies using soluble metal salt solutions have demonstrated the toxic effects of metals on nitrogenase activity and nodulation (Huang et al., 1974; McIlveen and Cole, 1974; Vigue et al., 1981). The toxicity of metals to nodule formation by red clover was ranked in the order Cd > Co > Cu > Zn by McIlveen and Cole (1974). The order of decreasing toxicity to symbiotic N_2-fixation is similar and has been ranked as Cd > Ni > Cu > Zn > Pb (McGrath et al., 1988). Raising soil pH or cation exchange capacity can, however, reduce metal toxicity to the nodulation process (McIlveen and Cole, 1974).

A 50% reduction in the rate of symbiotic N_2-fixation by white clover has been measured in sludged soils with Zn or Cd levels above the current maximum limits (Table 2) compared with farmyard manure treated or unsludged soil with background metal levels (McGrath et al., 1988, Stark and Lee, 1988). However, Stark and Lee (1988) also reported that the rate of N_2-fixation by clover was reduced by 15% in soil with metal levels that were below the current maximum limits, but this was only just significant at the 5% probability level compared with the control. Later studies (Kwan and Smith, 1990) at the same experimental site (Table 2) found no detrimental effect of sludge application on N_2-fixation. Under these conditions, effects of seasonal variation in environmental conditions on activity were more important than any metal related effects. Interestingly, results indicated that the rate of N_2-fixation by clover was significantly larger on the most metal-contaminated plot, which had received the largest addition of sewage sludge compared with the unsludged control. This was probably explained by the elevated organic matter content of this plot compared with the other lower rates of sludge addition and the control.

The apparent contradiction in the results from these studies may be explained because potentially detrimental effects reported in the earlier work were only transient. Once established, symbiotic N_2-fixation was not influenced by the level of soil contamination. This is supported by Smith et al.(1990) who screened soils from historically sludged pastures containing clover, for the presence of effective *Rhizobium*. Results showed that effective *Rhizobium* were present irrespective of the concentrations of heavy metals in the soil (Table 2).

Thus potential effects of soil metal contamination on *Rhizobium* are apparently minimised once infection and nodulation has occurred. This is because rhizobial bacteroids within root nodules are protected to some extent from soil metals because they are surrounded by host plant tissue (McGrath et al., 1988). It could be argued, therefore, that survival of effective rhizobial strains in soil may be of limited agronomic importance because legume crops can be supplied with commercial *Rhizobium* inoculants.

However, Obbard et al. (1990) reported that the rate of N_2-fixation by clover was reduced at an experimental site (established in 1968) with an acid sandy loam soil (pH 5.0–5.6) that had been supplied with large doses of monometallic sludges, compared with the control that had only received uncontaminated sludge. The toxicity of the metals was ranked in decreasing order as Zn > Cu > Ni, although N_2-fixation was similar or increased relative to the control in certain plots with elevated concentrations of Cr and Ni. High- and low-metal plots were compared for N_2-fixation activity (Table 2), but concentrations in both exceeded the current maximum permissible limits set for arable soil at pH 5.0–5.5 (Table 1). In addition, most probable number studies for the high-metal soils (Smith et al., 1990) indicated that Zn and Cu reduced the size of the rhizobial population to a greater extent than did Ni and Cr, which had no effect. However, the pH of soil from the Zn and Cu plots was especially low at pH 5.0 and 5.1, respectively, which would be prohibitive to sludge application on such a light soil (Department of the Environment, 1989). Detrimental effects of heavy metals applied in sludge were therefore probably explained by a combination of low soil pH and light soil texture, which resulted in increased metal availability.

A number of studies (Rother et al., 1982a, 1983; Obbard and Jones, 1989) have reported no effect of elevated metal concentrations in heavily polluted mine soils on heterotrophic N_2-fixation and symbiotic N_2-fixation by clover. These investigations are of particular interest because soil metal concentrations were considerably larger (Table 2) than the maximum metal limit values set for sludged agricultural soil and soil organic matter contents represented residual background levels. It was concluded, therefore, that microorganisms can adapt to high soil–metal conditions without loss or reduction of N_2-fixing ability. Though this may be the case for free-living heterotrophic microorganisms, these studies do not necessarily assess effects of metals on symbiotic *Rhizobium* directly. This is because host plants were growing on the soils tested thus offering protection to the bacteroids in root nodules. Effects of metals on survival and effectiveness of native *Rhizobium* strains can only be determined in the absence of the host plant. Alternatively, where the host is present, soils can be incubated with the host removed to determine the survival of indigenous *Rhizobium*, which is free-living in the soil. The development of resistance by microorganisms to increased soil metal concentrations is discussed later. However, Rother et al. (1982a) suggested an alternative explanation for the apparent resistance of N_2-fixing microorganisms at mine sites was that metals present in the soils examined were unavailable or in nontoxic forms being of geological origin.

The potential inhibition of symbiotic N_2-fixation by metals applied in sewage sludge may be due to direct toxicity (McIlveen and Cole, 1974) or may also be in response to an adaption or a selection of the microbial population to the elevated metal concentration. For example, Reddy et al. (1983) reported a decline in populations of *Rhizobium japonicum* in sludged soil and attributed this to increased soil metal concentrations. More recently, McGrath et al. (1988) found the strain of *Rhizobium leguminosarum* biovar *trifolii* in the Woburn soil formed root nodules on clover, but that these were ineffective at N_2-fixation. This was also reflected in the growth response of clover. The inability of *Rhizobium* strains to fix N_2 in the high-metal soil reduced yield by 25%, compared with the low-metal soil with effective *Rhizobium*. The survival of only ineffective strains of *Rhizobium* in sludged soil was later confirmed by Giller et al. (1989) by strain isolation and plant infection tests. Here, effective strains of *Rhizobium* did not survive, or became ineffective at N_2-fixation in the soil during a 2-month incubation period with inoculum doses of $\leq 10^7$ cells/pot. It could be argued, however, that strains not adapted or resistant to elevated soil metal conditions would not be expected to survive when introduced to such soils.

Smith et al. (1990) screened a number of soils from historically sewage sludge-amended arable sites with elevated heavy metal concentrations (Table

2), for the presence of effective strains of *Rhizobium* capable of symbiotic N_2-fixation with clover. These sites are of particular interest because organic matter levels have declined substantially through the intensive cultivation of the soil. Concentrations of Zn, Cu, Ni, and Cr in soils from most sites were within the maximum limits and the host plant was absent from all sites. However, Cd levels increased up to 4.5 times the maximum soil concentration limit for this element. Plant infection tests showed that *Rhizobium* were not present in the sludged soils and no ineffective resistant strains were isolated. However, a free soil assay showed *Rhizobium* could survive in soil at a Cd concentration of 3.1 mg/kg even though the plant infection test was negative. This probably indicates the presence of a small rhizobial population in the soil that was unable to nodulate the host plant in the infection test due to dilution (1:100 soil/water) prior to inoculation. The use of both test methods, and the contrasting metal concentration profiles examined showed that Cd could be important in determining the presence of *Rhizobium* in soils.

In other studies with soil from three experimentally sludged sites (established in 1978), Smith et al. (1990) found that *Rhizobium* were generally absent from soil with concentrations of Zn, Cu and Cd that exceeded the current permissible levels for these metals (Table 2). The sites were established on

(1) a sandy loam with low pH in the range of 5.1–5.4;
(2) clay of pH 4.9–6.0; and
(3) a calcareous loam of pH 7.4–7.6.

No indigenous clover was established on the sites. Interestingly, the clay soil did support effective *Rhizobium* on one sludged plot with the smallest Cd concentration (4.7 mg/kg) and largest levels of the other elements. Samples of the calcareous loam soil with metal levels approaching the maximum permissible concentrations (although the Cd concentration of one plot actually exceeded the maximum limit) all contained effective *Rhizobium*. These results show that the current maximum permissible concentrations for heavy metals in sludge-amended agricultural soil may protect *Rhizobium*. However, potentially detrimental effects of elevated metal concentrations in soil are modulated by soil physicochemical characteristics.

Concentration profiles of PTEs in sewage sludge-treated soil with elevated levels of Cd relative to the other elements are typical of soils that have received highly contaminated sludge in the past. Industrial effluent, which combines with domestic sewage and road run-off at the sewage treatment works, has been a major contributor of contaminants, such as Cd, to sewage sludge. However, trade effluent controls implemented by the water industry and changes in the nature of large manufacturing industries has significantly improved sludge quality. In particular, concentrations of Cd have declined markedly such that Cu and Zn derived largely from domestic sources and road run-off have become the main limiting PTEs to sludge recycling in agriculture. For example, sludge applied historically to soil at Woburn in the late 1950s contained up to 160 mg/kg of Cd (McGrath 1984). However, the

50-percentile concentration for Cd in sludge applied to agricultural soil in the U.K. was 8.5 mg/kg in 1982–1983 (Sleeman, 1984) and has declined significantly since then.

Using the plant infection approach to determine the presence of *Rhizobium* in soil, Smith (1991) showed that only 50% of inoculations were successful for an arable soil with large concentrations of Cd, but only moderate levels of the other PTEs (Table 2). This indicated that the population of *Rhizobium* was reduced, but not completely eliminated as was previously reported (Smith et al., 1990). In contrast, effective *Rhizobium* were present in 90% of soil samples taken from a field site with elevated levels of Cu and Zn, but with little Cd (Table 2). As with previous work, no ineffective strains were isolated. Thus, metals supplied in sewage sludge to agricultural soil determine the survival of *Rhizobium* through direct toxicity rather than having a genetic effect. The presence of Rhizobium is, therefore, largely site specific depending on the relative and absolute concentrations of individual PTEs that are a function of the quality of the sludge applied. Consequently, the model of Woburn, and other historical sites that were supplied with highly contaminated sewage sludges is unlikely to reflect the long-term effect of modern sludge application practices on symbiotic N_2-fixation or other soil microbial processes.

C. Mineralization and Nitrogen Transformations

Decomposition and mineral cycling processes are of considerable agronomic importance because they influence nutrient availability in soil and can satisfy a significant proportion of crop nutrient requirements. For example, Cooke (1980) indicated that soils containing residues of leguminous crops, grassland or organic manures could release over 200 kg N/ha/year. Mineralization of soil organic N can contribute as much as 70% of the total N uptake of winter cereals (Christian et al., 1985). Degradation and nutrient release from residual soil organic matter, crop residues, and organic amendments are entirely dependent on microbially mediated processes. Nitrogen is particularly unique in this respect because soil reserves of this major nutrient element are almost entirely in organic forms (Broadbent, 1986).

Mineralization of sewage sludge in soil has been studied frequently in relation to the fertilizer value of the material (Parker and Sommers, 1983; Chae and Tabatabai, 1986; Garau et al., 1986). However, comparatively little information is available on the effects of heavy metals in sewage sludge on soil mineral transformation processes. Soluble metal salts have been applied to soils often as a basis for assessing effects of heavy meals in sewage sludge on soil N transformations and results of these studies are summarized by Doelman (1986). Organic N mineralization and nitrification processes were inhibited by soil metal concentrations of approximately 100 mg/kg for Cu, Ni, and Zn; 100–500 mg/kg for Pb and Cr; 10–100 mg/kg for Cd; and 1–10 mg/kg for Hg. Denitrification is also inhibited by metal additions to soil, and Bollay and Barabasz (1979) showed metal toxicity was ranked in the order Cu > Cd > Zn > Pb. However, effects of soluble metal salts on N transformations are strongly governed by soil properties: high cation exchange capacity (Wilson, 1977a) and high pH (Quraishi and Cornfield, 1973) reduce toxicity.

Several incubation studies show that metals applied in sewage sludge have no apparent effect on the mineralization of organic carbon (Cornfield et al., 1976), N mineralization (Premi and Cornfield, 1971), or nitrification (Premi and Cornfield, 1969). Sludges used here came from sewage treatment works receiving combined domestic and industrial effluent. However, Wilson (1977b) reported that sludge derived primarily from an industrial source significantly reduced nitrification when incorporated into soil whereas domestic sludge had little or no affect. Differences in effects of sludge application on nitrification may be explained because the industrial sludge used by Wilson (1977b) generally contained much larger concentrations of PTEs compared with the other sewage sludges examined. For example, the Zn content of the industrial sludge was up to 6 times greater (at 9,240 mg/kg Zn) than that in the other studies, although the Cu concentration was significantly smaller. The 50-percentile concentration of Zn in sewage sludges utilized on agricultural land in the U.K. is in the region of 1000 mg/kg (Sleeman, 1984). Interestingly, sewage sludge applied at Woburn contained approximately 3000 mg/kg of Zn (McGrath, 1984). In the U.K. at least 98% of all sludges applied to agricultural land contain less than 1100 mg/kg of Zn (Sleeman, 1984). Therefore, it is unlikely that application of sewage sludge operationally to agricultural soils would have a detrimental effect on nitrification.

More recently, Brookes et al. (1984) reported that rates of mineralization of native soil organic C and N were similar in soil from the historically sludged experimental site at Woburn with high-metal concentrations (Table 2) and in low-metal soil that had only received farmyard manure. However, the nitrification potential of the sludge-treated soil was reduced compared with the farmyard manure plots. In contrast, Smith et al. (1990) measured the rate of nitrification of an arable soil that had received large and repeated applications of sewage sludge for 100 years. Results indicated there was no detrimental effect of large concentrations of heavy metals in soil (Table 2) on N transformations and that the rate of nitrate production could be explained as a simple function of soil organic matter content. In general, mineralization processes (which probably represent the most important components of soil fertility) are less sensitive to the effects of heavy metals than is nitrification. This is explained because of the large diversity of microorganisms that can decompose organic matter. However, there is little evidence available that indicates that the specialized groups of nitrifying bacteria are affected detrimentally by sewage sludge application to soil at the current maximum limit concentrations for heavy metals.

The adaptation of mineralizing and nitrifying microorganisms to high concentrations of heavy metals has been demonstrated by Rother et al. (1982b). Nitrogen mineralization in heavily polluted mine soils was not affected by large metal salt additions. However, nitrification was more sensitive to the effects of heavy metals than was mineralization, but tolerant populations of nitrifying microorganisms were present in the soils. The results also indicated that nitrifiers in noncontaminated soils supplied with metal salts eventually developed tolerance but that the rate of activity was reduced.

D. Vesicular–Arbuscular Mycorrhizae

Vesicular-arbuscular mycorrhizae (VAM) are fungus–root associations that form between fungal root endophytes and plants, including most agricultural crops. Crop plants benefit from this association due to the increased supply of P and other ions such as Cu, Zn, Si, Br, and Cl (Gilden and Tinker, 1983b), which the fungus acquires from the soil beyond the zone of depletion at the root surface, and transports back to the root (Rhodes and Gerdemann, 1980). Improved P status of legumes through root mycorrhizal infection increases growth, nodulation, and N_2-fixation by the crop (Barea and Azcón-Aguilar, 1983). Mycorrhizae also increase crop yields due to improved plant resistance to soil-moisture stress, modifications in plant hormonal balance, and enhanced resistance to plant pathogens (Dehne, 1982; Hayman, 1983).

Little research has been carried out to assess the effects of heavy metals from sewage sludge on VAM. Exposure of VAM to elevated soil heavy metal concentrations can significantly reduce the level of host plant infection and P uptake (Gilden and Tinker, 1983a, b). However, Koomen et al. (1990) demonstrated that metal tolerant strains of VAM persisted in sludge-treated soil at Woburn and could form normal associations with the roots of crop plants although the rate of root infection was reduced compared with farmyard manure plots. In addition, sludged soils often contain much available P, which is known to inhibit or reduce VAM infection (Pera et al., 1982) such that potentially reduced infection resulting from elevated metal levels is unlikely to affect plant growth (Koomen et al., 1990).

Vesicular–arbuscular mycorrhizae can enhance the uptake of heavy metals by the host and, therefore, increase the phytotoxic effects of metals on plants (Killham and Firestone, 1983). However, Gilden and Tinker (1983a) showed that strains of VAM vary in tolerance to heavy metals and that increasing tolerance reduced phytotoxicity of metals to crop plants. Dueck et al. (1986) reported that VAM infection of grass, grown in soil heavily contaminated with Zn, alleviated the detrimental effects of Zn on plant growth. The reason for this is not clear but could be explained because metals accumulate in fruiting bodies of mycorrhizal fungi, and there is other evidence that fungal mycelium acts as a filter for heavy metals (Moore, 1988).

E. Enzyme Activities

Enzymes produced by soil microorganisms catalyze biochemical processes involved in nutrient cycling in soil and may provide an index of total microbial activity. These enzymes may be exocellular, secreted by microorganisms and plant roots, occurring freely in the soil, or endocellular and active only in living cells.

Information available on the effects of sewage sludge applied to soil on enzyme activities is often contradictory and both beneficial and detrimental responses have been reported. For example, sewage sludge application may reduce the activity of dehydrogenase (Reddy et al., 1987; Stark and Lee, 1988 Reddy and Faza, 1989), urease (Frankenberger et al., 1983; Reddy et al., 1987), and phosphatase (Reddy et al., 1987). Tabatabai (1977) reported that the effects of metal salts on urease activity in order of decreasing inhibition were $Ag \geq Hg > Cu > Cd > Zn > Sn > Mn$. In contrast, Achberger and Öhlinger (1988) determined that dehydrogenase and protease activity increased in soil supplied with metal contaminated sewage sludge. There was no effect of sewage sludge application on β-glucosidase and urease activity, whereas phosphatase declined only at large rates of applied sludge. Brookes et al. (1984) also reported that phosphatase activity was not affected by elevated soil metal concentrations due to sewage sludge application. Phosphatase measurements may be a poor indicator of microbial activity in cropped soil because it is an exocellular enzyme system and plant roots also contribute the enzyme to the soil pool (Stark and Lee, 1988).

Soil properties can have an important influence on the effects of heavy metals on soil enzyme activity. For example, Doelman and Haanstra (1979a) showed that increasing Pb concentration reduced dehydrogenase activity in sandy soil, but had no effect in clay and peat soils with large cation exchange capacities. In addition, inhibitary effects of heavy metals on urease activity declined with increasing soil organic matter content (Tabatabai, 1977).

F. Soil-Borne Plant Pathogenic Microorganisms

The addition of organic matter to soil can suppress plant diseases caused by soil-borne pathogens (Henis, 1986). However, antagonistic effects of sewage sludge application on plant pathogenic microorganisms are poorly understood. Lumsden et al. (1983) supplied composted sewage sludge to several crop species and showed that Aphanomyces root rot of peas, Rhizoctonia root rot of bean, cotton, and radish, Sclerotinia drop of lettuce, Fusarium wilt of cucumber, and Phytophthora crown rot of pepper were significantly decreased by compost addition. Other diseases, such as Pythium damping-off of pea and bean, Fusarium root rot of pea, and Thielaviopsis root rot of bean and cotton, were not affected or were increased by compost addition. Disease suppression was generally enhanced by increasing the time between soil amendment and planting. Increased soil microbial activity and introduction of microbes that compete with soil pathogens are possible mechanisms which may explain the lower incidence of plant diseases following application of organic matter to soil (Meyer and Linderman, 1986).

G. Soil Structural Stability

Soil structural conditions are important if yield responses of agricultural crops to, for example, fertilizer inputs are to be optimized (Smith et al., 1989). Sewage sludge application improves soil physical conditions by increasing the stability of soil aggregates (Metzger and Yaron, 1987). Guidi (1981) indicated that improvements in physical properties of sludge-amended soils were largely mediated by microbial processes. Lynch and Bragg (1985) also showed that microorganisms had a dominant role in the improvement of aggregation following the addition of organic materials to soil. It follows, therefore, that soil aggregate stability increases with microbial biomass (Lynch, 1981). Changes in soil structure after sludge addition are characterized by a phase of aggregate formation followed by decreasing soil aggregate content to a level that is greater than that in unamended soil (Tisdall and Oades, 1982; Metzger et al., 1987). Metzger et al. (1987) indicated that this structural improvement was correlated with increased fungal activity. Cementing fungal carbohydrates and physical entanglement by mycelium appear to be important binding mechanisms in the formation of stable aggregates when sludge is applied to soil. Effects of sewage sludge on microbially mediated soil structural stability characteristics form an interesting contrast to those reported from biomass studies.

III. Resistance of Soil Microorganisms to Elevated Heavy Metal Concentrations in Soil

Doelman (1986) recently reviewed literature concerning the resistance of soil microorganisms to heavy metals. It was concluded that, in general, increasing soil metal levels may reduce the composition and diversity of the total soil microbial population, causing a shift towards resistant strains that become dominant in such soils. Fungi and actinomycetes are more resistant than bacteria (Maliszewska et al., 1985), although Gram-negative bacteria exhibit greater tolerance than do Gram-positive types (Duxbury and Bicknell, 1983). Increasing abundance of resistant strains under elevated metal conditions enables the continuation of essential processes that maintain soil fertility (Barkay et al., 1985).

Numerous studies (Doelman and Haanstra, 1979b; Olson and Thornton, 1982; Barkay et al., 1985; Yamamoto et al., 1985) show that resistant strains of microorganisms are present in soils containing only background metal levels . For example, Kelly and Reanney (1984) showed that 5% of the bacteria which they isolated from low Hg reference soils exhibited resistance to Hg. Raising soil metal concentrations above background levels increases the number of resistant microorganisms (Doelman, 1986) and the diversity of the tolerant population (Barkay et al., 1985). However, Giller et al. (1989) showed that *Rhizobium* strains that were resistant to heavy metals in sludged soil, but were ineffective at N_2-fixation, had similar plasmid profiles indicating a lack of genetic diversity in the population.

The process of resistance does not occur rapidly, but may develop over a period of 1–4 years (Doelman and Haanstra, 1979b, Barkay et al., 1985) following the application of metals to soil. However, mechanisms involved in the development of resistance to metal contamination are complex and poorly understood. The dissemination of genetic material encoding resistance may be partly involved in the adaption process (Kelly and Reanney, 1984). However, microorganisms can develop resistance without the transfer of resistant plasmids (Barkay et al., 1985; Duxbury and Bicknell, 1983; Yamamoto et al., 1985). For example, Yamamoto et al. (1985) suggested Cu tolerance in fungal populations was related to innate properties of certain strains already present in the soil. Barkay et al. (1985) proposed that bacterial resistance to Cd in sludged soil occurs through an accommodation mechanism. This is because resistant strains normally undergo an extended lag phase in the presence of Cd, followed by normal growth rates (Aiking et al., 1982). Other work (Olson and Thornton, 1982) suggests that soil bacterial populations can tolerate small inputs of Cd, up to several mg/kg, without showing any significant effect on numbers. In contrast, Williams and Wollum (1981) reported that small concentrations of Cd ($< 5\ \mu g/ml$) in culture media were toxic to soil bacteria, but that a tolerance mechanism operated at large levels of Cd ($> 5\ \mu g/ml$).

IV. Conclusions

Our knowledge of the effects of heavy metals in sewage sludge on soil microbial processes is insufficient to establish whether the innate fertility of agricultural soils is significantly affected by sludge application within the current maximum soil limits for metals. What is evident, however, is that the soil microbial population has a remarkable adaptive capacity to elevated PTE concentrations such that, in general, essential soil processes are maintained even in heavily contaminated soils. Nevertheless, there is evidence indicating that heavy metals in soil applied in sewage sludge may affect certain soil microbial processes but at levels that apparently exceed current maximum soil limit concentrations. Microbial processes performed by specific and specialized microorganisms, such as N_2-fixation and nitrification, are most sensitive to the effects of soil contamination with heavy metals resulting from sewage sludge application although the agronomic significance of this has yet to be established. Most of the published information on symbiotic N_2-fixation has dealt with clover. However, there is a need to expand current research to include other important leguminous arable and woodland crops that may also receive sewage sludge.

Research involving the application of soluble metal salts directly to soil, or in sewage sludges (Zibilske and Wagner, 1982) has often been used as a basis for establishing the effects of heavy metals in sewage sludge on microbial processes. However, the response of microorganisms to inorganic salt additions whether applied directly to soil or *via* spiked sludges, is unlikely to reflect that obtained with sewage sludge applied operationally because heavy metals in the sludge are largely unavailable due to the formation of insoluble organometal complexes (Stover et al., 1976; Davis and Carlton-Smith, 1981; Kiekens, 1984). For example, Chang and Broadbent (1982) showed that N transformations were severely inhibited by Cr (III) salt additions to soil. In contrast, Wiseman and Zibilske (1988) incorporated sewage sludge containing 38,220 mg/kg Cr (III) in soil to give similar concentrations of Cr as Chang and Broadbent (1982) and found no effect on C and N mineralization. In addition, sludged soils contain numerous heavy metals at varying concentrations and often have a large organic matter content. Therefore, results from field studies are often confounded because metal effects on microorganisms could be attributed to a combination of factors in the soil or applied sewage sludge (Barkay et al., 1985). An alternative approach may be to use prepared sludges with controlled metal contents following the method of Davis and Carlton-Smith (1981). This could enable a systematic long-term study of the effects of heavy metals in sewage sludge on soil microbial processes by providing contrasting soil metal concentrations at similar organic matter inputs. Such a study would assist the interpretation of results from current research using historically sludged soils.

References

Agbim, N.N., B.R. Sabey, and D.C. Markstrom. 1977. Land application of sewage sludge: V. Carbon dioxide production as influenced by sewage sludge and wood waste mixtures. *J. Env. Qual.* 6:446–451.

Aichberger, K., and R. Öhlinger. 1988. Effects of sewage sludge and waste compost on some enzymatic activities tested in a field experiment. Poster paper presented at *EC/EWPCA symposium*, September 1988, Amsterdam.

Aiking, H., K. Kok, H. van Herrikhuizen, and J. Van'Riet. 1982. Adaptation to cadmium by *Klebsiella aerogenes* growing in continuous culture proceeds mainly via formation of cadmium sulphide. *Appl. Environ. Microbiol.* 44:938–944.

Avnimelech, Y. 1986. Organic residues in modern agriculture. *In* Y. Chen and Y. Avnimelech (Eds.). *The Role of Organic Matter in Modern Agriculture.* Martinus Nijhoff Publishers, Dordrecht, The Netherlands, pp. 1–10

Barea, J.M., and C. Azcón-Aguilar. 1983. Mycorrhizas and their significance in nodulating nitrogen-fixing plants. *Adv. Agron.* 36:1–54.

Barkay, T., S.C. Tripp, and B.H. Olson. 1985. Effect of metal-rich sewage sludge application on the bacterial communities of grassland. *App. Environ. Microbiol.* 49:333–337.

Bhuiya, M.R.H., and A.H. Cornfield. 1972. Effects of addition of 1000 ppm Cu, Ni, Pb, and Zn on carbon dioxide release during incubation of soil alone and after treatment with straw. *Env. Pollution* 3:173–177.

Bollay, J.M., and W. Barabasz. 1979. Effect of heavy metals on the denitrification process in soil. *J. Env. Qual.* 8:196–201.

Broadbent, F.E. 1986. Effects of organic matter on nitrogen and phosphorus supply to plants. *In* Y. Chen and Y. Avnimelech (Eds.). *The Role of Organic Matter in Modern Agriculture.* Martinus Nijhoff Publishers, Dordrecht, The Netherlands, pp. 13–27.

Brookes, P.C., C.E. Heijnen, S.P McGrath, and E.D. Vance. 1986a. Soil microbial estimates in soils contaminated with metals. *Soil Biol. Biochem.* 18:383–388.

Brookes, P.C., and S.P. McGrath. 1984. Effects of metal toxicity on the size of the soil microbial biomass. *J. Soil Sci.* 35:341–346.

Brookes, P.C., and S.P. McGrath. 1987. Adenylate energy charge in metal contaminated soil. *Soil Biol. Biochem.* 19:219–220.

Brookes, P.C., S.P. McGrath, and C.E. Heijnen. 1986b. Metal residues in soils previously treated with sewage sludge and their effects on growth and nitrogen fixation by blue–green algae. *Soil Biol. Biochem.* 18:345–355.

Brookes, P.C., S.P. McGrath, D.A. Klein, and E.T. Elliott. 1984. Effects of heavy metals on microbial activity and biomass in field soils treated with sewage sludge. Environmental Contamination (International Conference, London, July 1984) CEP Ltd, Edinburgh, U.K., pp. 574–583.

Carlton-Smith, C.H. 1987. Effects of metals in sludge-treated soils on crops. *Environment Technical Report TR 251.* WRc Medmenham, Marlow, U.K.

Carlton-Smith, C.H., and J.H. Stark. 1987. Sites with a history of sludge deposition. Interim report of field trials April 1983–March 1986 (SDA 9166 SLD). WRc Report No. DoE 1376-M. WRc. Medmenham, Marlow, U.K.

Chae, Y.M., and M.A. Tabatabai. 1986. Mineralisation of nitrogen in soils amended with organic wastes. *J. Env. Qual.* 15:193–198.

Chaney R.L., J.M. Kelly, and R.C. Strickland. 1978. Influence of cadmium and zinc on carbon dioxide evolution from litter and soil from a black oak forest. *J. Env. Qual.* 7:115–119.

Chang F.H., and F.E. Broadbent. 1982. Influence of trace metals on some soil nitrogen transformations. *J. Env. Qual.* 11:1–4.

Chaussod, R., S.K. Gupta, J.E. Hall, B. Pommel, and J.H. Williams. 1985. Nitrogen and phosphorus value of sewage sludges. Commission of the European Communities, Revision of Document Nr SL/82/82-XII/ENV/35/82.

Christian, D.G., R. Crees, and R.J. Dowdell. 1985. Yield and uptake of fertilizer nitrogen by direct-drilled winter barley growing on a chalk soil. *Soil Use Manage.* 1:74–79.

Coker, E.G., and C.H. Carlton-Smith. 1986. Phosphorus in sewage sludges as a fertiliser. *Waste Manage. Res.* 4:303–319.

Coker, E.G., J.E. Hall, C.H. Carlton-Smith, and R.D. Davis. 1987a. Field investigations into the manurial value of lagoon-matured digested sewage sludge. *J. Agric. Sci., Camb.* 109:467–478.

Coker, E.G., J.E. Hall, C.H. Carlton-Smith, and R.D. Davis. 1987b. Field investigations into the manurial value of liquid undigested sewage sludge when applied to grassland. *J. Agric. Sci., Camb.* 109:479–494.

Cooke, G.W. 1980. *Fertilising For Maximum Yield.* 3d ed. Granada, London, U.K.

Coppola, S., S. Dumontet, M. Portonio, G. Basile, and P. Marino. 1988. Effect of cadmium-bearing sewage sludge on crop plants and microorganisms in two different soils. *Agric. Ecosystems and Environ.* 20:181–194.

Cornfield, A.H., P.H.T. Beckett, and R.D. Davis. 1976. Effect of sewage sludge on mineralisation of organic carbon. *Nature* 260:518–520.

Davis, R.D., and C.H. Carlton-Smith. 1981. The preparation of sewage sludges of controlled metal content for experimental purposes. *Env. Pollution (Series B)* 2:167–177.

Dehne, H.W. 1982. Interaction between vesicular arbuscular mycorrhizal fungi and plant pathogens. *Phytopathology* 72:1115–1119.

Department of the Environment. 1989. *The Use of Sewage Sludge in Agriculture. A National Code of Practice.* HMSO, London, U.K.

Doelman, P. 1986. Resistance of soil microbial communities to heavy metals. *In FEMS Symposium, No 33. Microbial Communities in Soil.* Copenhagen, August 4–8, 1985. Elsevier Applied Science Publishers Ltd, London, U.K., pp. 369–398.

Doelman, P., and L. Haanstra. 1979a. Effect of lead on soil respiration and dehydrogenase activity. *Soil Biol. Biochem.* 11:475–479.

Doelman, P., and L. Haanstra. 1979b. Effects of lead on the soil bacterial microflora. *Soil Biol. Biochem.* 11:487–491.

Doelman, P., and L. Haanstra. 1984. Short-term and long-term effects of cadmium, chromium, copper, lead, and zinc on soil microbial respiration in relation to abiotic soil factors. *Plant Soil* 79:317–327.

Domsch, K.H. 1961. Bodenatmung: Sammelbericht über Methoden und Ergebnisse. *Zentralblatt für Bakteriologie, Parasitenkunde, und Infektionskrankheiten* Abteilung 2 (116):33–78.

Dueck, T.H.A., P. Visser, W.H.O. Ernst, and H. Schat. 1986. Vesicular–arbuscular mycorrhizae decrease zinc toxicity to grasses growing in zinc-polluted soil. *Soil Biol. Biochem.* 18:331–333.

Duxbury, T., and B. Bicknell. 1983. Metal-tolerant bacterial populations from natural and metal-polluted soils. *Soil Biol. Biochem.* 15:243–250.

Eiland, F. 1985. Determination of adenosine triphosphate (ATP) and adenylate-energy charge (AEC) in soil and use of adenine nucleotides as measures of soil-microbial-biomass and activity. PhD thesis, Royal Veterinary and Agricultural University, Copenhagen, Denmark.

Frankenberger, W.T. Jr., J.B. Johanson, and C.O. Nelson. 1983. Urease activity in sewage sludge-amended soils. *Soil Biol. Biochem.* 15:543–549.

Garau, M.A., M.T. Felipér, and M.C. Ruiz de Villa. 1986. Nitrogen mineralization of sewage sludges in soil. *J. Env. Qual.* 15:225–228.

Gilden, A., and P.B. Tinker. 1983a. Interactions of vesicular-arbuscular mycorrhizal infection and heavy metals in plants. I. The effects of heavy metals on the development of vesicular-arbuscular mycorrhizas. *New Phytologist* 95:247–261.

Gilden, A., and P.B. Tinker. 1983b Interactions of vesicular–arbuscular mycorrhizal infections and heavy metals in plants. II. The effects of infection on uptake of copper. *New Phytologist* 95:263–268.

Giller, K.E., S.P. McGrath, and P.R. Hirsch. 1989. Absence of nitrogen fixation in clover grown on soil subject to long-term contamination with heavy metals is due to survival of only ineffective *Rhizobium. Soil Biol. Biochem.* 21:41–484.

Guidi, G. 1981. Relationships between organic matter of sewage sludge and physico-chemical properties of soil. *In* P. L'Hermite and H. Ott (Eds.). *Treatment and Use of Sewage Sludge.* Reidel Publishing Company, Dordrecht, The Netherlands, pp. 530–544.

Hayman, D.S. 1983. The physiology of vesicular arbuscular endo-mycorrhizal symbiosis. *Can. J. Bot.* 61:944–963.

Henis, Y. 1986. Soil microorganisms, soil organic matter and soil fertility. *In* Y. Chen and Y. Avnimelech (Eds.). *The Role of Organic Matter in Modern Agriculture.* Martinus Nijhoff Publishers, The Netherlands, pp. 159–168.

Henriksson, E. 1971. Algal nitrogen fixation in temperate regions. *In* T.A. Lie and E.G. Mulder (Eds.). *Biological Nitrogen Fixation in Natural and Agricultural Habitats. Plant Soil,* Special Volume, pp. 415–419.

Huang, E.Y., F.A. Bazzazz, and L.N. Vanderhoef. 1974. The inhibition of soybean metabolism by cadmium and lead. *Plant Physiol.* 54:122–124.

Jenkinson, D.S., and N.J. Ladd. 1981. Microbial biomass in soil: Measurement and turnover. *In* E.A. Paul and J.N. Ladd (Eds.). *Soil Biochemistry* Vol. 5. Marcel Dekker, New York, pp. 415–471.

Jones, K.C., J.P. Obbard, P. Ineson, N. Kichenside, and J.H. Stark. 1987. Microbial biomass and microbially mediated N-transformation processes in historically sludged and metal contaminated grassland soils. WRc unpublished report, WRc Medmeham, Marlow, U.K.

Kelly, J.W., and D.C. Reanney. 1984. Mercury resistance among soil bacteria, ecology and transferability of genes encoding resistance. *Soil Biol. Biochem.* 16:1–8.

Kiekens, L. 1984. Behaviour of heavy metals in soils. *In* S. Berglund, R.D. Davis, and P. L'Hermite (Eds.). *Utilization of Sewage Sludge on Land: Rates of Application and Long-term Effects of Metals.* Reidel Publishing Company, Dordrecht, The Netherlands, pp. 126–134.

Killham, K., and M.K. Firestone. 1983. Vesicular–arbuscular mycorrhizal mediation of grass response to acidic and heavy metal depositions. *Plant Soil* 72:39–48.

Koomen, I., S.P. McGrath, K.E. Giller. 1990. Mycorrhizal infection of clover is delayed in soils contaminated with heavy metals from past sewage sludge applications. *Soil Biol. Biochem.* 22:871–873.

Kwan, K.H.M., and S.R. Smith. 1990. Sites with a long history of sludge disposal. A study of microbial activity in soil contaminated with heavy metals resulting from the application of sewage sludge. WRc Report No. DoE 2508-M. WRc Medmenham, Marlow, U.K.

Letunova, S.V., M.M. Umarov, G.A. Niyazova, and I. Melekhin. 1985. Nitrogen-fixation activity as a possible criterion for determining permissible concentrations of heavy-metals in soils. *Soviet Soil Sci.* 17:88–92.

Lighthart, B., J. Batham, and V.V. Volk. 1983. Microbial respiration and chemical speciation in metal-amended soils. *J. Env. Qual.* 12:543–548.

Lumsden, R.D., J.A. Lewis, and P.D. Millner. 1983. Effect of composted sewage sludge on several soil-borne pathogens and diseases. *Phytopathology* 73:1543–1548.

Lynch J.M. 1981. Promotion and inhibition of soil aggregate stabilisation by microorganisms. *J. Gen. Microbiol.* 126:371–375.

Lynch, J.M., and E. Bragg. 1985. microorganisms and soil aggregate stability. *Adv. Soil Sci.* 2:133–171.

Maliszewska, W., S. Dec, H. Wierzbicku, and A. Wozniakowsku. 1985. The influence of various heavy metal compounds on the development and activity of soil microorganisms. *Env. Pollution* 37:195–215.

McGrath, S.P. 1984. Metal concentrations in sludges and soil from a long-term field trial. *J. Agric. Sci., Camb.* 84:25–35.

McGrath, S.P., P.C. Brookes, and K.E. Giller. 1988. Effects of potentially toxic metals in soil derived from past applications of sewage sludge on nitrogen fixation by *Trifolium repens* L. *Soil Biol. Biochem.* 20:415–424.

McIlveen, W.D., and H. Cole Jr. 1974. Influence of heavy metals on nodulation of red clover. *Phytopathology* 64:583.

Metzger, L., D. Levanon, and U. Mingelgrin. 1987. The effect of sewage sludge on soil structural stability: Microbiological aspects. *Soil Sci. Soc. Am. J.* 51:346–351.

Metzger, L., and B. Yaron. 1987. Influence of sludge organic matter on soil physical properties. *Adv. Soil Sci.* 7:141–163.

Meyer, J.R., and R.G. Linderman. 1986. Response of subterranean clover to dual inoculation with vesicular-arbuscular mycorrhizal fungi and a plant growth-promoting bacterium, *Pseudomonas putida*. *Soil Biol. Biochem.* 18:185–190.

Moore, P.D. 1988. Essential elements from waste. *Nature* 333:706.

Obbard, J.P., and K.C. Jones. 1989. The presence of effective nitrogen fixing strains of rhizobia in heavily contaminated Welsh metal mine soils. Seventh International Conference on Heavy Metals in the Environment, Geneva.

Obbard, J.P., S.R. Smith, and K.C. Jones. 1990. Nitrogen fixation in sewage sludge amended soils: A review of recent studies. *Proceedings of EC Cost 681 Seminar on Effects of Organics in Sewage Sludge on Soil Fertility, Plants and Animals.* FAL Braunschweig, FRG, 6–8 June 1990.

Öhlinger, R., H. Supersperg, and P. Cepuder. 1990. Effects of different contaminated sewage sludges on some microbial properties. *Proceedings of EC Cost 681 Seminar on Effects of Organics in Sewage Sludge on Soil Fertility, Plants and Animals.* FAL Braunschweig, FRG, 6–8 June 1990.

Olson, B.H., and I. Thornton. 1982. The resistance patterns to metals of bacterial populations in contaminated land. *J. Soil Sci.* 33:271–277.

Pagliai, M., G. Guidi, M. La Marca, M. Giachetti, and G. Lucamente. 1981. Effects of sewage sludges and composts on soil porosity and aggregation. *J. Env. Qual.* 10:556–561.

Parker, C.F., and L.E. Sommers. 1983. Mineralisation of nitrogen in sewage sludges. *J. Env. Qual.* 12:150–156.

Pera, A., M. Giovannetti, G. Vallini, and M. de Bertoldi. 1982. Land application of sludge: Effects on soil microflora. *In:* G. Catroux, P. L'Hermite and E. Suess (Eds.). *The Influence of Sewage Sludge Application on Physical and Biological Properties of Soils.* Reidel Publishing Company, Dordrecht, The Netherlands, pp. 141–166.

Premi, P.R., and A.H. Cornfield. 1969. Incubation study of nitrification of digested sewage sludge added to soil. *Soil Biol. Biochem.* 1:1–4.

Premi, P.R., and A.H. Cornfield. 1971. Incubation study of nitrogen mineralisation in soil treated with dried sewage sludge. *Env. Pollution* 2:1–5.

Quraishi, M.S.I., and A.H. Cornfield. 1973. Incubation study of nitrogen mineralisation and nitrification in relation to soil pH and level of copper (II) addition. *Env. Pollution* 4:159–163.

Reddy, G.B., A. Faza, and R. Bennett, Jr. 1987. Activity of enzymes in rhizosphere and non-rhizosphere soils amended with sludge. *Soil Biol. Biochem.* 19:203–205.

Reddy, G.B., C.N. Cheng, and S.J. Dunn. 1983. Survival of *Rhizobium japonicum* in soil-sludge environment. *Soil Biol. Biochem.* 15:343–345.

Reddy, G.B., and A. Faza. 1989. Dehydrogenase activity in sludge amended soil. *Soil Biol. Biochem.* 21:327.

Rhodes, L.H., and J.W. Gerdemann. 1980. Nutrient translocation in vesicular–arbuscular mycorrhizae. *In* C.B. Cook, P.W. Pappars, and E.D. Rudolph (Eds.). *Cellular Interactions in Symbiosis and Parasitism.* Ohio State University Press, Columbus, Ohio, pp 173–195.

Robson, M.J., A.J. Parsons, and T.E. Williams. 1989. Herbage production: Grasses and legumes. *In* W. Holmes (Ed.). *Grass: Its Production and Utilisation.* Blackwell, Oxford, U.K., pp. 7–88.

Rother, J.A., J.W. Millbank, and I. Thornton. 1982a. Seasonal fluctuations in nitrogen fixation (acetylene reduction) by free-living bacteria in soils contaminated with cadmium, lead, and zinc. *J. Soil Sci.* 33:101–113.

Rother, J.A., J.W. Millbank, and I. Thornton. 1982b. Effects of heavy-metal additions on ammonification and nitrification in soils contaminated with cadmium, lead, and zinc. *Plant Soil* 69:239–258.

Rother, J.A., J.W. Millbank, and I. Thornton. 1983. Nitrogen fixation by white clover (*Trifolium repens*) in grassland on soils contaminated with cadmium, lead, and zinc. *J. Soil Sci.* 34:127–136.

Seaker, E.M., and W.E. Sopper. 1988. Municipal sludge for minespoil reclamation: I Effects on microbial populations and activity. *J. Env. Qual.* 17:591–597.

Skujins, J., H.O. Nohrstedt and S. Odén. 1986. Development of a sensitive biological method for the determination of a low-level toxic contamination in soils. *Swed. J. Agric. Sci.* 16:113–188.

Sleeman, P. 1984. Determination of pollutants in effluents (MPC 4332 C). Detailed analysis of the trace element contents of U.K. sewage sludges. WRc Report No. 280-S. WRc Medmenham, Marlow, U.K.

Smith, S.R., J.E. Hall, and P. Hadley. 1989. Composting sewage sludge wastes in relation to their suitability for use as fertiliser materials for vegetable crop production. International Symposium of Compost Recycling of Wastes, Athens, 4–7 October.

Smith, S.R., J.P. Obbard, K.H.M., Kwan, and K.C. Jones. 1990. Symbiotic N_2-fixation and microbial activity in soils contaminated with heavy metals resulting from long-term sewage sludge application. WRc Research Report FR 0128. WRc Medmenham, Marlow, U.K.

Smith, S.R. 1991. Sites with a long history of sludge disposal: Phase II. Relationships between soil physico-chemical characteristics and soil contamination with potentially toxic elements and the presence of effective *Rhizobium leguminosarum* biovar *trifolii* in soil supplied with long-term applications of sewage sludge. Interim Report. WRc Report No. DoE 2752-M. WRc Medmenham, Marlow, U.K.

Stadelmann, X., and O.J. Furrer. 1982. Influence of sewage sludge application on organic matter content, microorganisms, and microbial activities of a sandy loam soil. *In* G. Latroux, P. L'Hermite, and E. Suess (Eds.). *The Influence of Sewage Sludge Application on Physical and Biological Properties of Soils.* Reidel Publishing Company, Dordrecht, The Netherlands, pp. 141–164.

Stark, J.H., and D.H. Lee. 1988. Sites with a history of sludge deposition. Final report on rehabilitation field trials and studies relating to soil microbial biomass (LDS 9166 SLD). Final report to the Department of the Environment. WRc Report No. DoE 1768-M. WRc Medmenham, Marlow, U.K.

Stover, R.C., L.E. Sommers, and D.J. Silviera. 1976. Evaluation of metals in wastewater sludge. *J. Water Pollution Control Fed* 48:2165–2175.

Stroo, H.F., and E.M. Jencks. 1985. Effect of sewage sludge on microbial activity in an old, abandoned minesoil. *J. Env. Qual.* 14:301–304.

Tabatabai, M.A. 1977. Effects of trace elements on urease activity in soils. *Soil Biol. Biochem.* 9:9–13.

Tisdall, J.M., and J.M. Oades. 1982. Organic matter and water-stable aggregates in soils. *J. Soil Sci.* 33:141–163.

Tomati, U., A. Grappelli, and E. Galli. 1985. Soil microorganisms and long-term fertility. *In* J.H. Williams, G. Guidi, and P. L'Hermite (Eds.). *Long-term Effects of Sewage Sludge and Farm Slurries Applications.* Elsevier Applied Science Publishers, London, U.K., pp. 14–22.

Vigue, G.T., I.L. Pepper, and D.F. Bezkiek. 1981. The effect of cadmium on nodulation and nitrogen fixation by dry beans. *J. Env. Qual.* 10:87–90.

Wei, Q.F., B. Lowery, and A.E. Peterson. 1985. Effect of sludge application on physical properties of a silty clay loam soil. *J. Env. Qual.* 14:178–180.

Williams, S.E., and A.G. Wollum. 1981. Effect of cadmium on soil bacteria and actinomycetes. *J. Env. Qual.* 10:142–144.

Wilson, D.O. 1977a. Nitrification in three soils amended with zinc sulphate. *Soil Biol. Biochem.* 9:277–280.

Wilson, D.O. 1977b. Nitrification in soil treated with domestic and industrial sewage sludge. *Env. Pollution* 12:73–82.

Wiseman, J.T., and L.M. Zibilske. 1988. Effects of sludge application sequence on carbon and nitrogen mineralisation in soil. *J. Env. Qual.* 17:334–339.

Witty, J.F., P.J. Keay, P.J. Froggatt, and P.J. Dart. 1979. Algal nitrogen fixation on temperate arable fields—the Broadbalk experiment. *Plant Soil* 52:157–164.

Yamamoto, H., K. Tatsuyama, and T Uchiwa. 1985. Fungal flora of soil polluted with copper. *Soil Biol. Biochem.* 17:785–790.

Zibilske, L.M., and G. Wagner. 1982. Bacterial growth and fungal genera distribution in soil amended with sewage sludge containing cadmium, chromium and copper. *Soil Sci.* 134:370.

Estimating Soil Water Retention from Soil Physical Properties and Characteristics

W.J. Rawls, T.J. Gish, and D.L. Brakensiek

I. Introduction

Soil and water are two fundamental resources of our agricultural environment. With the advent of new agricultural technologies, tremendous quantities of water and soluble chemicals are applied to soils each year. This enhanced use, or in some cases abuse, has endangered our soil and water resources. In addition, there are increased incidents of groundwater contamination, erosion, and contamination of surface waters from point and nonpoint sources. In response to these environmental concerns there has been an increased effort to develop improved agricultural management practices.

The potential response of soil water properties is a key indicator of the

Advances in Soil Science, Volume 16

impact of agricultural management on the movement of water and chemicals through the soil. Unfortunately soil water characteristics are difficult to measure in the field, hence methods are needed to relate easily measured soil physical properties to soil water retention characteristics (International Workshop on Indirect Methods For Estimating the Hydraulic Properties of Unsaturated Soils, 1989, University of California, Riverside). There are many potential user groups for the information generated by such an effort; hydrologists need estimates of soil water content and retention for partitioning rainfall into runoff and infiltration components; agronomists need the same data for input into crop yield simulation models used to establish irrigation schedules; and meterologists use surface soil moisture for partitioning net radiation into latent and sensible heat components.

Unfortunately, the soil water retention relationship, characterized by soil water content and potential, are not unique and are affected by a number of environmental and soil factors. Temperature affects soil water retention; however, it is generally not critical in moist soils, nor is it in sandy or loamy soils, which are typical in agricultural settings (Taylor, 1958; Hopmans and Dane, 1985, 1986). Irregular pore geometry and discontinuity, and variations in texture and mineralogy are the primary soil properties influencing soil water retention. Variability in these soil factors greatly increases the uncertainty of predicting soil water retention values. As a result, soil scientists and agricultural engineers have been developing methods for defining the hydraulic characteristics of soil using mathematical models and/or soil physical properties (Carsel and Parish, 1988). This information may then be used to generate large-scale probability estimates of potential runoff or the potential fate of agricultural chemicals. The reason for relating soil water retention characteristics to soil physical properties is that direct measurments are very time consuming, costly, and frequently impractical due to the high degree of spatial and temporal variability. Coefficients of variation (cv) for commonly measured saturated and unsaturated hydraulic conductivity often exceed 100% (Nielsen et al., 1973; Gumma, 1978; Cassel, 1983; Babalola, 1978; Libardi et al., 1980). On the other hand, texture, soil bulk density, and organic matter exhibit coefficients of variation of $<30\%$ (Courtin et al., 1983; Cassel and Bauer, 1975).

Vereecken (1988) subdivided empirical soil water retention models into four categories based on their functional relationships (1) exponential; (2) power, (3) cosine hyperbolic, (4) and error function. Most authors used a power function relationship to characterize soil water retention (Visser, 1966; Gregson et al., 1987; Brooks and Corey, 1964; Campbell, 1974; van Genuchten, 1980). The models most frequently being used are the power function models proposed by Brooks and Corey (1964), Campbell (1974), and van Genuchten (1980). The model proposed by van Genuchten (1980) permits a representation of the total soil water retention curve, whereas, Campbell (1974) and Brooks and Corey (1964) do not. Table 1 shows the mathematical interrelationship of these models, along with assumed correspondences between the model parameters as indicated by Rawls and Brakensiek (1985).

Table 1. Soil water retention models

Hydraulic soil characteristic	Parameters	Parameter correspondence
	A. Brooks–Corey (1964)	
Soil water retention $\frac{\theta - \theta_r}{\theta_s - \theta_r} = \left(\frac{h_b}{h}\right)^{\lambda}$	θ_s = saturated water content distribution λ = pore size index h_b = bubbling pressure or air entry pressure θ_r = residual water content ϕ = total porosity	$\lambda = \lambda$ $h_b = h_b$ $\theta_r = \theta_r$ $\theta_s = \phi$
	B. Campbell (1974)	
Soil water retention $\frac{\theta}{\theta_s} = \left(\frac{H_b}{h}\right)^{1/b}$	θ_s = saturated water content H_b = bubbling pressure or air entry pressure b = constant	$\theta_s = \phi$ $H_b = h_b$ $b = \frac{1}{\lambda}$
	C. van Genuchten (1980)	
Soil water retention $\frac{\theta - \theta_r}{\theta_s - \theta} = \left(\frac{1}{1 + (\alpha h)^n}\right)^m$	θ_s = saturated water content θ_r = residual water content α = constant n = constant m = constant	$\theta_s = \phi$ $\theta_r = \theta_r$ $\alpha = (h_b)^{-1}$ $n = \lambda + 1$ $m = 1 - \frac{1}{\lambda + 1}$

θ = Volumetric water content.

h = Capillary suction (cm).

ϕ = Porosity.

Source: Rawls and Brakensiek, 1985.

Past research relating water retention characteristics to soil physical properties and characteristics has taken the approach of estimating specific water retention values or estimating parameters for water retention models. Unfortunately, most of these studies have been conducted using limited data sets that severely limit their usefulness. This review was performed to focus future soil water retention research better by: (1) identifying soil physical properties and characteristics that have been used to describe soil water retention; (2) identifying methods for relating soil physical properties and characteristics to soil water retention; and (3) identifying data bases that can be used for developing and testing various relationships, relating soil properties to water retention characteristics.

II. Soil Physical Properties and Characteristics

The amount of water retained at high soil water potentials (> −10 KPa) depends primarily on the pore size distribution, and is thus strongly affected by soil structure, bulk density, and porosity. As the soil dries out (< −150 KPa), water adsorption becomes critical, and those soil properties that affect specific surface become important such as texture, organic matter content, and clay mineralogy. Soil physical properties used to describe soil water retention can be grouped into four categories (Table 2): (1) soil particle size properties; (2) hydraulic characteristics; (3) morphological properties; and (4) chemical properties. The soil properties and models selected to estimate soil water retention will depend on which soil properties were measured and how much variation was exhibitied by the measured properties. As a result, a comprehensive analysis identifying the importance of specific soil properties on soil water retention is not presently possible.

A. Particle Size Properties

To estimate soil water retention the soil particle size classes researchers most frequently use are sand, silt, and clay percentages (Table 3) (McCuen et al., 1981). This demonstates that the three major particle size classes were statisti-

Table 2. Properties and characteristic found to affect moisture retention

Particle size properties	Hydraulic characteristics	Morphological properties	Chemical properties
Sand	−33 KPa water	bulk density	CEC
Silt	−1500 KPa water	organic carbon	SAR
Clay	Reference moisture retention curve	organic matter	$CaCO_3$
		porosity	Iron
Fine sand		horizon	
Very coarse sand		structure	
Coarse sand		order	
Coarse fragments		color	
Particle size distribution		clay type	
Medium particle size		consistence	
Geometric mean particle size			
Standard deviation geometric mean particle size		pedality	
Water-stable aggregates			

CEC, Xr. SAR, soil adsorption ratio.

cally significant in describing soil water retention. A fourth particle size class that was shown to be beneficial for sandy soils was the fine sand percentage (Table 3). Perhaps, the principal reason more particle sizes are not used may be due to the cost and time involved in data collection.

Particle size distributions and particle size indexes of medium-sized particles, along with their geometric mean particle size and standard deviation, also appear to be useful in estimating soil water retention (Bloemen, 1980; Campbell and Shizawa, 1990; Vereecken, 1990). However, as was mentioned previously, they must be calculated for more than the standard three major particle size classes. Additionally, coarse fragments (>2 mm) have been shown to be important when they comprise more than 10% of the soil volume (Peterson et al., 1968a,b).

B. Hydraulic Characteristics

Field-scale soil heterogeneity results in water retention values having high variability (Table 4). As a result, several authors (Rawls, et al., 1983; Meng et al., 1987) have shown significant improvement in predicting soil water retention by including one or more measured soil water retention values. The measured values in conjuction with other soil physical properties, greatly improve predictions.

C. Morphological Properties

Organic carbon or organic matter (Biswas and Ali, 1969) and soil bulk density (Hill and Summer, 1967) have been commonly used to help describe soil water retention (Table 3). These properties are commonly measured and have a strong effect on soil structure and soil surface area. Usually water retention at all matric potentials is augmented with increases in organic carbon. Bulk density has a greater affect on the water retention at matric potentials > -33KPa. Normally, when bulk density increases, water retention decreases.

In soils containing a large percentage of clay ($>10\%$), clay mineralogy is another variable that is important in predicting soil water retention (Baumer and Brasher, 1982). The 2:1 layer silicates, especially montmorillionite, have the greatest effect, whereas the 1:1 nonexpandable clays, (kaolinite) have the least impact. In addition, factors, such as the cation exchange capacity, which have an inference to clay mineralogy, may also be important (Baumer and Brasher, 1982; Kutilek, 1973).

Soil morphology, specifically aggregation, has a distinct effect on the pore size distribution by dominating the soil water characteristic curve as the soil approaches saturation. As soil water content decreases, the water films around aggregates and soil grains become the dominating soil characteristic influencing water retention. Consequently, in natural soils, aggregation tends to dominate soil water retention only at high water contents, whereas texture

Table 3. Summary of authors and properties used to describe moisture retention characteristics

Reference	Year	Sand (2–0.005 mm)	Silt (0.05–0.002 mm)	Clay (<0.002 mm)	Fine sand (0.25–0.1 mm)	Very fine sand (0.1–0.05) mm)	Coarse sand	Very coarse sand (2.0–1.0 mm)	Coarse fragments (72 mm)	Particle size distribution 9 szs.	Organic carbon, Organic matter	Bulk density	Porosity	−33 KPa water	−1500 KPa water	CEC	SAR	$CaCO_3$
Ahuja et al.	1985											X		X	X			
Arya and Paris	1981									X		X						
Baumer and Brasher	1982	X		X							X		X			X	X	
Beke and MacCormick	1985	X	X	X							X							
Bloemen	1977											X						
Borg	1982		X	X							X	X						
Clapp and Hornberger	1978																	
Campbell and Shizawa	1990										X							
Cosby et al.	1984	X		X														
Dane and Puckett	1990	X		X	X							X						
DeJong et al.	1983	X	X	X							X							
DeJong and Loebel	1982	X	X	X														
Ghosh	1980	X	X	X									X					
Gupta and Larson	1979	X	X	X									X					
Hall et al.	1977	X	X	X		X					X	X						
Hartge	1969									X	X		X					
Haverkamp and Parlange	1986									X	X	X						
Hirschi and Moore	1980			X								X						
Madankumar	1985	X	X	X								X						
McBride and MacKintosh	1984		X	X									X					
McKeague	1987											X			X			
Meng et al.	1987	X	X	X				X			X			X	X			X

Mungare et al.	1983	X	X	X							X		X		
Oosterveld and Chang	1980	X		X								X			
Pachepsky et al.	1990	X		X								X			
Peterson et al. a,b	1968		X	X					X		X	X			
Puckett et al.	1985	X		X	X							X			
Rajakai and Varallyay	1990			X							X	X			
Rawls et al.	1983	X		X							X	X		X	X
Reeve et al.	1973								X			X			
Riley	1979			X											
Salter and Willams	1965a,b		X								X				
	1967			X	X										
Satterwhite	1980	X	X	X											
Saxton et al.	1986	X		X			X					X			
Shayewich and Zwarch	1968		X	X	X						X				
Vereecken	1990	X	X	X						X	X	X			
Williams et al.	1990		X		X						X	X			
Wösten and van Genuchten	1988		X	X							X	X			

Abbreviations as in Table 2.

Table 3. (*Cont.*)

Reference	Year	Iron	Medium particle size	Geometric mean particle size	Standard deviation geometric mean particle size	Particle size index	Water-stable aggregates (72.5 mm)	Texture	Horizon	Structure	Order	Clay type	Amporphous material	Consistence	Color	Reference moisture tension curve	Pedality
Ahuja et al.	1985															X	
Arya and Paris	1981																
Baumer and Brasher	1982											X					
Beke and MacCormick	1985																
Bloemen	1977		X			X											
Borg	1982								X								
Clapp and Hornberger	1978							X									
Campbell and Shizawa	1990			X	X												
Cosby et al.	1984							X									
Dane and Puckett	1990																
DeJong et al.	1983							X	X	X					X		
DeJong and Loebel	1982																
Ghosh	1980																
Gupta and Larson	1979																
Hall et al.	1977								X								
Hartge	1969																
Haverkamp and Parlange	1986																
Hirschi and Moore	1980																
Madankumar	1985																
McBride and MacKintosh	1984							X		X							
McKeague	1987												X	X			
Meng et al.	1987																

Mungare et al.	1983			X						
Oosterveld and Chang	1980									
Pachepsky et al.	1990				X					
Peterson et al. a,b	1968					X	X			
Puckett et al.	1985					X				
Rajakai and Varallyay	1990									
Rawls et al.	1983				X					
Reeve et al.	1973					X				
Riley	1979					X				
Salter and Willams	1965a,b									
	1967									
Satterwhite	1980									
Saxton et al.	1986									
Shayewich and Zwarch	1968									
Vereecken	1990	X	X				X			
Williams et al.	1990							X	X	X
Wösten and van Genuchten	1988					X				

Table 4. Water retention properties classified by soil moisture

Texture class	Sample size	Total porosity (ϕ) cm^3/cm^3	Residual water content (θ_r) cm^3/cm^3	Brooks–Corey Parameters: Bubbling pressure (h_b) Arithmetic cm	Bubbling pressure (h_b) Geometric** cm	Pore size distribution (λ) Arithmetic	Pore size distribution (λ) Geometric**	Water retained at −33 KPa cm^3/cm^3	Water retained at −1500 KPa cm^3/cm^3
Sand	762	0.437*	0.020	15.98	7.26	0.694	0.592	0.091	0.033
		(0.374–0.500)	(0.001–0.039)	(0.24–31.72)	(1.36–38.74)	(0.298–1.090)	(0.334–1.051)	(0.018–0.164)	(0.007–0.059)
Loamy	338	0.437	0.035	20.58	8.69	0.553	0.474	0.125	0.055
sand		(0.368–0.506)	(0.003–0.067)	(−4.04–45.20)	(1.80–41.85)	(0.234–0.872)	(0.271–0.827)	(0.060–0.190)	(0.019–0.091)
Sandy	666	0.453	0.041	30.20	14.66	0.378	0.322	0.207	0.095
loam		(0.351–0.555)	(−0.024–0.106)	(−3.61–64.01)	(3.45–62.24)	(0.140–0.616)	(0.186–0.558)	(0.126–0.288)	(0.031–0.159)
Loam	383	0.463	0.027	40.12	11.15	0.252	0.220	0.270	0.117
		(0.375–0.551)	(−0.020–0.074)	(−20.07–100.3)	(1.63–76.40)	(0.086–0.418)	(0.137–0.355)	(0.195–0.345)	(0.069–0.165)
Silt	1206	0.501	0.015	50.87	20.76	0.234	0.211	0.330	0.133
loam		(0.420–0.582)	(−0.028–0.058)	(−7.68–109.4)	(3.58–120.4)	(0.105–0.363)	(0.136–0.326)	(0.258–0.402)	(0.078–0.188)
Sandy	498	0.398	0.068	59.41	28.08	0.319	0.250	0.255	0.148
clay loam		(0.332–0.464)	(−0.001–0.137)	(−4.62–123.4)	(5.57–141.5)	(0.079–0.559)	(0.125–0.502)	(0.186–0.324)	(0.085–0.211)
Clay loam	366	0.464	0.75	56.43	25.89	0.242	0.194	0.318	0.197
		(0.409–0.519)	(−0.024–0.174)	(−11.44–124.3)	(5.80–115.7)	(0.70–0.414)	(0.100–0.377)	(0.250–0.386)	(0.115–0.279)
Silty	686	0.471	0.040	70.33	32.56	0.177	0.151	0.366	0.208
clay loam		(0.418–0.524)	(−0.038–0.118)	(−3.26–143.9)	(6.68–158.7)	(0.039–0.315)	(0.090–0.253)	(0.304–0.428)	(0.138–0.278)
Sandy clay	45	0.430	0.109	79.48	29.17	0.223	0.168	0.339	0.239
		(0.370–0.490)	(0.013–0.205)	(−20.15–179.1)	(4.96–171.6)	(0.048–0.398)	(0.078–0.364)	(0.245–0.433)	(0.162–0.316)
Silty clay	127	0.479	0.056	76.54	34.19	0.150	0.127	0.387	0.250
		(0.425–0.533)	(−0.024–0.136)	(−6.47–159.6)	(7.04–166.2)	(0.040–0.260)	(0.074–0.219)	(0.332–0.442)	(0.193–0.307)
Clay	291	0.475	0.090	85.60	37.30	0.165	0.131	0.396	0.272
		(0.427–0.523)	(−0.015–0.195)	(−4.92–176.1)	(7.43–187.2)	(0.037–0.293)	(0.068–0.253)	(0.326–0.466)	(0.208–0.336)

*X.
**X.
Source: Rawls et al., 1982.

and clay mineralogy (surface-area-dependent) become more important at low water contents. For example, Gardner (1968) found that the soil water content at −1500 KPa was highly correlated with the soil surface area. A number of researchers (Williams et al., 1990; Croney and Coleman, 1954; Sharma and Uehara, 1968, have demonstrated that the presence or absence of soil structure has a dramatic impact on the shape of the water retention curve. Unfortunately, morphological data has not been used very extensively in estimating soil water retention.

Peterson et al. (1968b) used textural and structural data from each soil horizon to assess the effect of structure on water retention characteristics of a Pennsylvania silt loam. He found that soil water retention in the surface horizon was not significantly affected by coarse fragments, whereas coarse fragments in the C horizon did have a significant effect. He also found that soil structure affected water retention more at −33 KPa than at −1500 KPa, and the degree of soil structure did not appear to dominate water retention as both weak- and moderate-to-strong structural units had a negligible effect on soil water content held at −33 KPa and −1500 KPa.

Also, Peterson (1968b) found that cultivation and gleying had no impact on soil water retention. Wosten et al. (1986) also found that the depth of soil horizon affects water retention. Various authors (Bhushan et al., 1973; Salter and Haworth, 1961; Chibber, 1964; Abrol and Palta, 1969; Amemiya, 1965) have found that tillage and aggregation affected water retention.

D. Chemical Properties

Chemical properties of the soil may also be important, especially if they affect the integrity of the soil aggregates or clay tactoids. Sodium adsorption ratio (SAR) appears to be an important field-scale parameter for predicting soil water retention as high Na levels disperse clay tactoids (Baumer and Brasher, 1982). Also, sodium hyposulfide-soluble iron oxides and cation exchange capacity have been shown to affect soil water retention (El Ashkar et al., 1956; Pachepsky, 1989).

III. Approaches for Estimating Soil Water Retention from Soil Physical Properties and Characteristics

Two approaches have been used for estimating soil water retention characteristics from soil properties. The first approach estimates soil water retention values from soil physical properties and characteristics using regression analysis. The second approach estimates parameters for water retention models from soil physical properties and characteristics using regression analysis.

Table 5. Summary of authors and methods used to describe moisture retention characteristics

Reference	Year	Moisture characteristic curve	−33 KPa moisture content	−1500 KPa moisture content	Moisture content at other tensions	Brooks–Corey parameters	Campbell parameters	van Genuchten parameters
Ahuja et al.	1985	X						
Arya and Paris	1981	X						
Carsel and Parrish	1988							X
Baumer and Brasher	1982		X	X	X			
Beke and MacCormick	1985		X	X				
Bloemen	1977					X		
Borg	1982						X	
Clapp and Horberger	1978						X	
Campbell and Shizawa	1990						X	
Cosby et al.	1984						X	
Dane and Puckett	1990		X	X	X			
DeJong et al.	1983		X	X				
DeJong and Loebel	1982		X	X				
Ghosh	1980						X	
Gupta and Larson	1979		X	X	X			
Hall et al.	1977		X	X	X			
Hartge	1969	X						
Haverkamp and Parlange	1986	X						
Hirschi and More	1980						X	
Madankumar	1985						X	
McBride and MacIntosh	1984	X		X				
Meng et al.	1987		X	X	X			
Mungare et al.	1983							
Oosterveld and Chang	1980		X	X				

Peterson et al.	1968a		X	X				
Puckett et al.	1985	X	X	X				
Rajakai and Varallyay	1990		X	X	X			
Rawls et al.	1973		X	X	X	X		
Reeve et al.	1973		X	X				
Riley	1979							
Salter and Williams	1965a,b		X	X				
	1967							
Satterwhite	1980		X	X				
Saxton et al.	1986							
Shayewich and Zwarch	1968		X	X				
Vereecken	1990							X
Williams et al.	1990						X	
Wösten and van Genuchten	1998							X

A. Estimation of Specific Points on the Soil Water Retention Curve

The most popular method for estimating soil water retention at specific water potentials has been to relate various soil properties (Table 2) to soil water retention at specific water potentials using regression analysis (Table 5).

The two most frequently estimated water contents are those corresponding to soil water potentials of −33 and −1500 KPa, primarily because they are commonly measured. Table 6 summarizes the soil water held at −33 and −1500 KPa for the USDA soil texture classes for the United States, British, and Australian data sets, which clearly demonstrate the effect of texture on water retention. To increase the accuracy of the regression equations using physical soil properties, Rawls and Brakensiek (1982) introduced the water content held at −1500 KPa or at both −33 KPa and −1500 KPa. Adding these variables, which require more costly and/or time-consuming laboratory procedures, increased the explained variation from 76% to 95%. In general, the water content held at −33 KPa was a more significant variable for estimating water retention at matric potentials between 0 and −33 KPa, and the water content held at −1500 KPa was a more significant variable for estimating water retention at water potentials between −33 and −1500 KPa.

Another method was proposed by Ahuja et al. (1985) who used linear extrapolation between the water contents held at −33 KPa and −1500 KPa on a log–log graph. Using data from Oklahoma, Ahuja et al. report that their procedure yields values of comparable accuracy to the equations that included the water held −33 and −1500 KPa.

Table 6. 33- and 1500-KPa water content for soil textures

	−1500 KPa Water content (% vol)			−33 KPa Water content (% vol)		
USDA Texture	Rawls et al. (1982)	Salter and Williams (1969)	Hall et al. (1977)	Rawls et al. (1982)	Salter and Williams (1969)	Hall et al. (1977)
Sand	3	4	3	9	9	8
Loamy sand	6	5	8	13	13	15
Sandy loam	10	7	11	20	20	21
Loam	12	12	11	27	27	—
Silt loam	13	12	13	33	35	26
Sandy clay loam	15	12	18	26	21	29
Clay loam	20	18	21	32	31	32
Silty clay loam	21	19	21	37	41	35
Sandy loam	24	18	—	34	26	—
Silty clay	25	22	26	39	41	35
Clay	27	26	30	40	37	41

Other methods include the development of soil water retention curves based on soil bulk density and particle size distribution (Arya and Paris, 1981; Ahuja et al., 1985; Hartge, 1969). Hartge (1969) developed a series of nomographs relating soil porosity and particle size distribution to a corresponding water retention curve.

Ahuja and associates (1985) proposed a scaling method to estimate the soil water retention curve from soil bulk density, water content held at −33 KPa, and a reference soil water retention curve. By using a mean soil textural water retention curve, such as that derived by Rawls et al. (1982) (Table 4), as a reference curve, the procedure produced results for the Oklahoma data set that were equivalent to the regression equations that included the water retained at −33 and −1500 KPa. These results imply that Ahuja's procedure can be applied with only soil texture class, soil water held at −33 KPa, and soil bulk density as input variables.

B. Estimation of Soil Water Retention Model Parameters

With increased interest in modeling, the need for a continuous function describing the soil water retention curve has become very important. Frequently water retention data is fitted to a water retention model (Table 1) and the model parameters subsequently related to physical soil properties using regression analysis (Table 5).

One approach to relating the model parameters to soil properties is to develop average parameter values as a function of soil texture classes. Such an approach was used to develop parameters values for the models of Brooks and Corey (1964) (Table 4) and Campbell (1974) (Table 7), (Rawls et al., 1982; Clapp and Hornberger, 1978; De Jong, 1982). As seen in Table 7 there is considerable variation in parameter estimations among the researchers, which can be a result of the data base used to develop the parameters.

The second approach for relating the water retention parameters to soil physical properties is to fit water retention models to data and then use regression analysis to relate the parameters to soil physical properties. Examples of this approach are given in Table 8 for the Brooks–Corey model (Rawls and Brakensiek, 1985; Bloemen, 1977) and the van Genuchten model (Vereecken, 1988). Using the correspondence between model parameters given in Table 1 the equations in Table 8 can be used with the Brooks–Corey, Campbell, and van Genuchten water retention models.

Another more recent approach introduced by Tyler and Wheatcraft (1990) uses fractal mathematics and scaled similarities to show that the empirical constant in the Arya–Paris (1981) model is equivalent to the fractal dimension of the tortous fractal pore. This information can be used to predict soil water retention from measured particle size distributions. Also, Mishra et al. (1989) developed a statistical procedure based on the Arya–Paris (1981) model

Table 7. Representative Campbell parameters for USDA soil texture

Texture	b			h_b			ϕ		
	Rawls et al. (1982)	Clapp and Hornberger (1978)	DeJong (1982)	Rawls et al. (1982)	Clapp and Hornberger (1978)	DeJong (1982)	Rawls et al. (1982)	Clapp and Hornberger (1978)	DeJong (1982)
Sand	1.44	4.05	3.21	15.98	12.1	8.39	0.44	0.40	0.40
Loamy Sand	1.81	4.38	4.09	20.58	9.0	12.68	0.44	0.41	0.39
Sandy Loam	2.64	4.90	4.78	30.20	21.8	26.03	0.45	0.44	0.39
Loam	4.51	5.39	6.60	40.12	47.8	26.63	0.46	0.45	0.49
Silt Loam	4.74	5.30	7.34	50.87	78.6	19.41	0.50	0.49	0.48
Silty Clay Loam	3.13	7.75	7.95	59.41	35.6	13.26	0.40	0.48	0.57
Sand Clay Loam	4.13	7.12	—	56.43	29.9	—	0.46	0.42	—
Clay Loam	5.65	8.52	9.84	70.33	63.0	10.21	0.46	0.48	0.50
Sandy Clay	4.48	10.40	—	79.48	15.3	—	0.43	0.43	—
Silty Clay	7.87	10.40	10.26	76.54	49.0	16.98	0.48	0.49	0.53
Clay	6.06	11.40	12.71	85.60	40.5	12.24	0.48	0.48	0.53

$$\frac{\theta}{\phi} = \left(\frac{h_b}{h}\right)^{1/b}$$

ϕ, toal porosity; h_b, bubbling pressure; b, constant.

Table 8. Estimation equations for water retention model

A. Brooks–Corey bubbling pressure (cm)

h_b: Brooks–Corey bubbling pressure (cm)

$h_b = e[5.340 + 0.185(C) - 2.484(\phi) - 0.002 (C)^2 - 0.044(S)(\phi) - 0.617(C)(\phi) + 0.001 (S)^2(\phi)^2 - 0.009(C^2)(\phi^2) - 0.00001 (S^2)(C) + 0.009(C^2)(\phi) - 0.0007(S^2)(\phi) + 0.00000(C^2)(S) + 0.500(\phi^2)C]$

λ: Brooks–Corey pore size distribution index

$\lambda = e[-0.784 + 0.018(S) - 1.062)(\phi) - 0.00005(S^2) - 0.003(C^2) + 1.111(\phi^2) - 0.031(S)(\phi) + 0.0003(S^2)(\phi^2) - 0.006(C^2)(\phi^2) - 0.000002(S^2)(C) + 0.008(C^2)(\phi) - 0.007)(\phi^2)(C)]$

θ_r: Brooks–Corey residual water content (vol. fraction)

$\theta_r = -0.018 + 0.0009(S) + 0.005)(C) + 0.029)(\phi) - 0.0002(C)^2 - 0.001(S)(\phi) - 0.0002(C^2) (\phi^2) + 0.0003(C^2)(\phi) - 0.002)(\phi^2)(C)$

B. Brooks–Corey Parameters[b]

$h_b = 2914\ f^{0.79}\ Md^{-0.96}$

$\lambda = 1.512(e^{0.3f} - 1)$

f = grain size distribution index

Md = median grain size (mm)

C. van Genuchten Parameters[c]

$\theta_r = 0.015 + 0.005(C) + 0.014(Ca)$

$\alpha = 10^{(-2.486+0.025(S)-0.351(Ca)-2.617(bd)-0.023(C)}$

$n = 10^{(0.053-0.009(S)-0.013(C)+0.00015(S)2)}$

$m = 1 - 1/n$

C, % clay; S, % sand; bd, bulk density (g/cm^3); Ca, % carbon; ϕ, porposity (vol. fraction).
Source: [a]Rawls and Brakensiek, 1985, [b]Bloemen, 1977, [c]Vereecken, 1988.

to estimate water retention parameters and the uncertainty from measured particle size data.

IV. Water Retention Data Bases

In the review several comprehensive data bases surfaced that have usefulness in future research. These data bases were cited by Rawls et al. (1982) and Baumer (1990) for the U.S.A.; De Jong et al. (1983) for Canada; Williams et al. (1990) for Australia; Vereecken (1990) for Belgium; and Hall et al. (1977) for Great Britain. The variables included in each data set are summarized in Table 3.

V. Conclusions

Research relating soil physical properties and characteristics to water retention characteristics has been limited by the properties included in the data bases. Present data bases need to be expanded to include (1) detailed particle size analysis with at least five to nine size classes; (2) clay mineralogy, organic matter, bulk density; (3) morphological information such as structure, horizon, order, grade, clay type, color, and consistence; (4) aggregate size and stability; (5) chemical properties; and (6) a distribution of pore sizes, their frequency, and their continuity with depth. The empirical approach of relating soil properties to water retention characteristics is at present adequate; however, more research is needed to describe pore geometry and the interaction between soil water and soil matrix better. The effects of rainwater and water quality on water retention characteristics also needs to be explored.

References

Abrol, I.P. and J.P. Palta. 1969. A study of the effect of aggregate size and bulk density on moisture retention characteristics of selected soils, Agrochimica. 14:157–163.

Ahuja, L.R., J.W. Naney, and R.D. Williams. 1985. Estimating soil water characteristics from simpler properties or limited data. *Soil Sci. Soc.* 49:1100–1105.

Amemiya, M. 1965. The influence of aggregate size on soil moisture content-capillary conductivity relation. *Soil Sci. Soc. Amer. Proc.* 29:744–748.

Arya, L.M. and J.F. Paris. 1981. A physicoempirical model to predict the soil moisture characteristic from particle-size distribution and bulk density data. *Soil Sci. Soc. Am. J.* 45:1023–1030.

Babalola, O. 1978. Spatial variability of soil water proporties for a tropical soil of Nigeria. *Soil Sci.* 126:269–279.

Baumer, O.W. 1990. Prediction of soil hydraulic parameters. *Proceedings of International Workshop in Indirect Methods for Estimating the Hydraulic Properties of Unsaturated Soils.* USDA-ARS/University of California, Riverside, CA.

Baumer, O.W. and B.R. Brasher. 1982. Prediction of water content at selected suctions. *Am. Soc. Agri. Eng.* Paper 82–2590.

Beke, G.J. and M.J. MacCormick. 1985. Predicting volumetric water retention for subsoil materials from Colchester County, Nova Scotia. *Can. J. Soil. Sci.* 65:233–236.

Bhushan, L.S., S.B. Varade, and C.P. Gupta. 1973. Influence of tillage practices on clod size porosity and water retention. *Indian J. Agric. Sci.* 43:466–471.

Biswas, T.D. and M.H. Ali. 1969. Retention and availability of soil water as influenced by soil organic carbon. *Indian J. Agric. Sci.* 39:582–583.

Bloemen, G.W. 1977. Calculation of capillary conductivity and capillary rise from grain size distribution. I. Real and theoretical values of the exponent in a formula of Brooks and Corey for the calculation of hydraulic conductivities. *ICW Wageningen nota* no. 952. II. Assessment of the values of the exponent in a formula of Brooks and Corey for the calculation of hydraulic conductivity from grain size distribution. *ICW Wageningen nota* no. 962. III. Air entry pressure and saturated conductivity calculated from grain size distribution and median grain size. *ICW Wageningen nota* no. 990. IV. Capillary rise in soil types and soil profiles. *ICW Wageningen nota* no. 1013.

Bloemen, G.W. 1980. Calculation of hydraulic conductivities from texture and organic matter content. *Z. Pflanzenernaehr. Boidenkd.* 143:581–605.

Borg, H. 1982. Estimating soil hydraulic properties from texture data. PhD. Thesis, Washington State University, Pullman, WA.

Brooks, R.H. and A.T. Corey. 1964. Hydraulic properties of porous media. *Colorado State University Hydrol.* Paper No. 3, pp. 27.

Campbell, G.S. 1974. A simple method for determining unsaturated conductivity from moisture retention data. *Soil Sci.* 117:311–314.

Campbell, G.S. and Sho Shizawa. 1990. Prediction of hydraulic properties of soils using particle size distribution and bulk density data. *Proceedings of International Workshop in Indirect Methods for Estimating the Hydraulic Properties of Unsaturated Soils.* USDA-ARS/University of California, Riverside, CA.

Cassel, D.K. 1983. Spatial and temporal variability of soil physical properties following tillage of Norfolk loamy sand. *Soil Sci. Soc. Am. J.* 47:196–201.

Cassel, D.K. and A. Bauer. 1975. Spatial variability in soils below depth of tillage. Bulk density and fifteen atmosphere percentage. *Soil Sci. Am. Proc.* 39:247–250.

Carsel, R.F. and R.S. Parrish. 1988. Developing joint probability distributions of soil water retention characteristics. *Water Resources Res.* 24(5)755–769.

Chibber, R.K. 1964. Aggregate size distribution and water relationships amongst some typical Indian soils. *Bull. Batn. Inst. Sci. India* 26:148–156.

Clapp, R.B. and G.M. Hornberger. 1978. Empirical equations for some hydraulic properties. *Water Resourc. Res.* 15:601–604.

Cosby, B.J., G.M. Hornberger, R.B. Clapp, and T.R. Ginn. 1984. A statistical exploration of the relationships of soil moisture characteristics to the physical properties of soils. *Water Resourc. Res.* 20:682–690.

Courtin, P., M.C. Feller and K. Klinka. 1983. Internal variability in some properties of disturbed soils in S.W. British Columbia, Canada. *J. Soil Sci.* 63:529–539.

Croney, D. and J.D. Coleman. 1954. Soil structure in relation to soil suction. *J. Soil Sci.* 5:75–85.

Dane, J.H. and W.E. Puckett. 1990. Field soil hydraulic properties based on physical and mineralogical information. *Proceedings of International Workshop in Indirect Methods for Estimating the Hydraulic Properties of Unsaturated Soils.* USDA-ARS/University of California, Riverside, CA.

De Jong, R. 1982. Assessment of empirical parameters that describe soil water characteristics. *Can. Agric. Eng.* 24:65–70.

De Jong, R. 1983 . Soil water desorption curves estimated from limited data. *Can. J. Soil Sci.* 63:697–703.

De Jong, R., C.A., Campbell, and W. Nicholaichuk. 1983. Water retention equations and their relationship to soil organic matter and particle size distribution for disturbed samples. *Can J. Soil Sci.* 63:291–302.

De Jong, R. and K. Loebel. 1982. Empirical relations between soil components and water retention at 1/3 and 15 atmospheres. *Can. J. Soil Sci.* 62:343–350.

El Ashkar, M.A., G.B. Bodman, and D.B. Peters. 1956. Sodium hyposulfite-soluble iron oxides and water retention by soils. *Soil Sci. Soc. Amer. Proc.* 20:352–365.

Gardner, W.R. 1968. Availability and measurement of soil water. *In Water Deficits and Plant Growth*, Vol. 1. Academic Press, New York, pp. 107–135.

Ghosh, R.K. 1980. Estimation of soil-moisture characteristics from mechanical properties of soils. *Soil Sci.* 130:60–63.

Gregson, K., D.J. Hector, and M. McGowan. 1987. A one parameter model for the soil water characteristic. *J. Soil Sci.* 38:483–486.

Gumma, G .A. 1978. Spatial variability of *in situ* available water. Ph.D. Dissertation. Univ. of Arizona, Tucson, pp. 140.

Gupta, S.C. and W.E. Larson. 1979. Estimating soil water retention characteristic from particle size distribution, organic matter percent and bulk density. *Water Reourc. Res.* 15:1633–1635.

Hall, D.G.M., M.J. Reeve, A.J. Thomasson, and V.F. Wright. 1977. Water retention, porosity, and density of field soils. *Tech. Monograph no. 9.* Rothamsted Exp. Stn., Harpenden, England.

Hartge, K.H. 1969 . Die Ermittlung der Wasserspannungskurve aus der Kornungssummenkurve und dem Gesamtporenvolumen. *Zeitschr. für Kulturtechn. und Flurbereinigung* 10 (Heft 1):20–29.

Haverkamp R. and J.Y. Parlange. 1986. Predicting water retention curve from particle size distribution, organic matter and bulk density. *Water Resourc. Res.* 15(6):1633–1635.

Hill, J.N.S. and M.E. Summer. 1967. Effect of bulk density on moisture characteristics of soils. *Soil Sci.* 103:234–8.

Hirschi, M.C. and I.D. Moore. 1980. Estimating soil hydraulic properties from soil texture. *Am. Soc. Agric. Eng.* Paper 80–2523.

Hopmans, J.W. and J.H. Dane. 1985. Effect of temperature-dependent hydraulic properties on soil water movement. *Soil Sci. Soc. Am. J.* 49(1):51–57

Hopmans, J.W. and J.H. Dane. 1986. Temperature dependence of soil hydraulic properties. *Soil Sci. Soc. Am. J.* 50:4–9.

International Workshop on Indirect Methods for Estimating the Hydraulic Properties of Unsaturated Soils. 1989. University of California, Riverside, California.

Kutilek, M. 1973. The influence of soil minerals and exchangeable cations on soil moisture potential. *In* X. Hadas et al. (Eds.). *Physical Aspects of Soil Water and Salts in Ecosystems.* Springer-Verlag, New York.

Libardi, P.C., K. Reinhardt, D.R. Nielsen, and J.W. Biggar. 1980. Simple field methods for estimating soil hydraulics conductivity. *Soil Sci. Soc. Am. J.* 44:3–6.

Madankumar, N. 1985. Prediction of soil moisture characteristics from mechanical analysis and bulk density data. *Agric. Water Manage.* 10:305–312.

McBride, R.A. and E.E. MacIntosh. 1984. Soil survey interpolations from water retention data: I. Development and validation of a water retention model. *Soil Sci. Soc. Am. J.* 48:1338–1343.

McCuen, R.H., W.J. Rawls, and D.L. Brakensiek. 1981. Statistical analysis of the Brooks–Corey and the Green–Ampt parameters across soil texture. *Water Resources Research* 17(4):1005–1013.

McKeague, J.A. 1987. Estimating air porosity and available water capacity from soil morphology. *Soil Sci. Soc. Am. J.* 51:148–152.

Meng, T.P., H.M. Taylor, D.W. Fryrear, and J.F. Gomez. 1987. Models to predict water retention in semiarid soils. *Soil Sci. Soc. Am. J.* 51:1563–1565.

Mishra, S., J.C. Parker, and N. Singhal. 1989. Estimation of soil hydraulic properties and their uncertainty from particle size distribution. *J. Hydrol.* 108:1–18.

Mungare, T.S., A.K. Shingte, and K.S. Pharande. 1983. Relationship of available water capacity with some of the physical properties of soils. *J. Maharashtra Agric. Univ.* 8(1):9–13.

Nielsen, D.R., J.W. Biggar and K.T. Erh. 1973. Spatial variability of field measured soil water properties. *Hilgardia* 42:215–259.

Oosterveld, M. and C. Chang. 1980. Empirical relations between laboratory determinations of soil texture and moisture retention. *Can. Agric. Eng.* 22:149–151.

Pachepsky, Ya. A. 1989. The effect of the content of soil solutions and exchangeable cations on water retention and hydraulic conductivity of soils. *Pochvovedenie* No. 3:53–65, (in Russian).

Pachepsky, Ya. A., E.V. Mironenko, and R.A. Shcherbakov. 1990. Prediction and use of soil hydraulic properties. *Proceedings of International Workshop in Indirect Methods for Estimating the Hydraulic Properties of Unsaturated Soils.* USDA-ARS/ University of California, Riverside, CA.

Peterson, G.W., R.L. Cunningham, and R.P. Matelski. 1968a. Moisture characteristics of Pennsylvania soils. I. Moisture retention as related to texture. *Soil Sci. Soc. Am. Proc.* 32:271–275.

Peterson, G.W., R.L. Cunningham, and R.P. Matelski. 1968b. Moisture characteristics of Pennsylvania soils: II. Soil factors affecting moisture retention within a textural class-silt loams. *Soil Sci. Soc. Am. Proc.* 32:866–870.

Puckett, W.E., J.H. Dane, and B.F. Hajek. 1985. Physical and mineralogical data to determine soil hydraulic properties. *Soil Sci. Soc. Am. J.* 49:831–836.

Rajkai, K. and G. Varallyay. 1990. Estimating soil water retention from simpler soil properties by regression techniques. *Proceedings of International Workshop in Indirect Methods for Estimating the Hydraulic Properties of Unsaturated Soils.* USDA-ARS/University of California, Riverside, CA.

Rawls, W.J. and D.L. Brakensiek. 1982. Estimating soil water retention from soil properties. *J. Irrig.* and *Drainage Div. ASCE* 108:166–171.

Rawls, W.J. and D.L. Brakensiek. 1985. Prediction of soil water properties for hydrologic modeling. *In* E.B. Jones and T.J. Ward (Eds.). *Watershed Management in the Eighties.* Proc. of Symp. sponsored by Comm. on Watershed Management, I & D Division, ASCE. ASCE Convention, Denver, CO, April 30–May 1, pp. 293–299.

Rawls, W.J., D.L. Brakensiek, and K.E. Saxton. 1982. Estimation of soil water properties. *Trans. ASAE* 25(5):1316–1220, 1328.

Rawls, W.J., D.L. Brakensiek, and B. Soni. 1983. Agricultural management effects on soil water processes: Part I Soil water retention and Green–Ampt parameters. *Transactions ASAE* 26(6):1747–1752.

Reeve, M.J., P.D. Smith, and A.J. Thomasson. 1973. The effect of density on water retention properties of field soils. *Soil Sci.* 24:335–367.

Riley, H.C.F. 1979. Relationship between soil moisture holding properties and soil texture, organic matter content, and bulk density. *Agric. Res. Exp.* 30:379–398.

Salter, P.J. and F. Haworth. 1961. The available water capacity of a sandy loam soil. I. The effects of farmyard manure and different primary cultivations. *J. Soil Sci.* 12:335–42.

Salter, P.J. and J.B. Williams. 1965a. The influence of texture on the moisture characteristics of soils. I. A critical comparison of techniques for determining the available water capacity and moisture characteristic curve of a soil. *J. Soil Sci.* 16:1–15.

Salter, P.J. and J.B. Williams. 1965b. The influence of texture on the moisture characteristics of soils. II. Available water capacity and moisture release characteristics. *J. Soil Sci.* 16:310–317.

Salter, P.J. and J.B. Williams. 1967. The influence of texture on the moisture characteristics of soils. IV. A method of estimating the available-water capacities of profiles in the field. *J. Soil Sci.* 18:174–81.

Salter, P.J. and J.B. Williams. 1969. The influence of texture on the moisture characteristics of soils. V. Relationships between particle-size composition and moisture contents at the upper and lower limits of available water. *J. Soil Sci.* 20:126–31.

Satterwhite, M.B. 1980. *Evaluating soil moisture and textural relationships using regression analysis ETL-0226*. U.S. Army Corps of Engineers, Fort Belvoir, VA, 31 p.

Saxton, K.E., W.J. Rawls, J.S. Romberger, and R.I. Papendick. 1986. Estimating generalized soil water characteristics from texture. *Soil Sci. Soc. Am. J.* 50(4):1031–1035.

Sharma, M.L. and G. Uehara. 1968. Influence of soil structure on water relations in low humic latosols. I. Water reflection. *Soil Sci. Soc. Amer. Proc.* 32:765–770.

Shayewich, D.F. and M.A. Zwarch. 1968. Relationships between soil physical constants and soil physical components of some Manitoba soils. *Can. J. Soil Sci.* 48:199–204.

Taylor, S.A. 1958. The activity of water in soils. *Soil Sci.* 86:83–90.

Tyler, S.W. and S.W. Wheatcraft. 1990. Fractal processes in soil water retention. *Water Reources Res.* 26:1047–1054.

van Genuchten, R. 1980. Predicting the hydraulic conductivity of unsaturated soils. *Soil Sci. Soc. Am. Proc.* 44:892–898

Vereecken, H. 1988. Pedotransfer functions for the generation of hydraulic properties of Belgian soils. PhD thesis. Katholieke Universiteit Leuven, Leuven, Belgium, pp. 254.

Vereecken, H. 1990. Derivation and validation of pedotransfer functions for soil hydraulic properties. *Proceedings of International Workshop in Indirect Methods for Estimating the Hydraulic Properties of Unsaturated Soils.* USDA-ARS/University of California, Riverside, CA.

Visser, W.C. 1966. Progress in the knowledge about the effect of soil moisture content on plant production. *Inst. Land Water management, Wageningen, Netherlands, Tech. Bull.* No. 45.

Williams, J., P. Ross, and K. Bristow. 1990. Prediction of the Campbell water retention function from texture structure, and organic matter. *Proceedings of International Workshop in Indirect Methods for Estimating the Hydraulic Properties of Unsaturated Soils.* USDA-ARS/University of California, Riverside, CA.

Wösten, J.H.M. and M. Th. van Genuchten. 1988. Using texture and other soil properties to predict the unsaturated soil hydraulic functions. *Soil Sci. Soc. Am. J.* 52:1762–1770.

Wösten, J.H.M., M.H. Banninl, J.J. de Gruijter, and J. Bouma. 1986. A procedure to identify different groups of hydraulic conductivity and moisture retention curves for soil horizens. *J. Hydrol* 86:133–145.

Index

MIX
Papier aus verantwortungsvollen Quellen
Paper from responsible sources
FSC® C105338

If you have any concerns about our products,
you can contact us on
ProductSafety@springernature.com

In case Publisher is established outside the EU,
the EU authorized representative is:
Springer Nature Customer Service Center GmbH
Europaplatz 3, 69115 Heidelberg, Germany

Printed by Libri Plureos GmbH
in Hamburg, Germany